CAP and the RURAL ENVIRONMENT in TRANSITION

A panorama of national perspectives

Floor Brouwer and Philip Lowe [editors]

Wageningen Pers

CIP-data Koninklijke Bibliotheek Den Haag

ISBN 90-74134-59-9 paperback

Subject headings:
Common Agricultural Policy
European Union
Environment

First published, 1998

© Wageningen Pers, Wageningen
The Netherlands, 1998

Printed in The Netherlands

TABLE OF CONTENTS

LIST OF CONTRIBUTORS

Siemen van Berkum, Agricultural Economics Research Institute, P.O. Box 29703, 2502 LS The Hague, The Netherlands

Alessandro Bordin, Department of Land and Agro-Forestry Systems, Faculty of Agriculture, Section of Economics, University of Padua, AGRIPOLIS, I- 35020 Legnaro (Padova), Italy

Jean-Marie Bouquiaux, Centrum voor Landbouweconomie, Ministerie van Middenstand en Landbouw, WTC 3, Simon Bolivarlaan 30, Verdieping 24, B - 1000 Brussel, Belgium

Floor Brouwer, Agricultural Economics Research Institute, P.O. Box 29703, 2502 LS The Hague, The Netherlands

João Castro Caldas, Department of Agrarian Economy and Rural Sociology, Higher Institute of Agronomy, Lisbon Technical University, Lisbon, Portugal

Helen Caraveli, Centre of Economics Research, Athens University of Economics and Business, 76 Patission Street, 104 34 Athens, Greece

Luca Cesaro, Department of Land and Agro-Forestry Systems, Faculty of Agriculture, Section of Economics, University of Padua, AGRIPOLIS, I- 35020 Legnaro (Padova), Italy

Brendan Flynn, Department of Government, University of Essex, Wivenhoe Park, Colchester CO4 3SQ, United Kingdom

Marielle Foguenne, Centrum voor Landbouweconomie, Ministerie van Middenstand en Landbouw, WTC 3, Simon Bolivarlaan 30, Verdieping 24, B - 1000 Brussel, Belgium

Paola Gatto, Department of Land and Agro-Forestry Systems, Faculty of Agriculture, Section of Economics, University of Padua, AGRIPOLIS, I- 35020 Padova, Italy

Lionel Hubbard, Department of Agricultural Economics and Food Marketing, University of Newcatle upon Tyne, Newcastle NEI 7RU, United Kingdom

Werner Kleinhanss, Institute of Farm Economics, Federal Agricultural Research Centre, Bundesallee 50, D-38116 Braunschweig, Germany

Ludwig Lauwers, Centrum voor Landbouweconomie, Ministerie van Middenstand en Landbouw, WTC 3, Simon Bolivarlaan 30, Verdieping 24, B - 1000 Brussel, Belgium

Michael Linddal, Statens Jordbrugs - og Fiskeriøkonomiske Institut (SJFI), Toftegårds Plads, Gammel Køge Landevey 1-3, DK - 2500 Valby (Copenhagen), Denmark

Philip Lowe, Centre for Rural Economy, Department of Agricultural Economics and Marketing, University of Newcastle upon Tyne, Newcastle NE1 7RU, United Kingdom

Maurizio Merlo, Department of Land and Agro-Forestry Systems, Faculty of Agriculture, Section of Economics, University of Padua, AGRIPOLIS, I- 35020 Padova, Italy

Asko Miettinen, Pohjois-Karjalan Maaseutukeskus, Koskitatu 11 C, P.O. Box 5, FIN - 80101 Joensuu, Finland

Andrew Moxey, Department of Agricultural Economics and Marketing, University of Newcastle upon Tyne, Newcastle NE1 7RU, United Kingdom

Pierre Rainelli, Institut National de la Recherche Agronomique (INRA), Centre de Rennes, Station d'Economie et Sociologie Rurales, 65 Rue de St. Brieuc, F-35042 Rennes Cedex, France

John Sumelius, Department of Economics and Management, University of Helsinki, P.O. Box 27, FIN-00014 Helsinki, Finland

José Sumpsi, Departamento Economia y Ciencias Sociales Agrarias, Escuela Técnica Superior de Ingenieros Agrónomos, Universidad Politécnica, Cuidad Universitaria s/n, 28040 Madrid, Spain

Consuelo Varela-Ortega, Departamento Economia y Ciencias Sociales Agrarias, Escuela Técnica Superior de Ingenieros Agrónomos, Universidad Politécnica, Cuidad Universitaria s/n, 28040 Madrid, Spain

Dominique Vermersch, Institut National de la Recherche Agronomique (INRA), Centre de Rennes, Station d'Economie et Sociologie Rurales, 65 Rue de St. Brieuc, F-35042 Rennes Cedex, France

Aage Walter-Jørgensen, Statens Jordbrugs- og Fiskeriøkonomiske Institut (SJFI), Toftegårds Plads, Gammel Køge Landevej 1-3, DK-2500 Valby (Copenhagen), Denmark

Neil Ward, Centre for Rural Economy, Department of Agricultural Economics and Marketing, University of Newcastle upon Tyne, Newcastle NE1 7RU, United Kingdom

Martin Whitby, Centre for Rural Economy, Department of Agricultural, Economics and Marketing, University of Newcastle upon Tyne, Newcastle NE1 7RU, United Kingdom

Michael Winter, Countryside and Community Research Unit, Department of Countryside and Landscape, Cheltenham and Gloucester College of Higher Education, Francis Close Hall, Cheltenham GL50 4AZ, United Kingdom

PREFACE

The European Commission initiated a research project entitled 'Thematic network on CAP and environment in the European Union'. The objective is to improve research methods to assess the effects of the Common Agricultural Policy (CAP) on the environment, nature and landscape, and to communicate the present state of understanding with representatives of national and regional governments, the European Commission, and interest groups. The emphasis in the project is on the post-1992 changes to the CAP, their impact on farming systems and consequences for the environment, nature and landscape in the European Union. The focus is on empirical assessments rather than on modelling or mathematical approaches. The book addresses the environmental effects of agricultural policy reform from the different national perspectives. In providing an overview of the role of the CAP in the rural environment in transition, it explores key linkages between agricultural policy and the physical environment, landscape and biodiversity.

The book has been prepared with financial support from the Commission of the European Communities, Agriculture and Fisheries (FAIR) specific RTD programme, FAIR3 - CT96 - 1793, Thematic Network on CAP and environment in the European Union. This support is gratefully acknowledged. It does not necessarily reflect its views and in no way anticipates the Commission's future policy in this area.

The book includes edited and revised versions of fourteen national reports which have been prepared as a collaborative effort of the partners involved in the project and complemented with additional contributions from experts in other countries. In addition, the first chapter reflects on the present state of knowledge across the European Union. First drafts of the chapters were discussed at a meeting, organized at Chania, Crete, September 25-27, 1997. Following that meeting, the individual contributions have been revised towards a panorama of national perspectives of CAP and the rural environment.

Editorial assistance was provided by Zayd Abdulla (LEI-DLO). His guidance in fine-tuning the book has been a great help. Patricia Grimminck and Helga van der Kooij prepared the manuscript for publication. The editors are grateful for their devotion to this project.

July, 1998
The Hague/Newcastle upon Tyne

1. CAP REFORM AND THE ENVIRONMENT

Floor Brouwer and Philip Lowe

1.1   Introduction

This introductory chapter reflects on the present state of knowledge regarding the effects of the current CAP on the quality of the physical environment, biodiversity and landscape across the European Union. Problems are identified for agricultural policy in both reducing harmful effects and creating beneficial effects on the environment. Also, certain research gaps regarding agricultural policy and the environment are identified.

Stating the problem

Agriculture not only produces food and fibre; it also shapes the rural environment. A managed environment that ensures the productivity of the land is of course essential to the maintenance of primary production. Increasingly, also, modern society values the environmental resources that arise as joint outputs with primary land use, including water supply, semi-natural habitats, wildlife, the historic pattern of land settlement and its associated cultural artefacts, rural landscapes and open spaces.

For these resources to be conserved and made available, the land must be managed and this mainly entails the continuity of certain farming practices. However, rapid changes in primary land use and technology have jeopardised the supply of these resources. The Common Agricultural Policy (CAP) has been criticized for helping to drive those changes and for not taking sufficient account of the environmental consequences, and in recent years European policymakers have begun to respond to such criticisms. Increasingly, a role for the CAP is also acknowledged in enhancing the beneficial effects of agriculture for landscape and biodiversity.

Agricultural policy, as originally formulated, was not given explicit responsibilities to enhance the role of agriculture in shaping the rural environment. The original objectives of the CAP were set out in Article 39 of the 1957 Treaty of Rome as:

-       to increase agricultural productivity by promoting technical progress and by ensuring the rational development of agricultural production and the optimal utilization of the factors of production, in particular labour;
-       thus to ensure a fair standard of living for the agricultural community, in particular by increasing the earnings of persons engaged in agriculture;
-       to stabilize markets;
-       to ensure stability of supplies;
-       to ensure that supplies reach the consumers at reasonable prices.

Not only did these objectives reflect the public priorities of the time, but the thought would not have occurred then that the pursuit of farming might in some ways not be in harmony with the maintenance of the countryside. From the 1970s onwards, however, public consciousness and scientific evidence grew in a number of European countries that certain contemporary developments in agriculture had damaging or undesirable side-effects. As common understandings emerged during the 1980s of the causes of these developments, including the role of agricultural policy, so also did public expectations rise of the contribution of the agricultural sector to achieving more sustainable conditions for the European environment. These concerns were at least partly addressed in the 1992 reform of the CAP by the provision to stimulate less intensive production methods. However, the reform was driven more by other considerations including the need to contain budgetary pressures, improve competitiveness of EU agriculture on the global market, curb surplus production and respond to the discussions on agricultural trade during the Uruguay round of the General Agreement on Tariffs and Trade (GATT).

Currently, the European Union (EU) is engaged in a further round of CAP reform as part of Agenda 2000 which sets out the objectives as those of increased competitiveness, high standards of food safety and quality, ensuring a fair standard of living for the agricultural community, the fuller integration of environmental goals, the creation of alternative job and income opportunities in rural areas and simplification of EU legislation and administration. These changes are seen as necessary to respond to contemporary public demands on agriculture and the countryside and to prepare the CAP for the imminent enlargement of the EU and for the reopening of world trade negotiations on agriculture. Any decisions made in the current round of CAP reform will be implemented from the year 2000 onwards. The proposals take forward the 1992 reform, not least in terms of the fuller integration of environmental goals. It is important therefore that the consequences of the 1992 reform be well and widely understood.

Some of the environmental problems of contemporary agriculture have also been addressed directly or indirectly in the development of EU environmental policy. Whereas the CAP measures provide payments to farmers subject to conditions on production, environmental policy tends to regulate the actions of producers. This has presented different strategies to solve environmental concerns. The distinction is being eroded, however, with agri-environmental policies that encourage farmers to introduce or continue agricultural production methods compatible with requirements on the protection of the environment and the maintenance of the countryside. They have not necessarily been targetted though to assist in achieving environmental legislation. Correlating the development and effects of agricultural and environmental policies over the past decade does allow us to question to what extent CAP reform has integrated environmental objectives.

Outline of the chapter

Some key linkages between agriculture, the physical environment, biodiversity and landscape are explored in Section 2, drawing wherever possible on EU-wide data and analysis. One of the objectives of the 1992 reform of the CAP was to achieve less intensive production methods and to contribute to the achievement of environmentally sustainable production methods. Section 3 therefore assembles evidence from the assessments made so far regarding the environmental consequences of CAP reform. Here, the various regimes will be reviewed, particularly focussing on those most affected by the 1992 reform. Finally, Section 4 explores the implications of the conclusions that can be drawn, for the development of research.

14

1.2   Linkages between agriculture, the physical environment, biodiversity and landscape

Both harmful and beneficial effects of agriculture on the environment need to be considered

The relationship between agriculture and the environment is complex, including both beneficial and benign effects as well as harmful ones. European agriculture is characterized by a broad geographical heterogeneity of production systems. Agri-environmental linkages are also complicated by the vulnerability and unpredictability of ecosystems. Intensive crop and livestock production systems exert pressures on the environment, leading to harmful effects in terms of the quality of soils, air and water (surface water and groundwater resources), biodiversity and landscape. Existing farming practices may also have positive effects in maintaining semi-natural areas and their associated habitats and wildlife, and features of the countryside that contribute to its landscape value (see also OECD (1997a) for an investigation of issues and policies related to the environmental benefits from agriculture).

Interactions between agriculture and the environment can be classified according to the following themes:

- *Soil quality* (in terms of erosion, nutrient supply, moisture balance, and compaction of soils due to the use of heavy machinery). The application of livestock manure in most cases improves soil quality. High supply levels, however, may saturate and degrade soils as well as cause water pollution. Soil erosion is also one of the most severe rural environmental problems in Mediterranean countries, and is partly linked to changing agricultural practices such as crop rotations without green cover crops during the winter period, the substitution of traditional arable fodder crops by maize for silage, and the farming of uncultivated land;

- *Water quality and water quantity* (leaching of nutrients and pesticides, water extraction and drainage). The excessive use of pesticides poses a widespread threat to the environment; the run-off or discharge of liquid slurry or other livestock wastes can cause severe pollution; high nitrate levels in groundwater may cause human health problems and eutrophication of surface water is a particular problem in certain parts of the European Union, causing algal blooms and fish-deaths. Water quality problems tend to be acute in regions with excess amounts of livestock manure from intensive livestock production. Water quantity problems arise in regions where water consumption exceeds critical levels in relation to the available water resources. For example, the share of water used for irrigation purposes in total national water consumption is over 80% in Spain compared with less than 10% in Austria. The sectors of Spanish agriculture that operate on international markets are highly dependent on irrigated land which, in total, accounts for about 80% of the country's agricultural exports (Varela-Ortega and Sumpsi, 1998);

- *Air quality* (emissions of ammonia and greenhouse gases). Emissions of ammonia contribute to acidification of soils and water, and agriculture is a major source of this type of pollution. Also, rising trends in greenhouse gas emissions may potentially alter global climatic conditions mainly due to the emissions of carbon dioxide from energy consumption, methane from cattle production and nitrous oxide from grazing livestock;

- *Biodiversity* (i.e. biological diversity including genetic, species and ecosystem diversity). The biodiversity of much of the EU is found on, or adjacent to, farmland, which accounts for more than 40% of the total land area in the EU, and is thereby considerably affected by agricultural management and practices. Work on birds, which have been studied in greater detail than other wildlife, suggests that, at a European scale, agricultural habitats have the highest overall species richness of any category of habitat (Tucker, 1997). Many modern farming practices, though, are inimical to wildlife. The intensification of agriculture has led to the widespread reduction or elimination of once-common species and habitats. However, about two-fifths of the EU's agricultural area remains under low intensity systems. Most of this is either grazing land under various systems of livestock management or permanent crops (olive trees, vines and fruit and nut trees) under traditional management. Typically these low intensity farming systems support semi-natural habitats and wildlife species of conservation importance. They may be threatened by either increased or decreased production pressures.
- *Landscape,* including preservation of landscapes by farming systems with high nature value. Marginalization of land used agriculturally, by which certain farmland areas cease to be viable under an existing land use and socio-economic structure, could eventually lead to abandonment, which limits management of semi-natural areas in Europe. Intensification of agriculture may lead to a general loss of landscape features such as hedges, ponds, field margins and woodlands, and the replacement of traditional farm buildings with industrialized structures.

The environmental problems of contemporary agriculture

During the past 20-30 years a series of environmental problems have arisen in relation to European agriculture, problems which can broadly be ascribed to the intensification and concentration of production. Intensification has involved a movement away from farming as a self-sustaining cycle, in which there was extensive re-use of animal and plant wastes and little import of additional nutrients, towards an industrial model of resource throughput in which the quantity of bought-in inputs is increased in order to increase yields. Agricultural pollution may become a major problem, because farmers use increasing quantities of potent inputs (particularly crop protection products and plant nutrients), if nutrients are applied surplus to the requirements of crops and animals, and where increasing quantities of potentially polluting farm wastes are produced. Intensive production systems, on the other hand, may be better equipped to meet the exacting standards laid down in environmental and other legislation and to respond to pressures to improve the environmental performance of the agricultural sector.

The modern arable and grazing systems which are now common in most of Europe are based on heavy applications of fertilizer (17 million tonnes of plant nutrients in fertilizer are used annually in the EU, Stanners and Bourdeau, 1995) and on the use of chemical crop protection products (ECPA, 1996 presents a recent survey). Many of these chemicals find their way into water courses where they may cause eutrophication and the elimination of sensitive aquatic species and into the groundwater, thus also contaminating human water supply systems. An additional

problem with irrigated farming in southern Europe is salinization linked to the over-use of aquifers and salt water incursion along the Mediterranean coastline. The intensive livestock production, which is found particularly across the Low Countries, and in Brittany, Galicia, Lombardy
and North Rhine-Westphalia, generates large quantities of liquid slurry which is a potent pollutant
both if it enters water courses and in the noxious fumes it emits, but also when spread on land
leads to nutrient leaching and gaseous emissions. The volume of animal production provides a
direct indication of the extent of production of global warming gases from agriculture. In particular, the population of ruminant livestock (85 million cattle and 120 million other grazing animals,
mainly sheep and goats) and its manures are responsible for the addition of 10 million tonnes of
methane, or twelve percent of annual EU emissions of this very powerful greenhouse gas, to the
atmosphere annually. Farm machinery also adds to agriculture's contribution to this global pollution problem. EU agriculture now has more than seven million tractors (compared with an EU
total of 144 million cars).

Alongside the intensification of production has gone the concentration (and specialization)
of production as some regions have emerged within the Single European Market with better advantages compared to others (through their superior soils or climate, better access to markets,
more rationalized farm structures and well-advanced integration of primary production with food
processing industry, or their better trained and supported farmers). The regions where production
has been concentrated may have lost wildlife, semi-natural habitats and traditional landscape features. Conversely, other, uncompetitive regions have suffered the withdrawal of capital and labour from agriculture. At its extreme this has led to land abandonment. In southern Europe many
regions with difficult terrain or poor soils have suffered extensively from the interlinked social
problem of rural depopulation and a series of environmental problems arising from the withdrawal of cultivation. These problems include the abandonment of terraces and water management systems, increased incidence of soil erosion, the invasion of scrub, increased risks of forest
fires and major floods, and the reversion of countryside to wilderness.

*Usage of crop protection products reduced ... but the consequences for the environment are un-
certain*

The total volume of crop protection products used in European agriculture fell by some 17% in
the first half of the 1990s (Table 1). Important factors contributing to the downward trend are innovations (e.g. new compounds requiring lower dosages and improved application technologies),
changes in farm management (e.g. Integrated Crop Management) and national mandatory reduction schemes (ECPA, 1996). Dry climatic conditions during the first half of the 1990s also contributed to the lower use of pesticides. This was particularly marked in Spain where consumption
of crop protection products went up in 1996 following the ending of several years of drought.

Overall reductions in the amount of crop protection products used do not necessarily mean
reduced environmental pressures. In many cases what is happening is that farmers are using less
of more potent products. This in turn detracts somewhat from the significance of aggregated data
concerning the tonnage of active ingredients used. However, Directive 91/414 concerning the
placing of plant protection products on the market is intended to establish common rules across
the Member States for the approval of such products and their active ingredients, including con-

sistent health and environmental criteria. The overall reduction at the European level, moreover, subsumes some divergent trends. The usage of crop protection products actually increased during the first half of the 1990s in Greece, Ireland and Portugal. More generally, the volumes sold in southern Europe fell at lower rates than in northern Europe. There were similar divergent trends intranationally. In Italy, for example, usage of agrochemicals increased in the lowland regions where production has increasingly concentrated in the past few decades, but fell elsewhere, thus reflecting the growing dichotomy between intensive production systems in the lowlands and extensification in the hills and mountainous areas (Borin et al., 1998).

Table 1   Volume of crop protection products sold in the European Union (tonnes of active ingredients)

Country	1985	1990	1995
Austria	-	4,247	3,231
Belgium	-	5,892 a)	4,572
Denmark	7,152	6,244	4,911
Finland	1,893	2,007	1,047
France	98,021	97,701	84,007
Germany	30,053 b)	29,883 b)	25,551
Greece	-	7,860 c)	8,525
Ireland	-	1,802	2,639
Italy	-	58,123 c)	48,190
Portugal	-	9,355 c)	9,712
The Netherlands	21,632 d)	18,835	10,923
Spain	39,134 d)	39,562	27,852
Sweden	3,660	2,344	1,224
United Kingdom	27,353	23,592	20,627
TOTAL	-	307,447	253,011

a) Situation in 1992; b) Situation in FRG; c) Situation in 1991; d) Situation in 1986.
Source: ECPA (1996).

Consumption of chemical fertilizers shows a declining trend

Fertilizer consumption increased during the 1960s and 1970s and, in combination with major advances in plant breeding, farm technology and land management, accounted for substantial increases in yields (EFMA, 1997). Total fertilizer consumption in most countries peaked around the mid to late 1980s, and has fallen since then (Table 2). Overall, at the EU-level, a reduction of almost 10% occurred between 1985 and 1990, but in Austria, Germany, Italy and the Netherlands reductions of twenty percent were achieved. In these countries, though, the rates of reduction levelled off considerably during the first half of the 1990s and were overtaken by another group of countries - Belgium, Denmark, Finland, Greece, Portugal and the United Kingdom - where con-

18

sumption had peaked a little later but which achieved reductions in excess of 10% in the period 1990-1995. Overall, nitrogen consumption in EU-15 fell by some 5% between 1990 and 1995.

The consumption of chemical fertilizers has been reduced in part through replacement by organic nutrients and the more efficient use of nutrients in general which has allowed yields to increase without a corresponding increase in chemical inputs. The major downturn in nitrogen consumption from the mid-1980s onwards is in part attributable to the imposition of milk quotas in 1984 and subsequent reductions in quota. The effect has varied though with the national significance of the dairy sector and its response to production quotas. As producers have sought to reduce their costs, the relative price of home-produced fodder compared to externally produced feed has been one important consideration. In the UK in the late 1980s, for example, dairy farmers were encouraged to reduce their dependency on bought-in feed and therefore to intensify their own grassland management using additional fertilizer (Lowe et al., 1997). In the Netherlands, in contrast, the consumption of nitrogen fertilizers has fallen markedly since the mid-1980s, mainly because dairy producers have increasingly come to rely on externally produced feed (Brouwer and Van Berkum, 1998).

In contrast to the effect of milk quotas, the 1992 reform of the CAP appears to have had little clear effect on fertilizer consumption, which has largely levelled off since then. One explanation is that the direct income supports given to farmers, by cushioning them from the drop in the prices they received, also reduced the pressures on them to cut their costs and therefore their use of inputs. A few countries - most notably France, Ireland and Spain - have resumed a strong upward growth in fertilizer consumption since the early 1990s, revealing (at least for major sectors and regions) a continuing tendency towards intensification.

Table 2   Nitrogen fertilizer consumption in EU15 countries (million kg N), 1970-1995

Country	1970	1980	1985	1990	1991	1992	1993	1994	1995
Austria	126	160	165	135	132	124	124	124	125
Belgium/Luxembourg	178	194	195	186	182	173	169	168	167
Denmark	289	374	382	396	370	333	325	316	291
Finland	169	197	202	205	177	174	171	183	183
France	1,453	2,147	2,408	2,492	2,569	2,154	2,222	2,308	2,392
Germany	1,642	2,303	2,286	1,788	1,720	1,680	1,613	1,787	1,769
Greece	201	333	450	407	391	393	328	346	316
Ireland	87	275	314	370	349	353	404	432	415
Italy	595	1,006	1,055	845	906	909	918	882	874
Netherlands	405	483	500	390	392	388	383	395	365
Portugal	77	137	137	150	140	127	129	128	126
Spain	578	902	962	1,064	999	819	935	929	1,037
Sweden	226	244	246	212	185	205	221	216	197
UK	801	1,240	1,568	1,515	1,365	1,219	1,248	1,339	1,328
EU-15	6,826	9,994	11,206	10,155	9,877	9,051	9,190	9,553	9,585

Source: 1970-1987 FAO; 1988-1995: EFMA.

in southern Europe, as well as grazing systems of moorlands and heaths in the uplands of the UK and Ireland. Relatively highly intensive grazing systems like the peat areas in the western part of the Netherlands, may also have a high nature conservation potential. The share of low-intensity farming systems is estimated to cover well over 40% of agricultural area of the countries included (Table 4).

Table 4   Estimated area (million ha) of farmland under low-intensity farming systems

Country	Land surface under agriculture	Agricultural area under low-intensity farming systems	Share of agricultural area under low-intensity farming systems (%)
Greece	9	6	61
Spain	31	25	82
France	31	8	25
Ireland	6	2	35
Italy	23	7	31
Portugal	5	3	60
United Kingdom	18	2	11
Total	122	52	43

Source: Bignal and McCracken (1996).

1.3   The effects of the CAP reform

The responsibility of the CAP

Many of these environmental impacts have been laid at the feet of the European Community, specifically its CAP (Baldock and Lowe, 1996), and it must be admitted that the CAP has been in place during what must have been the most widespread and rapid transformation of the rural environment in the whole of European history. To ascribe all responsibility for this transformation and its myriad changes to the CAP would, however, be naive.

There have been associated changes in rural, social and economic structure and in technology with which the CAP has interacted but which would have had profound consequences without the CAP. Arguably, one effect of the CAP has been to moderate some of the more detrimental, social and economic pressures and, given that many of the environmental benefits from rural land management depend upon the continuity of certain practices, it is possible that without the CAP there would have been even greater environmental losses. For example, the intensification and concentration in the pig and poultry sectors, with all of their attendant problems of disposal of waste products, have occurred without the CAP's commodity price supports, although they have been encouraged indirectly by a common and protected market 1). On a broader view, though, environmental problems associated with structural or technological changes in agriculture may point to gaps in policy, including at the Community-level, especially if other Community policies - such as the Single Market - are implicated.

Three broad areas of concern have been identified about the direct effects of the CAP: the level and efficiency of input use and the consequences for agricultural pollution; the rationalization of farm size and structure and the consequences for rural landscapes and habitats; and the maintenance and encouragement of farming in marginal areas. Below we consider each of these in turn.

It is generally agreed that high product prices paid under the CAP have encouraged a greater use of bought-in inputs than would otherwise have been the case. This has led to a less efficient use and hence a greater polluting surplus of chemical inputs (inorganic fertilizers and plant protection products); greater use of purchased feed and thus an encouragement to overstocking; and greater reliance on bought-in fertilizer, leading to even bigger surpluses of organic manures to be disposed of. The empirical evidence confirms that high price supports under the CAP have been associated with big increases in the use of plant protection products, inorganic fertilizers and surpluses of animal manures, though there are considerable variations between farms and regions (Brouwer and Van Berkum, 1996). It is generally assumed that lower agricultural supports should lead to environmental improvements, either by encouraging more efficient use of inputs or a shift to more extensive systems. The effects may not be marked, though, because of low price elasticities of demand for fertilizers and plant protection products.

A long-term objective of the CAP has been the improvement of the structure of farming. This has involved grants and technical aid to improve the age structure of the farming population, to modernise farms and to rationalise the size and structure of farm holdings. In a number of countries, these aids have been incorporated into national programmes that have orchestrated major investment in agricultural infrastructure (such as land drainage and new buildings) and the consolidation and reparcelling of land holdings. In countries such as the Netherlands and France, such programmes have had a dramatic environmental impact regionally by removing many traditional landscape features and micro-habitats (hedgerows, trees, small woodlands, wet areas) (Boisson and Buller, 1996). In other countries, similar consequences may have occurred but less due to CAP structural policies than to price supports. An investigation of structural changes in the UK provides 'at least casual support' for the argument that higher prices have provided an incentive to amalgamate farm holdings (Allanson and Moxey, 1996). It was also observed in the UK that a change in the occupancy of all or part of a farm is one of the major factors in landscape change leading, for example, to the removal of existing field boundaries and their associated hedgerows and trees. Contrary to this, the abandonment of agricultural land linked to rural depopulation and the economic pressures on agriculture, are considered to be a major concern for rural areas in Finland (Miettinen, 1998). Agriculture is considered to be an important agent in main-

---

1)   Import tariffs in the past gave significant price advantages to animal feed concentrates based on imported cereal substitutes, which encouraged the development of intensive livestock production especially in the vicinity of ports where these substitutes entered the Community, e.g. the Netherlands and Belgium.

taining overall population levels in the rural areas of Finland, and the structure of rural areas may be weakened by a decline of agricultural incomes.

Structural polices and price support may also have helped sustain farming in marginal regions, with undoubted environmental benefits, but in some cases losses too. The main structural policy to support agriculture in marginal regions has been the Less Favoured Areas (LFAs) designation under EC Directive 75/268. LFAs contain a million holdings and cover some 55% of the agricultural area of the EU (Brouwer and Van Berkum, 1996), including most of the land under low intensity systems (Baldock and Beaufoy, 1993). Potentially therefore they are of major significance.

Although the LFA Scheme may have helped to sustain some farmers in low intensity farming systems, the form of support, by adding to the basic headage payments for livestock, has provided a further incentive to raise stocking densities, which may be detrimental to nature conservation. A similar or even greater effect may have occurred through the beef and sheep premium which, while likewise supporting farming in marginal areas, may also have encouraged overstocking and local overgrazing and hence damage to swards and soils (Baldock and Beaufoy, 1993).

Reforming the CAP from an environmental perspective

Efforts to reform the CAP from an environmental perspective have been aimed both to overcome the negative externalities associated with production supports and to incorporate positive environmental aims into the objectives of the CAP (Reus et al., 1995). The 1992 reform was aimed to stimulate less intensive production methods, primarily through the reform of the arable crop regime, as well as the beef and sheep regimes, and it was anticipated that this would have indirect environmental benefits. Provisions were also made for environmental protection by allowing for conditions to be placed on compensatory payments for price reductions. The set-aside scheme included provisions regarding the maintenance and use of the areas set-aside. The payment of beef premiums was subject to limits on the number of eligible male animals per farm and the stocking density, meant to encourage extensive production methods. As part of the 1992 reform, specific measures were incorporated centrally in the CAP to promote environmentally-beneficial farming, under the agri-environmental Regulation 2078/92. Aid was thereby made available to reduce agro-chemical inputs, to assist organic farming, to facilitate shifts to extensive forms of crop production or grassland management and to support production methods that protect the environment and maintain the countryside.

Agri-environmental measures

Implementation of Council Regulation 2078/92 began in 1993. Some Member States were able to develop new schemes rapidly or to adapt existing national measures, but some programmes were not approved until 1995. FEOGA expenditures under the regulation were less than ECU 500 million a year during the period 1993-1995. The figure for the budget year 1996, though, was almost ECU 1,400 million (Table 5), of which some ECU 300 million was allocated to so-called Objective 1 regions (regions whose development is lagging behind). In comparison, back in 1990 a sum of just ECU 10 million had been dispensed in cofinancing agri-environmental payments.

Clearly, there has been a massive leap in the resources allocated to agri-environmental measures although they still account for only 4% of the overall CAP budget.

Regulation 2078/92 has provided for the first time a common European framework for national policies in the agri-environment field. In consequence, several states, that had not previously done so, have introduced agri-environmental programmes. It is obligatory on member states to implement a national programme and to include within it all the individual categories of measures listed in Article 2, unless there is a clear reason why these should not apply.

Aid measures under regulation 2078/92, article 2

Subject to positive effects on the environment and the countryside, the scheme may include aid for farmers who undertake:

a)   to reduce substantially their use of fertilisers and/or plant protection products, or to keep the reductions already made, or to introduce or continue with organic farming methods;

b)   to change, by means other than those referred to in (a), to more extensive forms of crop, including forage, production, or to maintain extensive production methods introduced in the past, or to convert arable land into exentensive grassland;

c)   to reduce the proportion of sheep and cattle per forage area;

d)   to use other farming practices compatible with the requirements of protection of the environment and natural resources, as well as maintenance of the countryside and the landscape, or to rear animals of local breeds in danger of extinction;

e)   to ensure the upkeep of abandoned farmland or woodlands;

f)   to set aside farmland for at least 20 years with a view to its use for purposes connected with the environemnt, in particular for the establishment of biotope reserves or natural parks or for the protection of hydrological systems;

g)   to manage land for public access and leisure activities.

Source: European Commission, 1992.

The Regulation sets certain precedents for agricultural policy which may have long-term consequences. It has established the principle that farmers, for both environmental and production control benefits, should be paid to de-intensity production and to manage the countryside. The Regulation thus legitimizes non-productivity agriculture, particularly low intensity systems. In certain regions and farming systems, it also brings small-scale and/or part-time farmers within the scope of agricultural support policies. These are potentially major shifts which challenge the legemony of organisations that represent large-scale productivist agriculture.

The Regulation introduces subsidiarity into agricultural policy to a substantial extent. Inevitably, this is leading to considerable variation in national and regional responses to the Regulation. To the extent that this reflects the varied nature of European rural environments and the social values attached to them, this is a desirable outcome. But it also seems to reflect variations in resources and capacities regionally and nationally.

The distribution of the funds under the Regulation is highly skewed between Member States. Just three countries account for over 70% of the budget - Austria, Germany and Finland. Undoubtedly, in these three, there is strong popular and official interest in promoting environmentally friendly farming, but there are other factors at work also. Germany was a prominent par-

Table 5   FEOGA Guarantee Section, expenditures during the budget year 1996 on agri-environment programme (million ECU) by country a) b)

Member State	Objective 1 regions	Other regions	Total
Austria c)	60.0	481.0	541.0
Belgium	0.2	1.2	1.5
Denmark	-	5.8	5.8
Finland	-	256.6	256.6
France	0.0	118.9	118.9
Germany	100.9	130.8	231.7
Greece	1.5	-	1.5
Ireland	43.4	-	43.4
Italy	21.6	20.0	41.5
Luxembourg	-	0.0	0.0
Netherlands	0.4	7.2	7.6
Portugal	40.0	-	40.0
Spain	32.8	0.0	32.8
Sweden	-	43.4	43.4
United Kingdom	1.6	23.9	25.5
Total	302.4	1,086.7	1,391.2

a) The FEOGA budget year is 16 October 1995 - 15 October 1996; b) Agri-environment expenditures in Objective 1 regions represents 75% of the total co-financible expenditure and 50% of the total in other regions; c) Agri-environment expenditure for Austria in 1996 includes spending committed in 1995 (brought forward under special arrangement owing to a post-accession delay in approval and implementation of the programme).
Source: European Commission, cited in House of Commons Agriculture Committee (1997).

ticipant in the measures which were precursors to Regulation 2078/92 (the extensification, voluntary set-aside and ESA schemes under Regulation 797/85) and had a range of local schemes that could readily be absorbed and expanded within the new Regulation. Finland and Austria, in contrast, devised their national programmes in response to Regulation 2078/92 and the main reason these are on such a scale is that, besides their environmental objectives, they have been seen and promoted as income transfer mechanisms to compensate for farm income losses due to EU accession (Finland and Austria, with Sweden, joined the EU in 1995).

In Austria, around 70% of farmers participate, covering some 80% of the land used agriculturally (Kleinhanss, 1998b). The high acceptance can be explained partly by the generally low requirements and relatively high premiums. Compensatory payments for some programmes are rather small compared to the costs of farm adjustments. The considerable budget, however, has contributed to the support and stabilization of farm incomes.

Amongst the other countries that have taken much smaller shares of the agri-environment budget, there are some, such as Belgium Greece and Portugal, that are newcomers to agri-environmental policy. For others, notably Denmark and the Netherlands, the relatively low expenditures reflect the fact that the payments available under the Regulation are insufficient to attract the participation of many farmers where intensive farming predominates. The low level of expenditure in the UK reflects an approach to the implementation of the Regulation that is oriented to the solution of descrete and specific problems of environmental conservation in the farmed countryside through circumscribed and targetted measures, rather than general extensification schemes.

One feature of agri-environmental programmes compared with commodity support programmes is that they trend to be costly to administer, especially where specific local or regional schemes have to be devised, where farmers require active guidance and where there is a need for compliance monitoring and evaluation of the results of schemes (Whitby and Falconer, 1998). Figures for the United Kingdom indicate that administrative and monitoring costs amounted to almost half of total expenditures on agri-environmental measures in 1992/93, but came down to 27% in 1995/96 as the number of farmer-recipients grew (Lowe et al., 1998). Such costs are borne largely by national governments and are usually ignored in the debate about the cost-effectiveness of agri-environmental programmes.

Monitoring and evaluation of EU agri-environmental policy will prove complex because of the extremely diverse agendas of individual member states as well as the problems of auditing the behaviour of some millions of participants in these voluntary schemes with complicated management requirements. Not only is there considerable variability in the packages of measures that member states have adopted but they also have very diverse administrative structures internally with the consequence that consistency of policy implementation across the EU will be very difficult to achieve or judge.

A number of member states have put in place procedures for environmental monitoring to reveal the impact on the environment of agri-environment measures. These measures promise both general and specific environmental benefits. However, in most cases, it is much too early to expect to see the full environmental consequences. For a start, many national schemes did not get under way until after 1995. Second, while it is possible to refer to the take-up rates of various schemes and changes in farming practices, the consequent second-order effects in such areas as changes in enterprises, farming systems or technology and farm structure may be complex and protracted. Third, the response of the natural environment may be particularly prolonged. It may take several decades, for example, to increase the biodiversity of agriculturally improved grassland or to attenuate harmful levels of soluble phosphorus in soils. In these circumstances, national monitoring often relies on mathematical models that combine environmental survey data with information on participation levels and changes in agricultural practices, to predict and assess the likely environmental impacts.

Preliminary monitoring results and assessments are available for a number of countries. In Finland, for example, it is judged that the greatest influence of the national agri-environmental programme should be on the reduction of nutrient losses, and already, during the first two years, there has been a sharp decline in the use of phosphorus in fertilisers. The environmental awareness and attitudes of farmers have also improved which is seen as crucial in the adoption of better farm management practices, on which any improvements in the state of the environment depend (Miettinen, 1998). The institutional framework to implement programmes and the managerial skills of the agricultural sector are also crucial factors in realising agri-environment programmes. In Italy, for example, significant progress has only been made where the aid available has built on already existing expertise, including networks of organic farming associations, extension services for integrated pest management, or in areas where incentives have been traditionally provided for the maintenance of alpine pasture and meadows (Bordin et al., 1998). In a number of countries the real environmental value of programmes dominated by general extensification schemes is being questioned. In Germany, for example, the various programmes are generally skewed in favour of grassland extensification. While this may support land use in less favoured areas it does little to solve the environmental problems of intensive agricultural regions (Kleinhanss, 1998a).

The reform of the arable crops regime

The arable crops regime was the one most altered by the reform of 1992. Intervention prices of cereals were gradually reduced by around one-third. To compensate cereal producers for income losses from price reductions, direct payments were introduced on a per hectare basis. Large-scale producers would only be eligible to hectarage compensation if they set aside part of their land.

The primary objective was to depress production levels. It was also anticipated that lowering product prices should lead to diminished intensity of production including reduced application of agrochemicals. This is what happened in Germany at least. Surveys of farmers in Baden-Wuerttemberg, Rhineland-Palatinate and North-Rhine Westphalia found that, with the start of the CAP reform, half of the surveyed farmers had reduced the intensity of cereal production and many intended to cut back further on their inputs (Zeddies et al., 1994). The effect has not been as marked everywhere, though. In some countries the price cut was more or less offset by other concurrent factors: including devaluation of the currency in the UK; and abolition of a fertilizer levy in Austria.

More generally, it is difficult to distinguish the effects of the price cut from the general tendency towards lower input use brought about in part by technical progress and improvements in farm management. A recent paper on the effect of the arable crop regime on pesticide use indicates that a 20% price reduction would have reduced use by 5% and that the set-aside requirements have contributed just a few percent to the reduction in pesticides (Oskam, 1998). Non-rotational set-aside may actually necessitate the application of more herbicides to control weeds. Another study has attributed only 20% of the reduction in pesticide use in the European Union to the CAP reform measures (Oppenheimer Wolff & Donnelly, 1998).

The set-aside scheme

Set-aside has been the other means, besides price cuts, of reducing production levels. By diverting land out of production it was also expected to have beneficial environmental impacts. The first set-aside scheme was introduced in 1988 as a voluntary scheme. It was taken up strongly in Germany and particularly in Italy. Presently, more than half of the old 'five year' set-aside can be found in Italy, mainly in connection with the less productive and less intensive agricultural systems of the central and southern part of the country (Bordin et al., 1998). There, the scheme has been criticized for assisting land abandonment and its associated problems of soil erosion. This scheme has now expired, and the area under five-year set-aside in the EU has steadily reduced over the past couple of years from 1.6 million ha in 1993/94 to 0.8 million ha in 1995/96 and down to less than 0.4 million ha (1996/97) (Table 6).

With the reform of the arable crops regime in 1992, compensation to farmers through direct area payments was subject to set-aside requirements, except for those who produced less than 92 tonnes of cereals equivalent. The percentage to be set aside is decided each year by the Council of Ministers. Initially, this obligatory set-aside had to be on a rotational basis and in 1993/94, over 4.6 million ha were part of the rotational set-aside scheme. In the following year, a non-rotational option was offered. The regulations were also revised to increase the scope for non-food production on set-aside land which has led to substantial plantings of, for example, industrial oilseeds.

The set-aside requirement has been progressively cut in response to changes in the world grain markets where agricultural production and stocks have declined throughout most of the 1990s and prices have risen sufficiently to take some of the pressure off the CAP. This is in marked contrast to earlier expectations that set-aside would have to be increased year-on-year to peg production back in the face of technological advances and slippage (i.e. the tendency for a given area reduction to achieve a less than proportionate decrease in output as farmers reallocate inputs, particularly that of land of differing quality).

Table 6   Areas under the different set-aside schemes for arable land, marketing year 1996/97 (x 1,000 ha)

Member State	Five-year set-aside a) (voluntary)	Annual set-aside		Total
		total b) (obligatory)	of which industrial set-aside c)	
Belgium	0	19	3	19
Denmark	2	220	26	222
Germany	78	1,217	243	1,294
Greece	0	14	0	14
Spain	26	1,338	16	1,364
France	48	1,405	250	1,454
Ireland	0	24	1	23
Italy	193	230	40	423
Luxembourg	0	0	0	0
Netherlands	3	10	0	13
Austria	0	114	8	114
Portugal	0	72	0	72
Finland	0	168	0	168
Sweden	0	309	4	309
United Kingdom	15	490	70	505
Total	365	5,630	661	5,993

a) Regulation (EEC) 2328/91; b) Regulation (EEC) 1765/92 (rotational or non-rotational, voluntary or special set-aside); c) Regulations (EEC) 1765/92 and 334/93.

The environmental effects of set-aside depend on its location and management (OECD, 1997c). Due to the exemption of small producers, obligatory set-aside is concentrated in those countries and regions with large-scale farm structures. These regions tend to have the most intensively used soils and are thus the ones most likely to benefit from taking land, at least temporarily, out of cereal production and from the associated reductions in inputs, particularly of fertilizers. Conversely, though, any reduction in inputs has been lower in those regions where small farms are dominant.

National regulations lay down appropriate management measures for set-aside and these differ between Member States, for instance with respect to restrictions on the application of agrochemicals, livestock manure and irrigation. Non-rotational set-aside is generally considered to have the more favourable consequences for the environment and conservation. It allows for field margins and buffer zones to be targetted, to encourage wildlife and to protect water courses. Certain types of guidelines or restrictions can significantly enhance the potential benefit: for example, requirements for farmers to maintain a 'green cover' to avoid bare ground and its consequent problems of leaching and erosion; or to encourage ground-nesting birds by not cutting this cover during certain periods. Specifying appropriate prescriptions depends crucially on an adequate appreciation of the heterogeneity of both the business structure and environmental circumstances of the farm population.

Obligatory set-aside can also have negative consequences. Production is intensified on the remaining land. Rotational set-aside may involve bringing into cultivation marginal areas on a farm, such as wet or scrubby areas or field edges or pasture land, that have in the past provided semi-natural habitats for wildlife. The production of industrial crops on set-aside often means that areas which would not normally have fertilizers or pesticides applied, receive significant amounts of chemicals.

The arable crops regime and nitrogen pollution from livestock waste

One of the objectives of the reform of the arable crops regime was to enhance the competitive position of EU agriculture. In particular, with the cut in cereal prices the use of cereals produced in the EU has become more attractive compared to imported feed concentrates. Eventually, this may have an impact on the location and intensity of pig production, resulting in a more balanced use of minerals in the sector and across regions of the EU. As yet, though, no shift in the concentration of pig production from the most important producing regions is evident (Brouwer and Van Berkum, 1998). On the contrary, the large producers have benefited from the general reduction in feed costs leading to yet further expansion of intensive rearing in areas such as Brittany and the Netherlands (Rainelli and Vermersch, 1998; Brouwer and Van Berkum, 1998).

What has occurred is a significant rebalancing of the share of cereals in compound feed. Previously, this had fallen in the EU from 42% in 1980 to 29% in 1989. Since 1992, the share of cereals has recovered significantly at the expense of soybean and corn gluten. In France, between 1993 and 1996 the share of cereals in compound feed increased from 32% to 42% (Rainelli and Vermersch, 1998). The consequence is a lower protein diet for livestock which means lower levels of nitrogen excretion. The average protein content of the diets of fattening pigs in Brittany, for example, fell by 1-1.5% during 1993 alone. This corresponds to a reduction in the nitrogen content of pig manure of 0.2-0.4 kg N per animal (Dourmad et al., 1995).

The reform of the arable crops regime has thus significantly reduced the cost of lowering the protein content of compound feed. As a result, it has begun to alleviate the pollution burden in livestock areas and facilitate nitrogen pollution control, through enabling the intensive livestock sector to apply nutritional management measures with reduced nitrogen output, without compromising the profitability of farm holdings. The results, though, are very sensitive to the relative price relationship between protein and energy sources. Soybean prices play a very significant role in the nitrogen content of compound feed. Any beneficial effects of the CAP price cuts on nitrogen pollution control may therefore be jeopardized by the volatile soya market (Brouwer et al., 1998).

Impacts of the reform of the beef regime

As part of the 1992 reform package, the intervention price for beef was reduced by 15% in three stages. The price reduction was compensated through direct headage payments for fattening bullocks and suckler cows, which are subject to a maximum stocking rate. The stocking density limit was progressively lowered between 1993 and 1996 down to 2 livestock units (LU) per hectare of forage crops. An additional extensification premium is provided if a producer reduces the stocking rate below 1.4 LU per forage hectare. These measures were intended to reduce the beef surplus, particularly by reducing beef production in dairy herds, and to safeguard farmers' incomes while containing budgetary costs. Extensification was primarily intended as a means of cutting production and placing limits, through the stocking rate rules, on levels of support payments, but it also promised environmental benefits.

The impact of the reform has varied between farming systems and regions reflecting the diverse nature and circumstances of beef production in the EU. The effects have been differentiated between intensive and extensive systems. These systems tend to have distinct environmental connotations. On the one hand, intensive beef production, especially when it takes place alongside other intensive forms of agriculture such as dairying or arable farming, may have a negative environmental impact including problems of livestock waste disposal, nitrogen leaching and intensive grassland management and arable fodder production. On the other hand, extensive grazing by beef cattle is essential to the maintenance of large tracts of countryside, particularly in upland and mountainous areas but also in other areas where the grazing is unimproved (for example, wet pastures, grazing marshes and the dehesas). Typically, these areas have significant landscape and ecological value. Beef production is a vital element in the viability of many extensive farming systems in less favoured areas and is therefore an important factor in counteracting the abandonment of marginal land. The traditional and distinct beef breeds also play an important role in the cultural landscape of different regions.

The reform measures have significantly reduced the profitability of intensive beef producers. Not only was the compensation inadequate but also their stocking densities were typically significantly higher than the eligibility ceilings for the compensatory premia. In the Netherlands, for example, only about 40% of the total number of fattening bullocks were eligible for compensation. On the loamy soils of Flanders a stocking density of three cows per hectare of fodder crops is the norm (Bouquiaux et al., 1998). For most intensive beef producers, it has not been a viable option to extensify and many have ceased beef production altogether. For those that remain, further intensification has been encouraged by lower feed prices and the premium for fodder cereals under the arable crops regime, leading to increased environmental pressures.

In the Ebro valley in Catalonia where intensive fattening units are concentrated, at stocking rates that clearly surpass the maximum of 2 LU/ha, producers have nonetheless benefitted from the beef premiums by renting pasture land in the nearby Pyrenees. The environment has suffered thereby not only from the continued intensification of production in the Ebro valley but also through the effective abandonment of mountain pasture zones (Varela-Ortega and Sumpsi, 1998).

Except for such abuse, the beef premiums have benefited the less intensive systems. Even here the extensification effects have not necessarily been marked and the environmental consequences are not clear cut, for a number of reasons. First, many farms are unaffected by the stocking density rules. Small producers with 15 LUs or less are exempt from the density norm. In Portugal this means that 90% of cattle holdings are exempt (Caldas, 1998). For many other farms the stocking rate is too high to bite. Cattle farms with more than 2 LU/ha are uncommon in Finland. In a survey of beef farms in the UK, just 20% anticipated a change in herd management to take account of the 2 LU/ha stocking rate when it was introduced (Winter and Gaskell, 1997). Most farmers, in fact, have not had to reduce stocking levels to qualify. The current limits of 2 and 1.4 LU per ha have also been criticized for not reflecting the wide variation in carrying capacity of grazing land and for being set too high for environmentally beneficial management.

Impact of the reform of the sheepmeat and goatmeat regime

The sheepmeat and goatmeat regime had been introduced in 1980 and was revised as part of the 1992 Mac Sharry reform. The regime had encouraged a rapid growth in breeding sheep numbers within Europe which had increased by about a quarter between 1983 and 1989. The growth was very uneven. The establishment of a common market in sheepmeat had favoured the producers of northern Europe where sheep production is usually not a specialized activity and which offered opportunities for expansion or diversification in the face of falling or restricted markets for other livestock products (beef and cows' milk). In Ireland, for example, breeding sheep numbers more than doubled between 1983 and 1989. In consequence, the more specialized sheep and goat producers of southern Europe faced increasing competition from northern European exports (mainly from the UK and Ireland) and lower prices which contributed to significant falls in the number of holdings, increased flock sizes and even greater specialization. In Italy and France, there was an overall decline in sheep numbers (Ashworth and Caraveli, 1988).

The intention of the 1992 reform was to control the levels of production in order to contain the costs of support and to create a balanced market. Finite limits were set for the number of animals to be supported (with the Sheep Annual Premium) with quotas for individual producers based on the number of ewes for which they had claimed in a given reference year (chosen form 1989, 1990 or 1991). Allowance was made for quota to be transferred between producers in certain circumstances but with restrictions on interregional transfers to protect regions dependent upon sheep production.

Sheep and goat farming are associated with some of the most environmentally sensitive and peripheral areas of Europe. In 1992, 75% of all ewes on which the Sheep Annual Premium was claimed were located in the Less Favoured Areas (CEC, 1995). the sheepmeat and goatmeat regime has undoubtedly helped retain and sustain producers in these areas. The typically extensive grazing is crucial to the management of unimproved pastures, moorland and fragile land in hilly and mountainous areas. The grazing of the stock on semi-natural vegetation helps prevent scrub encrouachment and maintains floristic diversity (Bignal and McCracken, 1996). The variety of traditional breeds and production systems contributes to the cultural heritage of different regions.

The restructuring of production encouraged by the pre-1992 regime, however, also caused some environmental damage. The expansion of sheep numbers in the UK and Ireland was associated in many areas with grassland improvement or overgrazing leading to a loss of plant species diveristy and the fragmentation of habitats. In the UK, for example, much upland heather - which is an important and distinctive habitat - has been lost or degraded as a result of heavy grazing by sheep (Lowe et al., 1998). In Southern Europe, the damaging effects were more complex and varied. The declining number of producers often meant the abandonment of more marginal areas and of traditional practices such as transhumance. At the same time, the enlargement of flocks led to the overexploitation of other areas exacerbating such problems as soil erosion. In Greece, for example, the abandonment of alpine grasslands has gone hand-in-hand with the overstocking and overgrazing of the rangelands at middle altitudes (Caraveli, 1998).

Since the 1992 reform, the expansion in sheep numbers has stopped. In some Member States, including France and Italy, breeding sheep numbers have stabilised, while in the UK, Ireland and Spain they have declined. However, the environmental consequences are mixed and uncertain. Reduced grazing pressures are not necessarily leading to the revitalization of degraded moor and grassland. Some of the changes which had previously occurred - such as pasture improvement, land abandonment or soil erosion - are not readily reversible. It is also not clear that the reform has tackled the twin underlying problems of land abandonment of some areas alongside intensification of grazing in other areas.

Council Regulation 233/94 introduced the option for Member States to limit or withhold payments for producers if they do not comply with environmental conditions fixed by the Member States. The UK is the only country so far to have taken up the option. What it did was to extend the overgrazing criteria used in its Hill Livestock Compensatory Allowance Scheme to the annual ewe premium. The enforcement process has proved cumbersome and only a small number of farmers have had payments withheld for overgrazing. Nevertheless, it is considered to have discouraged farmers from expanding their herds (Ashworth and Caraveli, 1998).

1.4   Implications for research

Identification of research gaps

European agri-environmental research is flourishing. The research effort is still in its infancy and inevitably it is not yet systematic in its coverage. Individual studies have been conducted in response to specific issues or particular circumstances. Therefore, we need to be modest on what we know of linkages between the CAP and the environment. Compiling an overview reveals gaps as well as the potential for more synoptic approaches.

The partiality of the existing research effort can be characterized in the following terms:

First, there is a northern bias in the research coverage, reflecting the strength of northern European concerns. This bias emerges in
- the geographical coverage of studies (with many studies for countries such as Germany, the UK and the Netherlands, but much fewer for southern Europe);
- the sectors and systems studied (temperate rather than Mediterranean crops, intensive rather than extensive systems);
- the problems and issues addressed (with pollution particularly of water by pesticides and nitrates well covered but not air pollution, with landscape deterioration better covered than biodiversity and with little attention to such agri-environmental problems as soil erosion, desertification, flooding, fire hazards or salinization).

Second, there is a strong interest in agri-environmental measures compared to the effects of other elements of the CAP. This bias towards what is novel in policy development is understandable. The point should not be missed, however, that the Agri-environment Regulation was introduced as an accompanying measure to the CAP reform. It is a minor component of the CAP. Not only do the main commodity regimes have much greater consequences for the environment, but there is also evidence that commodity supports actively discourage take up of agri-environment measures. Thus beneficial effects of the Agri-environment Regulation may be swamped by the environmental impact of the rest of the CAP. There is a risk, though, that the novelty and the intricacies of implementing the Regulation will distract attention away from the bigger picture.

Third, little if any work is being done on the environmental effects of:
- certain commodity regimes, e.g. tobacco and sugar;
- the other accompanying measures (the early retirement and afforestation schemes);
- the horizontal socio-structural measures (e.g. LFAs);
- regional and rural policy;
- other measures such as incentives for alternative crops, quality and label policy, biomass production, farm diversification.

Fourth, there are biases also in the style of research, with a tendency for single country studies, for an orientation to specific policy measures and for single disciplinary approaches. These lead to narrow or partial analysis, including:
- a lack of comparative studies (except, that is, for agri-environmental policy). Such studies are needed, for example, to understand variations in the implementation of common policies;
- a lack of integrated studies, focussing on a region or a farming system or a rural community and seeing how the complete set of policy changes interact upon it. Not only does this deficiency detract from a holistic assessment of the CAP but it is also linked to a poor understanding of the agricultural sector in its wider socio-economic context. For example, to what extent do the pressures of the CAP reflect broader social and technological changes and to what extent are these directly impinging on farmers and rural communities?

- a lack of linkage between agricultural economic analysis and farming systems/agro-ecology analysis. There is thus a disjuncture in understanding between the effects of policy on farm decision making and the effects of farmers' actions on the environment.

Uniform responses or single environmental outcomes from a given change in policy are not foreseen. We therefore increasingly consider the importance of higher-order effects on linkages between the CAP and the environment, as well as the site-specificity of environmental conditions, the great variability of production systems across Europe (e.g. farm types, breeds, production systems, farm management), and the regionalization of issues involved.

Future research needs

* Need for good comparative studies at the sub-national level to reveal the interaction between EU rules, national policies, different farming systems, and different environmental circumstances.
* Need to identify optimal policy strategies (multiple roles for agriculture, multiple policy objectives, joint products) for achieving environmental policy targets. This would also require investigations on the role of various actors (farmers, consumers, industry) in the achievement of sustainable production methods and the search for partnerships in achieving policy targets (e.g. management agreements). The identification of cost-effective measures would include assessments of the costs involved of achieving certain environmental objectives (to farmers, FEOGA budget and transaction costs).
* Need for dynamic modelling studies of strategic options or alternative development trajectories for farms/farming systems, including possible shifts in production/land use due to economic change.

A possible analytical framework for future studies

Comparative, holistic and dynamic studies are required at various levels, from the global down to the farm level. Increasingly, the thrust of policy is towards integrated territorial measures. The need to accommodate variability in social and environmental circumstances calls for subsidiarity to be respected in the implementation of policy. There is therefore a clear need also to understand the articulation between different levels, i.e. the ways in which actions and decisions taken at one level constrain or inform those taken at another level.

A framework to guide such studies and analysis is set out in Figure 1. It links local conditions with national, EU and global rules, including farm decision making processes with national and EU-wide policy measures and global trading rules.

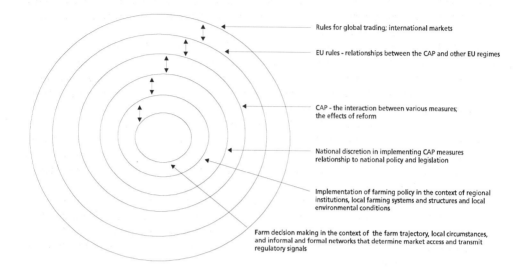

Figure 1   A possible analytical framework for future studies

References

Allanson, P.A. and A. Moxey (1996) *Agricultural land use change in England and Wales, 1892-1992;* Journal of Environmental Planning and Management, Vol. 39, pp. 243-254

Ashworth, S. and H. Caraveli (1998) *The EU sheepmeat and goatmeat regime and the natural environment;* Paper presented at the Workshop on CAP and the Environment in the EU; Wageningen, The Netherlands; February 5-8th

Baldock, D. and G. Beaufoy (1993) *Nature conservation and new directions in the EC Common Agricultural Policy: The potential role of EC policies in maintaining farming and management systems of high nature value in the Community;* London, Institute for European Environmental Policy

Baldock, D.M. and P.D. Lowe (1996) *The development of European agri-environmental policy;* In: M. Whitby (Ed.), The European experience of policies for the agricultural environment. Wallingford, Oxon, CAB International, pp. 8-25

Beaufoy, G. (1998) *Environmental considerations for the reform of the CAP olive-oil regime;* Paper presented at the Workshop on CAP and environment in the EU; Wageningen, the Netherlands; February 5-8, 1998

Bignal, E.M. and D.I. McCracken (1996) *Low-intensity farming systems in the conservation of the countryside;* Journal of Applied Ecology, Vol. 33, pp. 413-424

Boisson, J.-M. and H. Buller (1996) *Agri-environmental schemes in France;* In M. Whitby (Ed.), The European environment and CAP reform: Policies and prospects for conservation; Wallingford, Oxon, CAB International; pp. 105-130

Bordin, A., L. Cesaro, P. Gatto and M. Merlo (1998) *Italy;* Chapter 10 of this book

Bouquiaux, J.-M., M. Foguenne and L. Lauwers (1998) *Belgium;* Chapter 6 of this book

Brouwer, F.M. and S. van Berkum (1996) *CAP and environment in the European Union: Analysis of the effects of the CAP on the environment and assessment of existing environmental conditions in policy;* Wageningen, Wageningen Pers

Brouwer, F. and S. van Berkum (1998) *The Netherlands;* Chapter 7 of this book

Brouwer, F. and P. Hellegers (1997) *Nitrogen flows at farm level across European Union agriculture;* In: E. Romstad, J. Simonsen and A. Vatn (Eds.), Controlling mineral emissions in European agriculture: Economics, policies and the environment; Wallingford, CAB International, pp. 11-27

Brouwer, F., P. Hellegers, M. Hoogeveen and H. Luesink (1998) *Managing nitrogen pollution from intensive livestock production in the EU: Economic and environmental benefits of reducing nitrogen pollution by nutritional management in relation to the changing CAP regime and the Nitrates Directive;* The Hague, Agricultural Economics Research Institute (LEI-DLO) (forthcoming)

Caldas, J.C. (1998) *Portugal;* Chapter 12 of this book

Caraveli, H. (1998) *Greece;* Chapter 11 of this book

CEC (1995) Special report No. 3/95 on the implementation of the intervention measures provided for by the organization of the market in the sheepmeat and goatmeat sector; *Official Journal of the European Communities;* C285 Vol 38, 28 October 1995, p. 1

Dourmad, J.Y., C. Le Mouel and P. Rainelli (1995) *Réduction des rejets azotés des porcs par la voie alimentaire: évaluation économique et influence des changements de la politique agricole commune;* INRA Productions Animales, 8(2); pp. 135-144

ECPA (1996) *European Crop Protection: Trends in volumes sold, 1985-95;* Brussels, European Crop Protection Association

EFMA (1997) *Forecast of fertilizer consumption in EFMA countries to 2006/7;* Brussels, European Fertilizer Manufacturers Association

Hanley, N. (1997) *Externalities and the control of nitrate pollution;* In: Brouwer, F. and W. Kleinhanss (Eds.); The implementation of nitrate policies in Europe: Processes of change in environmental policy and agriculture. Kiel, Wissenschaftsverlag Vauk, pp. 11-22

Kleinhanss, W. (1998a) *Germany;* Chapter 2 of this book

Kleinhanss, W. (1998b) *Austria;* Chapter 13 of this book

Lowe, P., J. Clark, S. Seymour and N. Ward (1997) *Moralizing the environment: Countryside change, farming and pollution;* London, UCL Press.

Lowe, P., L. Hubbard, A. Moxey, N. Ward, M. Whitby and M. Winter (1998) *United Kingdom;* Chapter 5 of this book

Miettinen, A. (1998) *Finland;* Chapter 14 of this book

OECD (1997a) *Environmental benefits from agriculture: Issues and policies;* Paris, Organisation for Economic Co-operation and Development

2. GERMANY
Werner Kleinhanss

2.1   Introduction

Environmental protection is a major issue of policy debate in Germany. Agriculture is regarded as a main polluter of soil and water and partly responsible for the loss of biodiversity and land-scape values. It is argued that environmental damages in the past were mainly induced by intensi-fication and specialization of agricultural production, structural changes towards large-sized farms and land consolidation programmes (SRU, 1985). On the other hand, it is also well accepted that there are environmental benefits from agriculture and that these could be improved by extensifi-cation and a stronger orientation of agricultural policy towards environmental objectives.

The present chapter starts with a review of the level of pollution and of the links between agricultural development and environmental problems in the past. Emphasis is given to the envi-ronmental effects of market and price policy of the CAP reform and the accompanying measures. First evaluations of the environmental impacts of the Nitrate Directive and the German Fertilizer Decree will also be included.

2.2   Interactions between agriculture and environment

Nitrogen emissions and water pollution problems are a major issue of environmental policy. Some tentative assessments can also be made regarding the impact of greenhouse gases on global warming and the impacts of agriculture on biodiversity and landscape.

2.2.1   Pollution

Nitrogen

Risks resulting from nitrogen emissions include contamination of drinking water by nitrate, eutro-phication of surface water, changes in the species spectrum, the emergence of acidification and a contribution to global warming (SRU 1985, Wissenschaftlicher Beirat 1992 and 1993). Differ-ent levels of nitrogen surpluses have been reported for the German agricultural sector, based on different methods of calculation and base years. So-called farm gate balances applied at sector level come out with higher surpluses (144 kg N/ha in 1990/91 and 146 kg N/ha in 1991/92; Isermann 1995) than assessments based on the soil-surface balance method (Bach 1987, Wendland et al., 1993). Based on 1990/91 data, nitrogen surpluses of 113 kg/ha on average were estimated (Schleef and Kleinhanss, 1994). More recent estimations for the year 1995 show nitro-gen surpluses of 90 kg/ha on average (Niedersächsisches Umweltministerium, 1996). This signifi-cant change can be explained by (a) a decrease of mineral fertilizer use between 1988/89 and 1994/95 from 129.6 kg/ ha in the former FRG and 110.3 kg/ha in the former GDR to 103.2 kg/ha for the whole country; and (b) the drastic reduction of livestock in the 'New Laender' by about

70% for pigs and by about 50% for beef during the transformation of the agricultural sector after unification (Figure 1).

The supply of nitrogen from animal manure amounts to 87 kg/ha in 1993. Regional variations in nitrogen surpluses are mainly related to the concentration of livestock production varying from 0.6 LU/ha in the region of Braunschweig to 2.9 LU/ha in the regions of Münster and Weser-Ems where the highest surplus levels of more than 150 kg N/ha (Schleef and Kleinhanss, 1994). Maximum levels of nitrogen surpluses in the range of 300 kg/ha might be reached at the level of districts. Therefore environmental problems in Germany are often site specific.

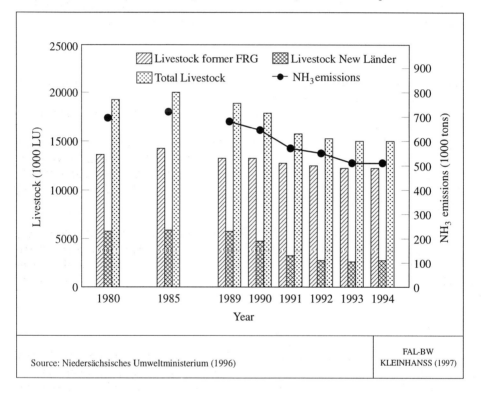

Source: Niedersächsisches Umweltministerium (1996)

FAL-BW
KLEINHANSS (1997)

Figure 1   Development of livestock and ammonia emissions from livestock production in Germany

Referring to Isermann (1994) about 10% of nitrogen surpluses are immobilized in soils, 38% are lost to the hydrosphere, about 25% by ammonia emissions and 27% by denitrification. The problem of water pollution from nitrogen is underlined by the fact, that:
1)   during the past 10 years, nitrate concentration in groundwater increased by 1 to 1.5 mg/l per year (Focken, 1988);

2) the limit of 50 mg NO$_3$/l for drinking water is exceeded by about 10% of all water sources (UBA, 1994) and several highly contamined groundwater pumping areas have had to be closed (Zeddies, 1995).

Estimates from Werner and Wodsack (1994) and Isermann (1994) show that nitrogen leaching from agriculture induces concentrations of 9.7 mg NO$_3$/l of groundwater on average. Agriculture also contributes about 44% of nitrogen pollution of surface waters (Isermann, 1995). Eutrophication problems of coastal waters are not yet solved and the objectives of the North Sea Conference held at The Hague to reduce the mineral load of surface water entering the sea by 50% will not be reached until the target year 2000 (Niedersächsisches Umweltministerium, 1996).

Phosphorus and potassium

Pollution of phosphorus and potassium from agriculture is less important with regard to environmental problems due to the following facts:
1) the contribution of agriculture to the phosphorus load of surface water was only 28% in 1987/89 (Isermann 1995);
2) phosphorus and potassium surpluses are relatively low; based on 1990/91 data, phosphorus surpluses were 42 kg/ha and those for potassium 76 kg/ha (Schleef and Kleinhanss, 1994);
3) the use of phosphorus from mineral fertilizer decreased rapidly from 66 kg/ha in 1989/90 to 34 kg/ha in 1993/94 and 23 kg/ha in 1995/96 (BML, 1996, 1997, UBA 1994). In 1995/96, 37.7 kg of potassium per hectare were used on average.

Special measures to reduce phosphorus and potassium pollution will be taken within the German Fertilizer Decree (1996). For highly saturated soils, the application of animal manure will only be allowed up to the level of crop uptake. First assessments by Schleef (1996) show that the use of animal manure will be much more restricted by this measure than by the limits on organic nitrogen.

Pesticides

In 1993/94, 22,246 tonnes of pesticides (active ingredients) were used in Germany, of which 80% was in agriculture (Höll and Von Meyer, 1996). Herbicides account for 56% of pesticides, with an average use of 0.9 kg/ha; fungicides 27%, with an average use of 0.5 kg/ha; and insecticides 4%, with an average use of 0.02 kg/ha. The intensity of pesticide use varies between regions; it is relatively high in regions with a high share of arable crops, horticultural crops and vineyards. Approximately 10% of groundwater pumping areas show quantities of pesticide residues exceeding the EU limit of 0.0005 mg per litre (Höll and Von Meyer, 1996). The most important sources of pollution are triazines (80% of contaminated samples; Steiner et al., 1996). Due to the ban on atrazine since the mid-1980s this problem might diminish in the near future.

2.2.2 Emissions of greenhouse gases

The problems of global warming are the subject of ongoing debate which was initiated by the UNCED conference in 1992. On the one hand, agriculture is declared as polluter, on the other hand its possible contribution to reduce emissions of greenhouse gases is mentioned (e.g. the transformation of conventional farming into organic, or the substitution of fossil energy by biomass). First assessments of greenhouse gases were made by the Enquete Kommission (1994) and strategies to reduce global warming were identified. Emissions of greenhouse gases are as follows:

1) Carbon dioxide (CO_2): 38.4 million tonnes in 1990/91 with 2.24 tonnes/ha of UAA emitted on average. The contribution of German agriculture to global CO_2 emissions is 3.9%. Emissions of CO_2 are mainly induced by the use of fossil fuels;

2) Ammonia (NH_3): 0.655 million tonnes in 1990, of which about 90% originated from animal production. Due to the drastic reduction of livestock in the New Laender, ammonia emissions were reduced to 0.525 million tonnes in 1994 (UBA, 1994);

3) Di-nitrogen oxide (N_2O): Of the total of 225,000 tonnes in Western Germany, about one third originates from agriculture. Average N_2O emissions are in the range of 2.5 to 3 kg/ha. Although this figure is very low the impacts of N_2O on global warming are important due to its conversion factor of 270 CO_2 equivalents. A controversial debate on the validity of the estimates and on strategies to reduce N_2O emissions is still going on.

Data on emissions of other greenhouse gases (NO_x and NMHC) for the German agricultural sector are not available.

The total emissions from agriculture contribute to about 15% of the global warming effect for which Germany is responsible.

2.2.3 Endangered species

Through intensification and land improvement, the removal of hedgerows and the destruction of landscape elements, agriculture contributes to losses of biodiversity. According to the Red List about 700 ferns and flowering plants and 30-50% of all animal species are threatened with extinction (Korneck and Sukopp, 1988). In the case of endangered plant species agriculture is identified as the main cause (70%) of their decline (Höll and Von Meyer, 1996). Typical plant species, especially in less favoured areas, are affected. The use of herbicides is the main cause of the reduction of plant species and particularly microbes.

According to the SRU (1985) these problems should be reduced by a general extensification of agriculture, by a set-aside of at least 5% of land for environmental purposes and a linked system of natural protected areas.

2.3 Impacts of the CAP and environmental policies

Assessments of the environmental impacts of agriculture have been made by the German Council of Environmental Policy (SRU, 1985), who ranked the salient problems as follows: loss of biodiversity; groundwater contamination by nitrates, phosphates and pesticides; soil compaction, wind and water erosion; surface water pollution mainly by nitrates and phosphates; residues in the food chain and emissions to the atmosphere.

The main causes of environmental problems have been identified as technical progress and agricultural policy favouring intensification and specialization (Zeddies, 1995). Landscape elements have been destroyed and low productive land transformed into a more productive one by land improvement and irrigation. The concentration of landless livestock production in the North-West is mainly an effect of GATT regulations allowing cheap feed imports (proteins and substitutes). Environmental and pollution problems exist in the following regions (SRU, 1985):

- intensively used arable areas in the northeastern part of Lower-Saxony, the northern part of North-Rhine-Westphalia, the east of Schleswig Holstein and parts of Bavaria and Baden-Wuerttemberg. Most of these regions are characterized by specialized cropping farms, crop rotations dominated by cereals, sugar beets and potatoes and a high level of mineral and pesticide use;
- areas with a high share of horticultural crops, fruits and vineyards and a low livestock density (Rhine valley, basins of the Havel and the Elbe);
- areas with a high livestock density (especially pigs and poultry), mainly located in the North of Germany (Münsterland and Weser-Ems) and characterized by high mineral surpluses, nitrate contamination of waters and air pollution.

For a significant proportion of farms, pollution is at unacceptable levels from an environmental viewpoint (Zeddies, 1995).

2.3.1   Market and price support measures

The use of mineral fertilizers and pesticides has gradually reduced since the late 1980s and the beginning of the 90s (see Section 2), due to decreasing producer prices, better advice and application techniques in fertilizer use, the set-aside of arable land and the implementation of extensification programmes (BML, 1997). During the reorganization of co-operative and state farms into private agricultural enterprises in the New Laender, fertilizer and pesticide use were reduced further due to liquidity shortages. Detailed evaluations of environmental impacts of the CAP reform for the whole sector are not available. The majority of analyses made on this subject focus on specific regions, farm types or environmental issues.

The arable regime

The main impacts of the relevant measures are summarized as follows:
- price reductions for cereals, oilseeds and protein crops induce lower fertilizer and pesticides use;
- compensation payments for cereals, oilseeds and protein crops based on regional average yields bring about lower regional concentrations of agricultural production in the medium-term and thus will break the tendency towards intensification on the best soils.

Calculations from Fuchs and Trunk (1995) show that for wheat the use of nitrogen fertilizer will be reduced by 13% and of pesticides by 10%, and those for maize by 30% and 20% respectively. CO_2 emissions will be reduced by one third. For rapeseed the predicted reductions were 65% for nitrogen fertilizer, 100% for pesticides and 50% for CO_2. Although these changes might be over-estimated, it is the case that extremely extensive systems of rapeseed production were introduced in some sandy soil regions, in response to the high area payments. In traditional regions of rape-seed production with good soils the intensity has been reduced by a simular magnitude as for ce-reals.

Surveys from Zeddies et al. (1994) on a sample of farms in Baden-Wuerttemberg, Rhine-land-Palatinate and North Rhine-Westphalia show that, with the start of the CAP reform, half of farms reduced the intensity of cereal production. The farmers' intentions to cut back on inputs with full implementation of the reforms are given in Table 1. The results from Kögl (1993) on extensification strategies show the same directions of adaptation.

Table 1   Proportion of farmers (%) reducing intensity for cereals and rapeseed in response to the CAP reform

	Cereals	Rapeseed
- lower use of nitrogen fertilizer	38	41
- reduction of fungicides	39	30
- reduction of herbicides	32	35
- reduction of insecticides	not applied	21

Environmental Impacts of set-aside depend on its share of UAA, location and if it is rotational or not. Under the national regulations it is not permitted to spread animal manure except on set-aside areas being used for non-food production. Therefore mineral fertilizers might be reduced in the same proportion as the share of set-aside. Reduction of pesticides might be lower because in the succeeding year more herbicides have to be applied to control weeds. The latter problem is one of the reasons that especially large farms prefer non-food production instead of fallow. With re-gard to biodiversity only non-rotational set-aside for at least 5 years might have positive effects. Due to the small producer scheme, obligatory set-aside is mainly observed in regions with large scale farm structures, primarily the arable regions in the North and East of Germany. In areas where small farms are dominant as in hilly areas and the Centre and the South of Germany, the share of set-aside and therefore the reduction of inputs is much lower. Due to equal premiums for cereals, oilseeds and protein crops, oilseed production is not more competitive in farms applying the small-producer scheme. Since compensatory payments are based on the cereal regime, they are much lower than within the oilseed regime. Therefore crop rotations will become less diversi-fied. Obligatory set-aside will also induce scarcity of the available land resource, which will con-tribute to an increase in land prices (Chatzis, 1996) and thereby to an increase of production inten-sity (Bauer et al., 1996).

In 1996 1.3 million ha of land were set aside (Table 2). Model calculations show that, due to the CAP reform (Becker and Kleinhanss, 1997), fertilizer use will be reduced by about 13%.

In contrast, restrictions on manure spreading will only reduce nitrogen surpluses by 2.1% on average (Becker and Kleinhanss, 1997).

Table 2   Set-aside in Germany in 1996

	Set-aside within the arable crops regime (ha)		Former set-aside within extensification programmes (ha)	Total (ha)	Set-aside on UAA (%)
	obligatory	voluntary			
Former FRG	418,878	148,341	56,829	624,048	5.4
New Laender	370,814	274,113	23,945	668,872	12.0
Germany	789,692	422,454	80,774	1,292,920	7.5

Source: BML, Agrarbericht der Bundesregierung (1997).

The beef and dairy regimes

Due to the CAP reform the following tendencies can be considered in the beef sector:
- no changes in intensive beef production systems, which are mainly based on maize silage;
- reduction of livestock on farms with low livestock density, to get additional premiums for extensive livestock systems (<1.4 LU/ha);
- increase of suckler cows especially in the New Laender induced by beef premiums and additional premiums for extensive livestock holding systems;
- reduction of heifer fattening which is excluded from compensation payments.

Table 3   Environmental impacts of grassland extensification a)

		Intensity of grassland		Rel. change (%)
		high	medium	
Purchase of fertilizer	kg/ha	85	42	-50
Grassland yield	qn/ha	94	83	-12
Livestock density	LU/ha	2.7	2.6	-4
Ammonia emission	g/kg of milk	5.9	5.4	0.8
Nitrogen losses	kg/farm	4,193	2,815	-23
Nitrogen surplus	kg/farm	1,909	718	-62
CH_4 emissions	g/kg of milk	21.7	20.3	-7
CO_2 emissions (100 years)	kg/kg of milk	0.3	0.62	-15

a) Assumption: constant number of 50 cattle and constant milk yields.
Source: adapted from Fuchs and Trunk (1995).

The extensification of beef production might induce positive environmental effects.

A different situation is found in the dairy sector which was not included in the Mac Sharry reform. Dairy production became more favourable due to lower feed prices and premiums for fodder maize under the cereals regime. Green maize production is thus favoured compared to other roughage crops. This will induce substitution of other roughage crops by maize, and therefore an intensification of arable fodder production. In some regions, problems of soil erosion will increase. On the other hand an extensification of grassland will be induced, which might be one reason that measures of grassland extensification within Regulation 2078/92 are well accepted.

Model calculations from Fuchs and Trunk (1995) show that the reduction of grassland intensity from a high to a medium level might contribute to a lowering of nitrogen losses, nitrogen surpluses and emissions of CO_2 (Table 3).

The increase in milk yields induced by lower fodder costs is also assessed by Fuchs and Trunk (1995). The projected reduction of pollution is considered to be lower than in the case of an extensification strategy. The authors conclude that the environmental side effects of the CAP reform can be positively judged for grassland farming and milk production.

2.3.2 Agri-environment policies

The official position concerning the environmental impacts of the CAP is that market and price policy measures give positive signals to encourage extensification. Specific programmes in the framework of Regulation 2078/92, however, are required to reorientate agriculture towards environmental concerns (BML, 1997). The programme named 'Förderung einer markt- und standortangepaßten Landbewirtschaftung' will be described below, including its economic and environmental impacts. First the impacts of the former extensification programmes are briefly reviewed.

The former extensification programmes

Structural and environmental policy are mainly the responsibility of the Laender. In the so-called Gemeinschaftsaufgabe the Federal State is involved in Bund-Laender programmes. In the early 1980s some Laender started with local programmes of landscape protection and nature conservation (Höll and Von Meyer, 1996). They have been expanded to other Laender in accordance with the EC Regulation 795/85, most of them being co-financed by the EU under Regulation 1780/87.

Extensification programmes were established in accordance with EC Regulation 2328/91. The so-called quantitative method was only applied in the beef sector where requirements are to achieve a reduction of beef fattening of at least 20%. During the period from 1989/90 to 1993/94, 82,000 LU were included in the programme (Zeddies, 1996). In the so-called technical method (extensive farming practise) another 25,000 LU were included, 33,000 ha of cereals without fertilizer and pesticides application and 391,000 ha for organic farming. In total 2.3% of UAA was included and 3.4% of farms took part in the programme (Frohberg, 1995).

The environmental impacts of the extensification programmes might be rather low because of limited take-up and the concentrated application of the programmes in the less polluted regions. It is a general experience that extensification measures are more profitable in regions with less favourable natural conditions than in the most productive regions (Frohberg, 1995). The market effects with regard to the reduction of surpluses were therefore less than expected (UBA, 1994).

In the beginning of the 90s some Laender established programmes aiming at a reduction of agricultural surpluses and the protection or improvement of cultural landscapes (MEKA (Baden-Wuerttemberg), KULAP (Bavaria), HEKUL (Hessen)). In 1993 they were integrated in the accompanying measures of the CAP reform (2078/92).

Accompanying measures (Regulation 2078/92)

The implementation of 2078/92 is organized both by the Federal State and the Laender. Programmes from the Laender include regional and site-specific measures. They are co-financed from the EU by 50%, and in the New Laender, due to their status as an Objective 1 region, by 75% (Mehl and Plankl, 1995; Plankl, 1996a; Plankl, 1996b). At the Federal State level support measures are based on common principles defined in the so-called Gemeinschaftsaufgabe zur Verbesserung der Agrarstruktur und des Kuestenschutzes (GAK). The programme is called Förderung einer markt- und standortangepaßten Landbewirtschaftung (msaL). It includes supporting measures for:
-    extensification of arable farming and permanent crops either without mineral fertilizer and pesticides or only without herbicides;
-    extensive grassland use including the transformation of arable land into extensive grassland;
-    organic farming.

All these measures aim at an extensification far beyond the standards of good agricultural practise. The programme is financed by 50% from the EU, 30% from the Federal State and 20% from the Laender; for the New Laender the shares of co-financing are 75/15/10% respectively.

The subsidy levels for the different measures are shown in Figure 2. Within the measures for arable and permanent crops as well as for organic farming, distinctions are made between introducing and maintaining extensive practices, with the latter attracting lower levels of subsidy. Extensification of grassland is only subsidized during the introduction of the measure. The Laender are free to alter the subsidy level upwards by up to 20% or downwards by as much as 40%, but in the case of changing cropland into extensive pasture by +/- 40%. The subsidy level is significantly lower than the maximum allowed premiums defined in Regulation 2078/92.

The participation of the Laender in the msaL varies considerably. Some of the Laender follow the common principles of extensification measures for arable and permanent crops, others apply the whole package and others do not participate at all. In the latter case there is only co-financing from the EU (and not from the Federal Government), but support levels are less restrictive and the Laender are free to pay so-called base premiums for maintaining cultural landscapes. They are paid within the Laender programmes of Bavaria (KULAP) and Saxonia. The base premiums are rather low; in Bavaria they are 40 DM/ha and in Saxonia it varies between 75 and 80 DM/ha.

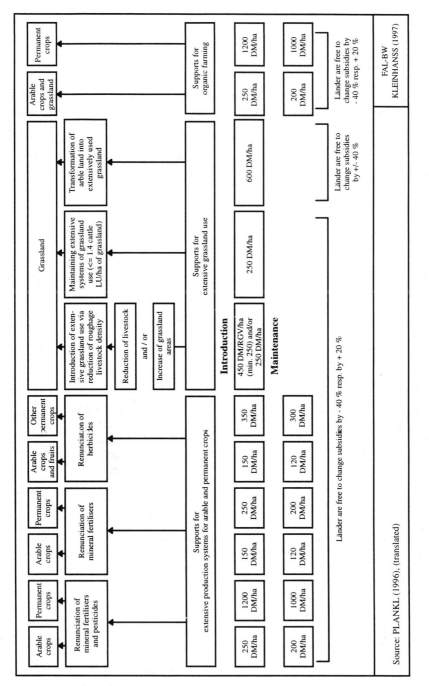

Figure 2 Support levels for measures included in the programmes Förderung einer markt- und standortangepaßten Landbewirtschatung

The acceptance

Data on participation rates in Germany are only available for the years 1993/94 and 1994/95 (BML, 1996 and 1997). In total 330,000 contracts were signed in 1994, which in total represent 4.96 million ha (29% of UAA) (Table 4). Almost 2.9 million ha were part of the support measure of maintaining cultural landscape. Another 1.21 million ha were included in grassland extensification, about 0.66 million ha in extensification of arable crops, 0.05 million ha in extensification of permanent crops and 113,000 ha in 'organic farming'. Under site-specific Laender programmes, compensatory payments for 'traditional land use schemes' were provided on 23,000 ha and for areas of high ecological value on 50,600 ha.

Table 4   Agri-environmental programmes in Germany a)

No.	Group of measures	Number of applications	Area or animals ha (or LUs)
1.	Permanent grassland and pasture	148,521	1,209,987
2.	Renucation of cattle livestock and sheeps (LU)	147	(3,604)
3.	Renunciation of agrochemicals on		
	- arable land	104,087	660,534
	- permanent crops and vineyards	50,294	50,219
4.	Environmental friendly production system based on EC Regulation 2092/91 (organic farming)	5,588	112,864
5.	Nature protected areas	10,480	14,051
6.	Endangered domestic animal species	1,727	(8,399)
7.	Endangered crops b)	-	-
8.	Long-term set aside (>= 20 years)	255	546
9.	Maintenance of abondonned areas	366	2,479
10.	Demonstration projects	20	-
11.	Environmentally oriented training projects	300	-
12.	Traditional land use systems	2,894	26,922
13.	Base supports for 'maintaining cultural landscape'	-	2,879,249
	Total c)	332,959	4,956,249

a) Based on payments for the year 1995; b) Measures are not yet included in the programme; c) Only areas included in the programmes.
Source: adapted from BML (1997).

Regional participation rates are relatively high in Laender with schemes for paying base premiums for maintaining cultural landscape e.g. Bavaria 86%, Baden-Wuerttemberg 52% and Saxonia 60% (BML 1996 - see Figures 3 and 4).

Extensive grassland measures are mainly applied in mountainous regions with a high share of grassland.

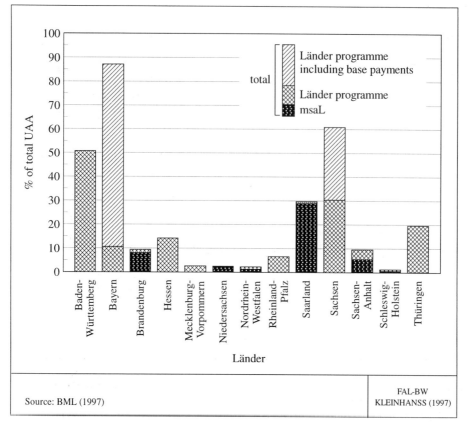

Source: BML (1997)

FAL-BW
KLEINHANSS (1997)

Figure 3   Regional application of 2078/92 measures

The high share of grassland extensification in the New Laender can be explained by the drastic reduction of livestock numbers, the replacement of intensive beef fattening by suckler cows, and the partially low grassland yields. These changes had already largely occurred and, therefore, participation in extensification programmes could happen without major changes in farming practice. The costs of adaptation might have been very low due to the low intensities that already prevailed.

The measures for extensive arable farming are applied to 4% of arable land; they are mainly applied in Baden-Wuerttemberg (27%), Saxonia (16.6%) and Bavaria (4.7%) (BML, 1995) where Laender programmes define comparatively low requirements.

Support for organic farming is paid on 112,864 ha (BML, 1997). In addition, about 350,000 ha are supported by the ongoing extensification programme under Regulation 2328/91. Projections from Jungehülsing (1997) show progressive replacement of that support so that by 1998 all measures will be based on Regulation 2078/92. The share of supported areas might be reduced by 10% because some Laender do not support continuing organic farming within 2078/92 if the former extensification programme has been applied. The share of organic farming on UAA is 3.4% in the Centre and South of Germany, 1.2% in the North and 3.2% in the New Laender.

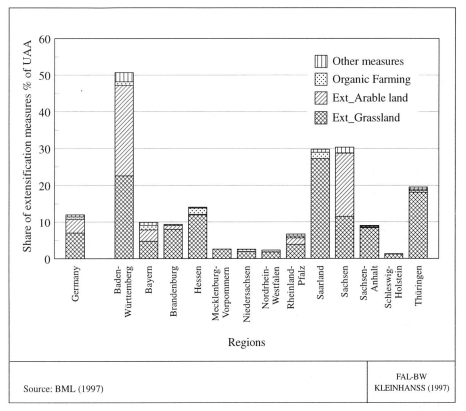

Source: BML (1997)

FAL-BW
KLEINHANSS (1997)

Figure 4   Regional application of agri-environmental measures (excluding areas for which base subsidies are paid)

It is worth mentioning that only 5.8% of the total area included in agri-environmental pro-grammes under 2078/92 were supported within the msaL with financial contribution from the Federal State. One of the reasons are budget constraints within the GAK. The budget is fixed at the Laender level and the budget can be used either for agri-environmental programmes or invest-ment supports. In some of the Laender the whole budget is totally used for investment support; therefore agri-environmental measures are not included in the GAK (see Figure 3). Due to the severe budget constraints on the Laender, the Laender programmes might be redefined in the fu-ture in order to bring them more in line with the msaL.

Concerning the potential appeal of agri-environmental measures, assessments are available for some Laender programmes. Assessments of the MEKA in Baden-Wuerttemberg (Zeddies, 1996) show that about 60% of all farms could participate, of whom 80% were actually signed up already in 1993 for a period of 5 years. Those measures are well accepted, where farms just reach the requirements or where the costs of adaptation are relatively low in relation to the supports. This is especially true for grassland extensification. Bruckmeier and Langkau (1996) were led to conclude that agri-environmental measures are skewed towards grassland extensification and are characterized by windfall profits.

A survey in North Rhine-Westphalia in 1993 based on a sample of 700 farms shows (Lett-mann, 1995) that 16% of the farmers there are willing to participate in extensification schemes for arable land, 24% for grassland and 4% for organic farming. The motives for participation vary: some farmers argue that environmental objectives are of primary importance while others only ask for higher support.

Environmental impacts

Assessments of environmental impacts of 2078/92 for the whole country are not yet available, although they are for some Laender programmes or certain measures (e.g. organic farming). A distinction needs to be made between the potential environmental benefits of specific measures and the acceptance of these measures. Generally, it can be concluded that there is a negative cor-relation between the environmental benefits or demands of some measures and the extent of their application. Site-specific Laender measures orientated towards landscape and nature protection are only applied on small areas. Organic farming and extensification of arable farming is intro-duced on almost 1 million ha, while highest participation is in grassland extensification and the maintenance of cultural landscape. In consequence, the SRU (1996) has criticized the limited ap-plication of agri-environmental measures, especially those with high environmental benefits.

Assessments of the MEKA programme show (Zeddies, 1996), that due to MEKA measures the intensity of arable land has been reduced by 14%, nitrogen losses by 10% and wind and water

Table 5   Classification of environmental effects of agri-environmental measures

Measures	Criteria of environmental concerns					
	soil structure	erosion	humus content	nutrient cycle	waste recycling	dynamics/ regulation of ecosystems
Extensification of arable and permanent crops: renunciation of:						
- mineral fertilizers and pesticides	+	+	+	+++	-	++++
- mineral fertilizers	+	+	+	+++	-	+++
- herbicides	0	0	0	++	-	+++
Organic farming	++	++	++	++++	-	++++
Extensification of grassland use						
- reduction of livestock	0	0	0	++	0	++++
- increase of grassland at constant livestock	0	0	0	++	0	++++
- maintaining extensive grassland use	0	0	0	++	0	++++

Remark: Positive signs indicate positive environmental effects.
Source: Geier et al. (1996).

erosion by 3%. About 6% of grassland became extensively used due to the reduction of livestock density or mineral fertilizer use.

A general evaluation of specific measures has been conducted by Geier et al. (1996). The measures were classified, taking into account their impacts on soil structure, erosion, humus content, nutrient cycle, waste recycling and the dynamics and regulation of ecosystems. The results of this evaluation are shown in Table 5. This classification shows high positive effects for all the measures with regard to the dynamics and regulation of ecosystems. There are also positive values for the criterion nutrient cycle, which indicates that pollution from fertilizers will be reduced.

Organic farming shows the greatest value on this item and also the highest values with regard to soil structure, humus and erosion.

Based on this classification the measures were ranked with regard to their ecological efficacy; the results are summarized below (Geier et al., 1996):

1. *Organic farming and renunciation of mineral fertilizers and pesticides on the whole farm (2328/91)* are judged to have the greatest ecological value. Assessments from Köpke and Haas (1997) show that compared to conventional farming the nitrogen content of groundwater under plots based on organic farming is reduced by 50%; flora and fauna are more diversified and pollution of greenhouse gases is much lower.

2. *Areas with high landscape values, maintenance of traditional farming and long-term set-aside for nature protection:* These measures have a relatively high ecological efficacy. They include areas for landscape and nature protection as well as the maintenance of dispersed fruit-trees cropping. The total area included in these measures is 117,000 ha.

3. *Grassland extensification:* Positive impacts on the protection of biological resources are given: they are high for those Laender measures which are specified for this target.

4. *Extensification of arable and permanent crops:* Depending on the requirements for special measures there is a relatively high variation in ecological efficacy. The introduction of green cover crops on arable land within the MEKA has low ecological value while the introduction/maintenance of diversified crop rotations (Laender programme of Thueringen) seems to be more effective.

5. *Maintaining cultural landscapes:* The ecological efficacy of this measure, which is mainly applied in Bavaria and Saxonia, is very low.

Economic impacts

As all the measures can be applied on a voluntary basis it can be assumed that farmers will apply for specific measures where costs of adaptation are less or equal to the returns through subsidies. Measures with high margins will be applied to a higher degree than those with low margins, high costs of adaptation or uncertain yield expectations. The latter is of importance in account of the 5 year contracts. In the case of organic farming there has been a high economic incentive for participation during the first years (Schulze-Pals, 1994). After 4 years of participation yield depressions were greater than predicted and due to the increasing supply there was a considerable drop in producer prices. The study of Nieberg (1997) shows, that organic farms orientated to beef and dairy products were not in a position to increase their income, while cropping farms became better off.

Economic assessments of MEKA measures by Zeddies (1996) show that subsidies for maintaining extensive grassland and dispersed fruit-tree cultivation induce windfalls profits of about 90%. On the other hand, subsidies for extensification of arable land are not much higher than the costs of adaptation. Subsidies given for the transformation of arable land into extensive grassland as well as for some regional measures aiming at nature protection seem to be too low.

It can be concluded, therefore, that most of the measures for arable and permanent crops as well as for organic farming do not have a significant income effect, with the exception of less favoured areas due to the low production intensities that already prevail. The measures are not very attractive in the most intensive arable regions and will therefore not contribute to solving environmental problems. Pollution problems of land-less livestock production and regions with high livestock densities will not be influenced by the available agri-environmental programmes. Positive income effects can be expected for all regional programmes with payments for maintaining cultural landscapes. The present system of payments might maintain the existing state of the environment, but will not induce a significant shift towards agricultural systems of a higher ecological value.

The SRU concluded that the increase in organic farming will have positive environmental effects. Programmes based on 2078/92 are in favour of supporting land use in less favoured areas, which is in line with the proposal of the SRU for an integrated land use with lower intensities. The problems are the short duration of the programmes, the modification of some measures due to budget constraints and the change of policy options, as well as the lack of transparency and the variability of the numerous Laender programmes.

2.3.3 Water protection policies

In the following, the impacts of the German Fertilizer Decree and of measures applied in water protected areas are described.

Fertilizer Decree

The Fertilizer Decree (DVO) was introduced in July 1996 (Düngeverordnung, 1996). It fulfils the requirements of the EU Nitrate Directive (Schleef, 1997). The following rules are of importance regarding environmental impacts:

- *Definition of vulnerable zones:* The whole country is defined as a vulnerable zone, therefore the DVO will be applied across the whole country;
- *Introduction of a code of good agricultural practices:* It is defined in collaboration with the Federal State and the Laender aiming at the use of best agricultural practices and to reduce pollution;
- *Restrictions on animal fertilizer use:* The maximum level of nitrogen application (excluding losses from the excretion level of a maximum of 30%) is 170 kg/ha on arable land and 210 kg/ha on grassland. The use of manure on soils saturated with phosphorus or potassium is only allowed up to the crop uptake. Further there are restrictions regarding when livestock manure can be applied. Farms of more than 10 ha are obliged to introduce a mineral

accounting system and to calculate mineral balances. Sanctions will be applied where the spreading or disposal of manure causes damage.

First evaluations show that about 12% of all farms will be affected by restrictions on manure use defined in the Nitrate Directive (Brouwer et al., 1997). A much larger number of farms might be affected by the restrictions defined for phosphorus and potassium (Schleef, 1997). It is not clear that sanctions are effective in the latter case. Problems of regional mineral surpluses might be solved for most of the regions by interregional transport and a better distribution of manure.

Model calculations from Schleef (1997) show that, in farms affected by the Directive, the nitrogen surpluses will be reduced by 4% in pig farms and 5.5% in poultry farms but by less than 1% in dairy farms. Assessments from Geier et al. (1996) show that the Fertilizer Decree will induce a reduction of nutrient surpluses, but the scale of the reductions depends on the adoption of effective control measures and accompanying support measures. From an ecological viewpoint the maximum levels for manure application seem to be set too high. The SRU argues that mineral balancing at farm level should be a crucial test. Taxation on the basis of balance surpluses (Bauer et al., 1996) might have greater positive environmental effects than taxation on mineral fertilizer.

Water protected areas

With regard to drinking water protection, command and control measures are introduced for water protected zones. Compensatory payments are provided to ease the costs of adaptation. Some of the measures of 2078/92 can be combined so that extensification effects will increase. Assessments based on the SCHALVO (Schutzgebiets- und Ausgleichs-Verordnung) scheme of Baden-Wuerttemberg show that, in addition to the obligatory SCHALVO measures a high proportion of farms apply for the MEKA measure grassland extensification (Baudoux et al., 1997). This might be so not only for the protected zone I where the use of mineral and organic fertilizers is forbidden, but also in other zones (II-IV) where additional extensification might be stimulated by 2078/92 measures. This is important because today 17.4% of the whole area of Baden-Wuerttemberg is designed as water protected zone I and the area will be extended to 28%. Calculations based on the RAUMIS model show that, on average, nitrogen surpluses could be reduced by an order of 60% either by a fertilizer levy of 100% or by command and control measures applied in water protected areas (Weingarten et al., 1995; Frohberg, 1995). Nitrate contents of more than 25 mg/l will be reduced drastically by both policies, but command and control measures in water protected zones are economically preferable.

2.3.4   Structural policies

Within structural policies only the compensatory allowance for Less Favoured Areas will be discussed. Of the total UAA (18.61 million ha) 9.426 million ha (50.6% of UAA) are declared as Less Favoured Areas, of which 4% as mountainous areas and 2% as small areas with specific natural handicaps (BML, 1996). In the year 1995 228,919 farms (about 41% of the total) applied for compensatory allowances; the total budget was 963.8 million DM. Some 60% of the budget is paid on the basis of livestock numbers, the remainder is based on the land area.

The environmental impacts can be summarized as follows (Bauer et al., 1996):

-    compensatory allowances support land use in Less Favoured Areas. Therefore the tendency towards concentration of agricultural production in favoured regions resulting from market and price policy will be reduced;
-    on the other hand, structural changes and increasing farm sizes will be retarded. Therefore land prices in Less Favoured Areas will increase, inducing higher intensities of land use.

The principal problem with this measure is that it is orientated to income support, but it is not specific enough to compensate for the economic burdens in the 'really' less favoured regions. Only 4% of the declared regions are mountainous areas, all the others are characterized by structural deficiencies. The SRU (1996) and Bauer et al. (1996) conclude that the measure should be specified in such a way that it supports agriculture only in those regions which would be abandoned under market conditions without supports. That would entail a considerable reduction in the areas included in the programme and the support being determined by the real economic burdens.

2.4   Concluding observations

Environmental protection is a major policy objective, both generally and specifically in agricultural policy. At first the preoccupations were water protection and reduction of air pollution. During the 1990s aspects of global warming have also become important.

Agriculture is seen as a main polluter of soils and water and to be partly responsible for the loss of biodiversity and landscape values. The main causes of environmental problems are identified as technical progress and agricultural policy favouring intensification and specialization. Environmental and pollution problems exist in some intensively used arable areas and in regions where horticulture, fruit growing and viticulture are concentrated or which have a high density of livestock. Referring to the assessment of the SRU (1996) the production systems of most German farms can be classified as posing environmental risks.

On the other hand, it is also well accepted that there are environmental benefits from agriculture and that these external effects could be improved by extensification and a general reorientation of agriculture and agricultural policy towards environmental objectives. This policy direction is also motivated by the opportunities to create new sources of agricultural income. Nevertheless, there is a heated debate over the degree to which positive external effects are for free.

The environmental effects of the market and price policy measures of the CAP reform are rather limited. Nevertheless, positive indications are given towards extensification. Specific programmes based on 2078/92 form the core of policies directed towards giving agriculture an environmental orientation. Through the responsibility of the Laender for environmental policy and the co-financing by the Federal State and the EU, both the Laender and the Federal State are involved in agri-environmental programmes. The Federal State is involved in the collaborative programme Förderung einer markt- und standortangepaßten Landbewirtschaftung (msaL). Only some Laender apply the common programme or part of the measures. Other Laender have introduced specific programmes with co-financing from the EU only. Apart from the competitive relationships between Federal and Laender policies, the following reasons might be important (a) that the maximum of the allowed support levels is significantly higher than the one defined in msaL and (b) that so-called base payments for maintaining cultural landscape are allowed.

Due to the different support levels, specification of and requirements for regional measures, natural conditions and intensity level, the regional participation rates are very different. Overall, 29% of UAA is included in agri-environmental programmes. Participation is very high in those Laender where compensatory payments are provided to maintain cultural landscape (Bavaria and Saxonia). The other major measures are grassland extensification, extensification of arable crops and organic farming.

The application of the existing programmes is skewed in favour of grassland extensification. In this case positive income effects are given. Poorer economic incentives are given for extensification measures for arable and permanent crops. Therefore the acceptance in the arable crop sector and in intensively used agricultural regions is rather low. This is also true for intensive livestock production. Therefore regional pollution problems induced by intensive crop production and high concentrations of livestock production might not be solved. The programmes mainly prevent an increase in pollution in low polluted regions. The SRU (1996) therefore criticizes the limited application of agri-environmental measures with high environmental benefits.

Some of the measures specifically support land use in less favoured areas, and therefore are in line with a move towards an integrated land use at lower intensities. Problems are seen to lie with the short term nature of the programmes and the lack of transparency and the variability in the numerous Laender programmes. Nevertheless, there is a consensus that agri-environmental programmes are a central part of agricultural policy and will become much more important in the future.

References

Bach, M. (1987) *Die potentielle Nitratbelastung des Sickerwassers durch die Landwirtschaft in der Bundesrepublik Deutschland;* Göttingen, Thesis

Baudoux, P., G. Kazenwedel and R. Doluschitz (1997) *Agrarumweltprogramme: Betriebliche Wirkungen und Einstellung von Landwirten;* Agrarwirtschaft 46, 4/5, 184-197

Bauer, S., J.P. Abresch and M. Steuernagel (1996) *Gesamtinstrumentarium zur Erreichung einer umweltverträglichen Raumnutzung;* Rat von Sachverständigen für Umweltfragen (eds), Materialien zur Umweltforschung, Band 26, Stuttgart, Metzler-Poeschel

Becker, H. and W. Kleinhanss (1997) *Assessment of CAP reform and fertilizer levies at regional level in the European Union;* In: Brouwer, F. and W. Kleinhanss (eds), The implementation of nitrate policies in Europe: Processes of change in environmental policy and agriculture; Kiel, Wissenschaftsverlag Vauk, pp. 193-206

BML (Bundesministerium für Ernährung, Landwirtschaft und Forsten) (1995) *Umsetzung der Verordnung (EWG) Nr. 2078/92 in der Bundesrepublik Deutschland*, Bericht an die Europäische Kommission für den Evaluierungsbericht der Kommission an den Rat und an das Europäische Parlament über die Umsetzung in der EU; Bonn

BML (Bundesministerium für Ernährung, Landwirtschaft und Forsten) (1996) *Agrarbericht der Bundesregierung 1996;* Bonn

BML (Bundesministerium für Ernährung, Landwirtschaft und Forsten) (1997) *Agrarbericht der Bundesregierung 1997;* Bonn

Brouwer, F., F. Godeschalk and P. Hellegers (1997) *Nitrogen balances at farmlevel in the European Union;* In: Brouwer, F. and W. Kleinhanss (eds); The implementation of nitrate policies in Europe: Processes of change in environmental policy and agriculture; Kiel, Wissenschaftsverlag Vauk; pp.79-90

Bruckmeier, K. and J. Langkau (1996) *Eine grüne GAP? Ökologisierung der Europäischen Agrarpolitik;* Köln, Institut für angewandte Umweltforschung KATALYSE

Chatzis, A. (1996) *Flächenbezogene Ausgleichszahlungen der EU-Agrarreform - Pachtmarktwirkungen und Quantifizeirung der Überwälzungseffekte;* Agrarwirtschaft, Sonderheft 154, Frankfurt, AgriMedia

Düngeverordnung (1996) *Verordnung über die Grundsätze der guten fachlichen Praxis beim Düngen;* (Düngeverordnung) vom 26. Januar 1996, Bundesgesetzblatt, Teil 1; pp. 118-121

Enquete-Kommission 'schutz der Erdatmosphäre' des Deutschen Bundestages (ed) (1994) *Schutz der Grünen Erde - Klimaschutz durch umweltgerechte Landwirtschaft und Erhalt der Wälder;* Bonn, Economica Verlag

Focken, M. (1988) *Wo liegen die derzeitigen Probleme und deren mögliche Ursachen? - Aus der Sicht der Wasserversorgung;* In: Sauberes Wasser und moderne Landbewirtschaftung - ein Gegensatz ? Frankfurt/M, DLG Kolloquium, Dezember 15, DLG

Frohberg, K. (1995) *Programs for stimulating extensive agriculture in Germany;* In: Hofreither, M. and S. Vogel (eds); The role of agricultural externalities in high income countries; Kiel, Wissenschaftsverlag Vauk, pp. 161-179

Fuchs, C. and W. Trunk (1995) *Auswirkungen der EU-Agrarreform auf die Umweltverträglichkeit der landwirtschaftlichen Produktion;* In: Grosskopf et al. (eds), Die Landwirtschaft nach der EU-Agrarreform, Schriften der Gesellschaft für Wirtschafts- und Sozialwissenschaften des Landbaues e.V, Vol 31; Münster-Hiltrup, Landwirtschaftsverlag, pp. 243-258

Geier, U., G. Urfei and J. Weis (1996) *Stand und Umsetzung einer umweltfreundlichen Bodennutzung in der Landwirtschaft* - Analyse der Empfehlungen des Schwäbisch Haller Agrarkolloquiums der Robert Bosch Stiftung. Schriftenreihe Institut für Ökologischen Landbau; Berlin, Verlag Dr. Köster

Höll, A. and H. von Meyer (1996) *The response of the Member States - Germany;* In: Whitby, M. (ed), The European Environment and CAP Reform - Policies and Prospects for Conservation; Wallingford, CAB International, pp. 70-85

Isermann, K. (1994) *Ammoniak-Emissionen der Landwirtschaft, ihre Auswirkungen auf die Umwelt und ursachenorientierte Lösungsansätze sowie Lösungsaussichten zur hinreichenden Minderung;* In: Studienprogramm Landwirtschaft der Enquete-Kommission 'Schutz der Erdatmosphäre' der Deutschen Bundestages; Bonn, Economica-Verlag

Isermann, K. (1995) *Stoffeinträge aus der Landbewirtschaftung in die Gewässer;* Stellungnahme als Sachverständiger zum Fragenkatalog des Ausschusses für Ernährung; Bonn, Landwirtschaft und Forsten des Deutschen Bundestages

Jungehülsing, J. (1997) *Entwicklung und Perspektiven des ökologischen Landbaues und dessen Rahmenbedingungen in Deutschland;* In: Nieberg, H. (ed), Ökologischer Landbau: Entwicklung, Wirtschaftlichkeit, Marktchancen und Umweltrelevanz, Landbauforschung Völkenrode, Sonderheft 175, Braunschweig, pp. 3-12

Kögl, H. (1993) *Wege zur Extensivierung der Landwirtschaft;* Landbauforschung Völkenrode, Sonderheft 142, Braunschweig

Köpke, U. and G. Haas (1997) *Umweltrelevanz des ökologischen Landbaues;* In: Nieberg, H. (ed), Ökologischer Landbau: Entwicklung, Wirtschaftlichkeit, Marktchancen und Umweltrelevanz, Landbauforschung Völkenrode, Sonderheft 175, Braunschweig, pp. 119-146

Korneck, F. and H. Sukopp (1988) *Rote Liste der in der Bundesrepublik Deutschland ausgestorbenen, verschollenen und gefährdeten Farn- und Blütenpflanzen und ihre Auswirkungen auf den Arten- und Biotopschutz,* Bundesforschungsanstalt für Naturschutz und Landschaftökologie; Bonn, Schriftenreihe für Vegetationskunde, Nr. 19, Bad-Godesberg

Lettmann, A. (1995) *Akzeptanz von Extensivierungsstrategien - Eine empirische Untersuchung in Nordrhein-Westfalen;* Bonn, Institut für Agrarpolitik, Marktforschung und Wirtschaftssoziologie der Universität Bonn, Forschungsbericht 20

Mehl, P. and R. Plankl (1995) *Doppelte Politikverflechtung als Bestimmungsfaktor der Agrarstrukturpolitik;* Untersuchung am Beispiel der Förderung umweltgerechter landwirtschaftlicher Produktionsverfahren in der Bundesrepublik Deutschland, Landbauforschung Völkenrode, 45, 4, pp. 218-232

Nieberg, H. (1997) *Wirtschaftliche Folgen der Umstellung auf ökologischen Landbau - empirische Ergebnisse von 107 Betrieben aus den alten Bundesländern;* In: Nieberg, H. (ed), Ökologischer Landbau: Entwicklung, Wirtschaftlichkeit, Marktchancen und Umweltrelevanz, Landbauforschung Völkenrode, Sonderheft 175, Braunschweig, pp. 57-74

Niedersächsisches Umweltministerium (1996) *Stickstoff-Minderungsprogramm,* Bericht der Arbeitsgruppe aus Vertretern der Umwelt- und der Agrarministerkonferenz; Hannover, September

Plankl, R. (1996a) *Synopse zu den Agrarumweltprogrammen der Länder in der Bundesrepublik Deutschland, Maßnahmen zur Förderung umweltgerechter und den natürlichen Lebensraum schützender landwirtschaftlicher Produktionsverfahren gemäß VO (EWG) 2078/92,* Bundesforschungsanstalt für Landwirtschaft, Institut für Strukturforschung, Arbeitsbericht 1/1996, Braunschweig

Plankl, R. (1996b) *Analyse des Finanzmitteleinsatzes für die Förderung umweltgerechter landwirtschaftlicher Produktionsverfahren in den Ländern der Bundesrepublik Deutschland;* Landbauforschung Völkenrode, 1, pp. 33-47

Schleef, K.-H. (1996) *Impacts of policy measures to reduce nitrogen surpluses from agricultural production - an assessment at farm level for the former Federal Republik of Germany,* Bundesforschungsanstalt für Landwirtschaft, Institut für Betriebswirtschaft, Arbeitsbericht 2/96, Braunschweig

Schleef, K.-H. (1997) *Impacts of the Nitrate Directive for Germany;* In: Brouwer, F. and W. Kleinhanss (eds); The implementation of nitrate policies in Europe: Processes of change in environmental policy and agriculture; Kiel, Wissenschaftsverlag Vauk, pp. 221-234

Schleef, K.-H. and W. Kleinhanss (1994) *Mineral balances in agriculture in the EU-Part I: the regional level*, Bundesforschungsanstalt für Landwirtschaft, Institut für Betriebswirtschaft, Arbeitsbericht 1/94, Braunschweig

Schulze - Pals, L. (1994) *Ökonomische Analyse der Umstellung auf ökologischen Landbau. Eine empirische Untersuchung des Umstellungsverlaufes im Rahmen des Extensivierungsprogramms*, Schriftenreihe des BMELF, Reihe A: Angewandte Wissenschaft, Heft 436; Münster: Landwirtschaftsverlag

SRU (Rat von Sachverständigen für Umweltfragen) (1985), *Umweltprobleme der Landwirtschaft (Sondergutachten);* Kohlhammer: Stuttgart and Mainz

SRU (Rat von Sachverständigen für Umweltfragen) (1996) *Konzepte einer dauerhaft-umweltverträglichen Nutzung ländlicher Räume (Sondergutachten),* Deutscher Bundestag, Drucksache 13/4109, Bonn, 14.03.1996

Steiner, M., H. Sprich, H. Lehn and G. Linckh (1996) *Einfluß der Land- und Forstbewirtschaftung auf die Ressource Wasser;* In: Linckh, G. et al.. (eds), Nachhaltige Land- und Forstwirtschaft - Expertisen; Berlin, Heidelberg, New York, etc., Springer, pp. 27-76

UBA (Umweltbundesamt) (1994) *Stoffliche Belastung der Gewässer durch die Landwirtschaft und Maßnahmen zu ihrer Verringerung*, Bericht 2/94; Berlin: Erich Schmidt Verlag

Weingarten, P., W. Henrichsmeyer and R. Meyer (1995) *Abschätzung der Auswirkungen von Vorsorgestrategien zum Grundwasserschutz im Bereich Landwirtschaft;* Agrarwirtschaft 44, pp.191-201

Wendland, F., H. Albert, M. Bach and R. Schmidt (1993) *Atlas zum Nitratstrom in der Bundes republik Deutschland;* Berlin, Heidelberg, New York, etc., Springer

Werner, W. and H.-P. Wodsack (1994) *Stickstoff- und Phosphoreintrag in die Fließgewässer Deutschlands unter besonderer Berücksichtigung des Eintragsgeschehens im Lockergesteinsbereich der ehemaligen DDR;* Schriftenreihe Agrarspectrum, 22, Frankfurt/M: Verlagsunion Agrar

Wissenschaftlicher Beirat beim BMLEF (1992) *Strategien für eine umweltverträgliche Landwirtschaft;* Schriftenreihe der Bundesministers für Ernährung, Landwirtschaft und Forsten, Reihe A: Angewandte Wissenschaft, Nr. 414; Münster-Hiltrup: Landwirtschaftsverlag

Wissenschaftlicher Beirat beim BMLEF (1993) *Reduzierung der Stickstoffemissionen der Landwirtschaft;* Schriftenreihe des Bundesministers für Ernährung, Landwirtschaft und Forsten, Reihe A: Angewandte Wissenschaft, Nr. 423; Münster-Hiltrup: Landwirtschaftsverlag

Zeddies, J. (1995) *Umweltgerechte Nutzung von Agrarlandschaften;* Berichte über Landwirtschaft, 73, pp. 204-241

Zeddies, J. (1996) *Analyse der laufenden und geplanten Programme (EU-, Bund-, Länderebene) zur Förderung umweltgerechter Produktionsverfahren - Modifikationen und Perspektiven;* In: Linckh, G. et al. (eds), Nachhaltige Land- und Forstwirtschaft - Expertisen; Berlin, Heidelberg, New York, etc.: Springer, pp. 655-700

Zeddies, J., C. Fuchs, W. Gamer, H. Schüle and B. Zimmermann (1994) *Verteilungswirkungen der künftigen EG-Agrarpolitik nach der Agrarreform unter besonderer Berücksichtigung der direkten Einkommenstransfers - dargestellt auf der Grundlage von Buchführungsergebnissen und Betriebsbefragungen;* In: Landwirtschaftliche Rentenbank (ed), Verteilungswirkungen der künftigen EU-Agrarpolitik nach der Agrarreform, Landwirtschaftliche Rentenbank, Schriftenreihe, Band 8, pp. 97-144

3. FRANCE

Pierre Rainelli and Dominique Vermersch

3.1 Introduction

Agriculture and the peasantry play a particular role in France. In spite of its post-war agricultural modernization, France remained in the early 1990s more or less impregnated by an agrarian tradition. A large share of the population identified themselves with the interests of the farmer, or at least the image of the traditional farming communities from which most had originated just one or two generations earlier. Farmers were seen as the traditional creators and protectors of the rural landscape. The role farmers have played in the provision of food during a period of scarcity (the Second World War) reminds many people of the crucial importance of agriculture.

Furthermore we have to keep in mind that France has both a relatively low population density and a substantial territorial imbalance. Nearly 20% of total population is concentrated in the Ile de France, whilst large parts of the territory are empty space. For instance, the population density is 30 inhabitants per square kilometre in the Massif Central, as opposed to 104 for France as a whole. Due to these conditions, the contribution of agriculture to balanced land-use planning and development is perceived as essential. This is considered to be of utmost importance in mountain and hill areas, in which the farming community generally accounts for a large share of the overall population.

The CAP appeared to offer French farmers an unquestionable opportunity to modernize through the provision of support for capital investments and a high price support system for agricultural products. Consequently, the CAP for a long time remained ignored in the public debate over the transformation of agricultural practices and their potential effects on the environment. More generally, concern for the agricultural environment focused on the policy to consolidate land holdings. This was aimed to contribute to substantial changes, not only to the agricultural landscape and to ecosystems, but also the broader agricultural environment. Pollution was predominantly perceived as an urban issue linked to point-source pollution by industry and urbanization. The only real exception to this otherwise general absence of confrontation between environmental organizations and farmers, is Brittany where the degradation of surface water quality let to the creation of a dynamic pressure group composed first of salmon and trout anglers, then of protectors of the environment (Association Eaux et Rivières).

The various factors summarized above explain why the agricultural community in France, the Ministry of Agriculture and the policymakers were rather reluctant in their efforts to introduce the early European agri-environmental measure, i.e. Article 19 of Council Regulation 797/ 85. Unlike the response by some other Member States, mainly the UK and Germany, France had a limited experience in domestic agri-environmental measures and was initially opposed to the linking of agriculture and environmental protection (Boisson and Buller, 1996). The Ministry of Agriculture feared to give its support to a policy that appeared to impose production limits on farmers, and to label them as simply gardeners of nature. This policy, therefore, was introduced very slowly. By 1991 only 22 ESA projects were approved by the National Committee for Agriculture and the Environment.

The same reasons also explain why CAP reform appeared as a threat towards agriculture itself and towards a balanced land use. The new supply control measures to limit production, and particularly the set-aside programme, were seen negatively as a move back to old agricultural practices where fallow was necessary to maintain soil fertility. In a country where the critical agricultural-environmental concern was land abandonment and the degeneration of cultivated areas to wilderness, such a measure was unthinkable. We have to keep in mind that the agri-environmental policy has been centred on retaining threshold densities of farmers in marginal areas to avoid the dereliction of the countryside through land abandonment, which is summarized by the French term 'désertification'.

Nowadays the debate on agriculture and the environment encompasses the loss of basic features of the landscape due to the accessory works accompanying land re-allocation, which pull out hedgerows, level embankments and slopes, and straighten and widen streams. Such redevelopment plus the spread of monoculture are adding to a new sense of the transformation of the landscape. The effects of agricultural intensification on aquifers and watercourses, in which the level of nitrates and pesticides is increasing, also raise questions on human health, and the response to the recent BSE crisis gives evidence of the public concern about the quality of food.

The new feature is that the debate occurs now within the agricultural community pitting, broadly speaking, 'productivity groups' against 'anti-productivity groups'. In the former are cereal growers and intensive livestock producers who are opposed to any type of environmental constraints limiting production by regulatory means, such as set-aside, or by economic incentives, such as the Polluter-Pays Principle. On the opposite side there are various groups that believe they can manage the environment through their agricultural practices, including small traditional holdings, modern extensified farms and organic producers. The majority of farmers, though, are preoccupied with getting acquainted with the new rules of the CAP and trying to adapt their behaviour to the changing context.

3.2   Impacts of the CAP

3.2.1   The arable regime

The arable crop regime covers cereals (except rice), oilseeds (rapeseed, sunflower seed, soybeans) and protein crops (field peas, beans and lupins). In order to compensate producers for the reduction in prices under the 1992 CAP reform, direct payments per hectare are made. In addition, land which is entering into the required set-aside to limit production receive a per-hectare payment equal to the cereals payment. From an environmental point of view this regime has both direct and indirect effects.

Concerning the direct effects, the eligible crop system has negative consequences in allowing farmers to consider maize either as a cereal or as a fodder crop according to their particular interest. Furthermore, the method for calculating per-hectare payments, which is partially based on regional yields, differentiates yields between irrigated and non-irrigated land. This mechanism is an incentive to develop maize at the expense of permanent grassland. At the national level permanent grassland decreased by 843,000 ha or 7.4% between 1990 and 1996. At the same time the acreage of maize increased by 168,000 ha (or 10.8%).

At the farm level, the acreage control programme has led to under-used production capacity with unemployed capital (machinery) and labour. This explains the interests of the French cereal growers in the non-food option for set-aside. In consequence, areas that normally would not have fertilizers or pesticides applied, receive significant amounts of chemicals. Moreover, cereal farmers are diversifying their farming system towards vegetables, horticulture and fruit. For example in Landes, a Département where cereals and maize largely dominate, the carrot acreage increased from 850 ha in 1992 to 2,000 ha in 1994, bringing Landes to the second place in carrot production. In this region, farmers were facing idle production capacity (capital and labour) because of the set-aside requirements, which was used in part to grow vegetables. Such a shift increases intensification and therefore, pollution. Lastly, direct payments are more important for irrigated land, since compensatory payments are based on average yield, regardless of whether irrigation has been used. This scheme involves a top premium part on water input which benefits irrigating farmers. They are entitled with property rights over water resources which are revealed through the payment scheme. Farmers' rights may prevail upon the rights of other categories of water users. This distribution of property rights does not reflect the social value of water in its alternative uses (Cohen et al., 1996).

Theoretically, lower prices may imply a significant reduction in input use, in terms of the derived demand on inputs, and a decrease in the yields. In order to analyse changes in production techniques due to the price cuts a specific tool has been built (Boussemart et al., 1996) linking together linear programming and an agronomic simulation model (EPIC i.e. Erosion Productivity Impact Calculator) adapted to the French conditions. It reveals the consequences of technical choices for nitrate pollution when the producer optimizes revenues. This bio-economic model has been applied in the region of La Beauce, where cereal specialized farms are dominant, and in the Toulouse agricultural plain, a less specialized region.

The model first indicates a change in the crop pattern with a greater specialization in winter cereal production in La Beauce, and an increase in durum wheat in the south-west, because of the specific subsidy existing for this crop. Secondly, there is a movement towards relatively less intensive techniques inducing a nitrate reduction of 5 to 8 kg per hectare according to the region. Thirdly, in terms of technology used, intensive dry and irrigated techniques decrease with the CAP reform. In La Beauce, for example, the share of irrigated area, which was over 19% before the reform, is projected to drop to 8% under the new regime. The effect is less clear in the Toulouse plain where the high level of subsidies to irrigated land induces an increase in the irrigated area from 18% to 33%. These theoretical projections, however, are not confirmed by the empirical trends given in Table 1.

The region Centre corresponds approximatively to La Beauce, and the Toulouse plain is located in Midi-Pyrénées. Although in both regions and in France overall, there has been a slight decrease in total cereals acreage, no clear trend is yet apparent. In Centre the wheat acreage has shown a steady increase, whilst for yields and the rate of nitrogen used there has been no declining trend.

If the yields do not decrease as foreseen the reason is a relative intensification of the production as the US experience has shown. When the land input is restricted, through set-aside, farmers

Table 1   *Acreage of cereals, wheat yields and nitrogen use per hectare 1990-1996 for regions Centre, Midi-Pyrénées and total of France*

	1990	1991	1992	1993	1994	1995	1996
CENTRE							
- cereals acreage (1,000 ha)	1,328.9	1,343.2	1,286.4	1,124.9	1,082.6	1,103.8	1,202.7
- wheat acreage (1,000 ha)	759.8	711.7	698.2	717.1	722.7	730.9	762.2
- wheat yield (ton/ha)	6.2	6.9	6.2	6.9	6.8	6.5	6.4
- nitrogen (kg/fertilizable ha)	126	109	110	111	123	123	n. a
MIDI-PYRÉNÉES							
- cereals acreage (1,000 ha)	773.9	785.3	794.7	704.3	633.5	660.8	701.3
- wheat acreage (1,000 ha)	262.3	216.3	249.8	180.6	168.8	207.5	224.7
- wheat yield (ton/ha)	4.3	5.2	5.2	4.5	5.0	5.4	5.8
- nitrogen (Kg/fertilizable ha)	70	70	72	72	70	70	n. a
FRANCE							
- cereals acreage (1,000 ha)	9,060.6	9,226.1	9,338.5	8,541.2	8,166.9	8,293.0	8,829.3
- wheat acreage (1,000 ha)	4,737.5	4,635.0	4,630.8	4,262.2	4,314.9	4,485.2	4,741.3
- wheat yield (ton/ha)	6.6	6.9	6.6	6.6	6.8	6.6	7.3
- nitrogen (kg/fertilizable ha)	95	89	92	87	90	90	n. a

Source: Statistique Agricole Annuelle.

increase the use of other inputs (capital, labour, chemicals ...) on the land remaining in production. Furthermore, detailed analysis shows the existence of technical and allocated inefficiencies. A shock, like the price cut, can induce farmers to move towards the frontier function. The reduction of technical inefficiencies which occurs involves a possible increase in yields which has been estimated at 6.4% in Aquitaine, 11.4% in Centre and 12.3% in Ile de France (Vermersch et al., 1993; Piot-Lepetit et al., 1997).

Concerning indirect effects, the most important is related to the development of intensive rearing. As a matter of fact the price cut for cereals, but also for oilseeds and pulses, led to a fall in feed costs. Consequently prices for livestock products, mainly pork and poultrymeat, decreased. These commodities, becoming more competitive, have expanded at the expense of beef. Using a simplified world trade model which allows the simulation of the consequences of the CAP reform, it has been demonstrated that pork and poultrymeat would increase in France by 8.5% from 1993 to 1996 (Léon and Quinqu, 1994).

Empirical evidence confirms the predictions of the model. From 1992/3 to 1995/6 the price reduction for pig feed compounds reached 8.7%, whilst the production of pork and poultrymeat increased by 9.3% (Baromètre Porc, 1996). The expansion of intensive rearing increases environmental pressures since this expansion affects primarily regions having the lowest cost, in fact those regions with the highest stocking rates already. From this point of view the case of Brittany is exemplary: in 1992 its total pork and poultry output accounted for 43% of French production; in 1995 this share reached 46% (Statistique Agricole Annuelle).

But the cut in cereal prices introduces important modifications in the formulation of feed rations since cereals, which become more competitive, substitute for soybean and corn gluten feed. The share of cereals in compound feed increased during the period between 1992/3 and 1995/6 from 32% to 42% (Baromètre Porc, 1996) giving a better balanced diet. More cereals at the expense of proteins mean less nitrogen in manure. A 1 to 1.5 point reduction in the protein content of feeds is estimated to result in a diminution of nitrogen excretion by pigs in the range of between 0.2 to 0.4 kg per pig. It has been shown that these results are very sensitive to the relationship between the prices of protein and energy (Dourmad et al., 1995).

3.2.2   The dairy regime

The introduction of controls on milk production, through milk quotas in 1984, was basically aimed at cutting support costs, imposed by earlier policies. The rationale for the introduction of quotas was to maintain producers' incomes while keeping costs within the constraint of the available budget. Milk quotas have been allocated to producers on the basis of their milk output. In only a few countries, namely the Netherlands and the United Kingdom, can individual quota rights be leased or rented. In France and Germany the redistribution of quota rights is handled administratively leading in appearance to a frozen situation. In fact, this system has had a powerful restructuring effect in changing the numbers and sizes of dairy farms.

In 1979 the number of milk producers was 520,000, two-thirds of whom were producing less than 50,000 litres on small holdings. By the end of 1993 the number was 177,000, indicating a dramatic concentration of production and a systematic ejection of the small holders. By 1990 the decline was running at 8% per year. Government aid programmes explain this reduction: over ten years they involved 150,000 producers, or 40% of 1983 suppliers (Perrier-Cornet, 1995). In consequence, the available quotas were distributed in mountainous areas or to young farmers.

This shift from small farms to larger farms is accompanied by an intensification with higher stocking rates and a more intensive farming system as shown in Table 2.

As indicated in Table 2, the smaller the dairy farms are the more extensive they are. For the specialized ones, when the stocking rate increases from 1.0 to 1.8 in LU per ha the share of agri-

Table 2   *Main features of dairy farms according to their size in 1993*

Size in wheat units a)	Specialized dairy farms				Mixed dairy farms			
	< 20 ha	20-40	40-75	> 75	< 20 ha	20-40	40-75	> 75
Average size in ha	20	38	58	100	21	40	61	108
Dairy cows per farm	13	27	42	73	8	17	29	51
Livestock Units per ha	1.0	1.4	1.6	1.8	.9	1.3	1.7	2.0
Grassland in % of agricultural area	91	88	82	79	61	43	30	23

a) Wheat unit refers to the economic size of farms and is based on the Standard Gross Margin.
Source: Perrier-Cornet, 1995.

cultural area in grassland decreases from 91% to 79%. For the mixed dairy farms the intensification is more significant and the share of pasture decreases more dramatically (from 61% to 23%). Such intensification implies an increase of protein consumed or more fertilizers and therefore more manure per hectare.

But a geographical mobility of quotas, through tradable rights, would have induced a greater concentration. Using a sample of dairy or predominantly dairy farms, a restricted cost function has been estimated for 1991 (Guyomard et al., 1996). Marginal cost functions and characteristics of the market equilibrium have then been derived. Regional dummy variables allow geographical aspects to be taken into account. Assuming that capital is fixed and farmers face different milk prices it is possible to estimate the effects on implementing a national leasing market of quota for the various French regions.

Table 3   Regional characteristics of a milk quota leasing market

Administrative Region	Number of Producers a)		Average Marginal Costs (FF/hectolitre)	Average Milk Price (FF/hectolitre)	Average Production Change per Farm (hectolitre)	Average Producer Welfare Gain(FF)
Champagne-Ardennes	61	(2,319)	91.57	189.60	-539.42	5,527.9
Picardie	39	(1,084)	117.74	201.33	-2,036.80	2,0096.1
Haute-Normandie	57	(3,275)	91.90	208.24	-95.88	2,097.4
Centre	14	(684)	95.41	192.66	-767.36	9,569.7
Basse-Normandie	158	(19,466)	89.53	204.50	123.78	1,504.3
Bourgogne	16	(820)	95.14	195.17	-368.19	3,249.1
Nord - Pas de Calais	69	(3,008)	74.67	194.55	289.17	1,119.8
Lorraine	119	(4,851)	90.42	199.55	127.88	3,023.4
Alsace	25	(1,073)	91.94	197.24	-113.09	984.8
Franche-Comté	154	(7,724)	86.32	199.22	118.28	2,191.6
Pays de la Loire	187	(21,162)	82.63	196.61	180.81	2,252.2
Bretagne	287	(26,866)	82.86	198.99	185.18	1,380.4
Poitou-Charentes	29	(1,923)	82.86	194.58	122.38	2,539.4
Aquitaine	42	(3,214)	103.50	191.27	-1,037.34	5,763.7
Midi-Pyrénées	57	(5,461)	98.38	194.80	-441.35	4,087.3
Limousin	37	(1,085)	98.18	188.62	-1,111.76	6,813.5
Rhône - Alpes	92	(11,757)	90.04	210.32	117.38	1,327.7
Auvergne	153	(11,779)	87.71	187.58	- 303.88	2,702.7
Languedoc - Roussillon	3	(172)	27.94	181.62	120.48	55.8

a) The first column gives the number of dairy farms in the sample; the second column gives the number of holdings represented by the sample.

Note: The detailed results presented in this Table have to be considered mainly as an illustration. They are subject to important caveats stemming from the use of FADN data and regional quota flows have thus to be interpreted with great caution, particularly if they are used for political purposes. The omission of small herds tends to underestimate the outflow of quota from regions with more small herds and smaller herd sizes if small size and high (marginal) cost production are positively correlated. Furthermore, the size distribution is likely to be different for the various regions. These factors tend to bias our results, in a direction and for a size which is difficult to evaluate.

Source: Guyomard et al., 1996.

Table 3 illustrates the regional shifts which would occur if there was a national milk quota leasing market. It appears that the regions which are specialized in dairy production (Basse-Normandie, Pays de la Loire and Brittany) would gain quotas. Brittany, where the number of producers is the largest, and the level of pollution originating from agriculture the highest, would have the largest production increase per farm. As noted by the authors, there is no clear difference between mountain and plain areas since two mountain regions (Midi-Pyrénées, Auvergne and Limousin) should experience a decrease. Nevertheless the weighted decrease is higher than the weighted increase.

3.2.3   The beef regime

The 1992 CAP reform resulted in significant reductions in intervention prices for beef compensated by a system of direct payments mainly based on the number of suckler cows and male beef animals. An extensification complement is given when the stocking rate is less than 1.4 per ha of fodder crop. Lastly there is a grassland premium, an agri-environmental measure adopted under EC Regulation 2078/92. It applies for the entire French agricultural area for holdings of at least 3 ha of usable agricultural area capable of supporting a minimum of 3 livestock units (LU) providing that a stocking rate of 1.0 is not exceeded. But this rate can rise to between 1.0 and 1.4 if pasture lands cover at least 75% of the total agricultural area.

Table 4 indicates that the beef regime is a semi-decoupled policy; the system of direct payments is mainly based on the number of livestock and therefore is related to the scale of beef production. The only exception to this is the grassland premium scheme. In comparison with a policy of completely decoupled support (an area based system) the present system disadvantages extensive cattle rearing. As a matter of fact the revenue of beef herds with less than 1 LU per ha is 15% below the national average, whilst it is 11% above the national average for herds with a stocking rate between 1.4 and 2.0. With a decoupled support the most extensive farms would have a 2.3%

Table 4   The French beef regime of direct payments in 1996

Type of premium Dispositions	
Suckler cow premium	a headage premium of FF 1,160 per cow per year (of which 200 F from French budget for the first 40 animals). Total amount FF 4,840 million (of which FF 659 million from French budget).
Male beef premium	a headage premium of FF 720 per animal per year. Total amount FF 2,100 million (entirely paid by the EU budget).
Extensification complement	a headage premium of FF 240 per animal per year. The stocking rate must be less than 1.4 LU per ha.
Grassland premium	a headage premium of FF 300 per ha per year. Total amount FF 1,430 million (half from French budget). The stocking rate must be less than 1 LU/ha or between 1 and 1.4 LU/ha if pasture lands include more than 75% of the total agricultural area.

increase in revenue, whilst the farms with 1.4 to 2.0 LU per ha would have a 1.3% decrease (Litvan, 1996).

Thus it is clear that the present beef regime favours intensive regions, mainly Brittany, at the expense of the most extensive such as Auvergne and Limousin. This system increases regional disparities, but also worsens the environmental problems of the most intensive regions. As previously indicated the arable crop regime increases the competitiveness of the intensively produced meat (by lowering costs of feed), over grass-based beef output. Intensive beef rearing supposes more silage maize and therefore greater use of chemicals.

3.2.4   The sheepmeat regime

Although lamb prices have stabilized since 1990, and have even increased recently through the consumer reaction to BSE, the general trend during the period 1980-1996 has been a decline. Nevertheless this price decline has been overcompensated by the increase in the sheep premia. The compensatory sheep premium, which is a headage payment, reaches FF 165 per ewe. The total amount in 1995 was FF 1,150 million from the EU budget. Moreover there is a particular premium for the farms located in the Less Favoured Areas which account for 60% of the national herd. The two sorts of premia represent about 30% of the value of the sheep output. The major difficulties for the French sheep sector stem from the competitiveness of UK and Irish production which represent over three quarters of total lamb meat imports.

Because low-intensity sheep production is practised in the uplands and the mountainous regions there are positive effects on the environment. From an ecological point of view this system helps to maintain the nature conservation value of these areas. High grazing pressure by livestock for a relatively short period is essential to prevent scrub encroachment and maintain floristic diversity (Bignal and Mc Cracken, 1996) especially when land is inaccessible to tractors. From a human and cultural point of view, the sheep regime helps avoid depopulation and reduces forest fire risks in remote areas.

It has been shown in the Northern Massif Central that sheep holdings represent 15% of the total number of holdings, but constitute only 10% of the total units of livestock (LUs) (Benoit et al., 1997). The discrepancy reflects the extensive way these holdings are farmed. Compared with beef production or dairy farming, the flocks of sheep exploit rough grazing and seasonal grazing which could be abandoned since they are less productive.

The sheep regime is favouring the use of these extensive areas. A panel of sheep farms, for which records cover the period 1987 to 1995 clearly shows the existence of two subperiods: before the CAP reform, and after. The pre-reform subperiod is characterized by an increase of the usable agricultural area by 2 ha per year. During the same time the degree of specialization in sheep increases at the expense of cattle. Since 1992 the yearly acreage increase becomes more significant reaching 2.7 ha. In parallel the use of compound feeds per ewe decreases from 145-150 kg in 1987/8 down to 130 kg in 1995 indicating a substitution between compound feed and grass.

3.2.5   Agri-environment policies

The EC Regulation 2078/92 is widely adopted in France, except, that is, for long term set-aside. The type of measures and the way they are implemented - whether as national, regional or targetted local programmes - are summarized in Table 5.

There are two nation-wide agri-environmental measures. The most important is the voluntary grassland premium which is aimed to protect land used for low density grazing. This premium also is to maintain low density farming systems. The conditions of eligibility have been defined above in the examination of the beef regime. Each year, since its introduction in 1993, over 100,000 applications have been made. Some 5.8 million ha were covered by the end of 1995. There is an important concentration of the applications in the Massif Central and, not surprisingly, 45% of the acreage is in mountain regions, 36% in the other Less Favoured Areas and only 18% is in lowland regions (CNASEA, 1996). It is clear that the grassland premium scheme operates in areas where a process of extensification has been in progress for years and it assists in the maintenance of farming in areas under threat of depopulation.

Table 5   *Agri-environmental measures adopted following EC Regulation 2078/92*

Type of programme	Objective	Measure	Annual payment in 1995
National	- To maintain extensive production	- Grassland premium	- FF 300/ha with a FF 30,000 ceiling per farm
	- To promote economic sustainability	- Sustainable farm development plans	
Regional	- Water quality protection (i) Source protection (ii) River protection (iii) Erosion protection	- Reduction of inputs - 20-year set-aside - Conversion to grassland	FF 3,000/ha ± 20% FF 2,500/ha ± 20%
	- Extensification by enlargement	- Reduction in stocking rate on original area	FF 1,500/livestock units (LUs) removed
	- Preservation of threatened breeds	- Subsidy per hectare according to crop type	FF 300/livestock units (LUs)
	- Nature protection - Conversion to organic farming	- 20-year set-aside - Subsidy per hectare according to crop type	up to FF 3,000/ha FF 700-4,700/ha
	- Training		up to FF 15,000 per person
Local	Sensitive ecosystems, land abandonment and countryside management	Subsidy in relation to constraints imposed	from FF 100 to 1,100/ha

Source: Boisson and Buller, 1996 and CNASEA, 1996.

71

'départements' are concerned: those where a large amount of land was freed, and those were the number of farmers taking early retirement was low.

For the first group the most representative département was Vendée which had the highest number of farmers leaving the land in 1992 and 1993 with a rate of 56% of eligible farmers, whereas the average French level was only 21% (Allaire and Daucé, 1995). Because of the amount of land freed up, takers where not found for all of it, especially in the least favoured areas. The main consequence is abandoned land resulting in degradation of the landscape and the environment. Vendée is representative of cattle rearing in the whole western France.

For the second group, the most representative département is Vaucluse where only 11% of eligible farmers opted for the early retirement scheme in 1992 and 1993. This 'département' is representative of fruit and vegetable areas in the Mediterranean region suffering from over-intensive agriculture. In Vaucluse 15% of freed land did not find a taker. The most affected sector is market gardening. In these areas unused land may help the environment because of reduced pollution from pesticides and fertilizers.

Concerning the Less Favoured Area Scheme its purpose is to allow a continuation of farming in these areas by ensuring a minimum population level or conserving the countryside. The financial incentives used are not intended to encourage conservation per se, but instead they are targeted at a more general positive effect on the environment and the countryside. LFAs are to affect three types of farm situations:
(i)   mountain areas handicapped because of high altitude or by steep slopes;
(ii)   areas in danger of depopulation;
(iii)   other areas affected by specific handicaps arising from permanent natural conditions which are unfavourable for farming.
The proportion of the French Utilized Agricultural Area (UAA) included in LFAs reaches 12.4 million ha (40%).

Even though the relationships between intensification of agriculture and nature conservation is not well understood, the consensus is that intensification reduces species diversity. Stocking densities and grazing patterns are from this point of view particularly important. Traditional management practices, such as late harvesting of meadows and arable crops, or the shepherding and seasonal movements of livestock create favourable conditions for grassland flora and nesting birds. Fertilizer use and overgrazing may lead to an impoverishment of sward diversity and even to soil erosion in Mediterranean regions. From this point of view, investment supports that lead to a higher degree of mechanization with efficient methods of harvesting or favouring irrigation may have negative impacts on the environment. The stocking density limits to obtain the compensatory allowances (0.2 Livestock Unit, LU, per hectare in mountains over 1,200 m, and 1 LU per fodder hectare in the other LFAs) may in particular circumstances involve an intensification of grassland management.

In fact, in many cases LFAs are in the process of being run down or abandoned, often because of the poor agricultural income which they are able to generate and the lack of successors to the present farmers. The investment aids in tourist and craft industries on farms are very useful to encourage the development of supplementary activities to farming, thus ensuring a better standard of living for farm families. By this way agricultural structures are strengthened helping to develop the social fabric of rural areas, to safeguard the environment and to preserve and maintain the countryside.

The headage and area payments can contribute significantly to the survival of low intensity farming in many areas. The statistical service of the French Ministry of Agriculture has estimated the effect of the CAP Reform on Gross Operating Surplus per farm at regional level considering LFAs (Blogowski and Boyer, 1994). The most important increases occur in the LFAs, mainly in mountainous areas (The Massif Central and the Pyrénées). If we consider Limousin, the poorest French region, we have a low intensity livestock system representative of the mountainous LFAs. The most important increase of Gross Operating Surplus occurs in mixed bovine systems (dairy and beef production, + 33%) in sheep systems (+ 22%), in beef production (+ 17%) and lastly in dairy systems (+ 9%). For the region as a whole the total increase is 16%, the highest. So we can see that breeding regions characterized by geographic handicaps improve their situation in normal conditions because of LFA payments.

The LFA scheme will have a general positive impact on the environment and landscape by avoiding land abandonment. But, it is an unsophisticated instrument which cannot discriminate between farms providing environmental values and farms which are damaging the natural resources because appropriate stocking levels are not maintained.

3.3   Environmental conditions in agricultural policy

Existing environmental policies of course have an impact on agricultural policies and on agricultural practices. It can be seen through both general and specific environmental regulations.

3.3.1   General requirements

Environmental protection and landscape take into account the incompatibility between some production activities and residential areas. It is a matter of land use and thus comes under town planning law. At the local level, the majority of the 36,000 French municipalities, do have a land use plan (Plan d'Occupation des Sols, POS). A land use plan sets out general rules for land use within the municipality and allows for specialization of activities upon the local territory: urban sites, areas of natural beauty, agricultural land, sensitive areas and areas of special interest from an ecological or aesthetic point of view. Although such plans do not influence agricultural policy, nevertheless they have a strong impact on farming systems, mainly in the neighbourhood of urbanized areas.

However, the most important conservation policy was the setting up of the National Parks in July 1960. National Parks are primarily designated because of their value as 'important sanctuaries of nature', but an important feature is that they cover land which is inhabited, largely privately owned and used economically. Therefore it is necessary for their management to take into account both their value for conservation and the indigenous 'human' element. The result is a compulsory regulation which distinguishes between two types of areas:

(i) a central area, characterized by wilderness, generally uninhabited, with strict protection. Sometimes, the core is a reserve in full where fauna and flora are protected for scientific reasons;

(ii) around this central area, there is a peripheral zone used economically by an indigenous population. It has implications for the objectives and operations of such designated areas, placing particularly emphasis on integrating rural conservation with other land-use, particularly agriculture which has to meet strict environmental conditions.

The seven National Parks cover 12,758 sq km which corresponds to 2.2% of the territory. But the Regional Parks cover 8% of the territory. These latter are generally situated in economically disadvantaged areas with relatively high percentage of employment in farming, low population density, and difficult natural conditions. In contrast to National Parks, uninhabited in their central areas, the Regional Parks aim for the maintenance of local communities and the reversal of rural decline. For this reason conservation is considered as a practice which requires active management rather than static protection of valued land. This involves an integrated approach to conservation and development issues for the greatest benefit of the environment.

Moreover, a more comprehensive environmental approach has been introduced through the Coastal Protection Act, the Mountain Act and the Landscape Act.

In order to safeguard threatened coasts and lake shores, a national government agency, the Coast and Lake Shore Conservancy Authority has been established in the spirit of the British National Trust (July 1975). Its sole task is to purchase land to maintain the natural inheritance and to preserve it for future generations. To preserve or re-establish traditional land use, the Conservancy does not manage the sites it owns. There are management agreements with local authorities (communities and 'départements') or with voluntary organizations or farmers. According to the state of the ecosystems, the management is oriented towards either simple conservation or restoration. In this way the Coastal Protection Act has an impact on local agriculture.

The Mountain Act of January 1985 is based on the same principles: there are specific problems for these areas and they require a comprehensive approach. Such an approach must take into account aesthetic aspects, but also the historical, economic and sociological elements that define a mountainous massif. The Mountain Act deals with 5,600 communities and aims at controlling the impact of economic development on landscape. It is interesting to recall that France around 1880 introduced a policy to limit the effects of over-farming in restoring plots of mountain and hill land. This over-farming, which was due to overpopulation, caused extremely harmful erosion in the most fragile areas, causing serious consequences, for areas far downstream.

In comparison with the two other Acts, the Landscape Act of January 1993 defines a more general policy. The underlying philosophy of this law is its extension from the protection of interesting natural features to more ordinary areas. Nowadays any programme of land development or land management has to take into account the countryside. These programmes include land consolidation since difficulties remain with the removal of hedgerows, the ploughing up of permanent pastures and the destruction of woodlands and scrub.

3.3.2   Specific requirements

Many specific environmental regulations deal indirectly with agricultural policy. The most significant relate to the recent Clean Air Act, the protection of catchment areas, the system of registered installations and the conditions of spreading sewage sludge and municipal waste on agricultural land.

The new Clean Air Act requires the incorporation into fuel of oxygenated compounds such as rapeseed methyl ester, ETBE (Ethyl Tertio Butyl Ether) produced from sugarbeet and wheat, and MTBE, which may be derived from oil. This decision is regarded by the farm unions as a key tool in reducing the need for set aside. But Article 7.4 of Regulation 1765/92 gives the option of keeping set-aside land in production rather than letting it idle. It defines eligibility as 'products not primarily intended for human or animal consumption' and detailed rules are laid down in Regulations 334/93 and 2595/93. Eligible crops are primarily cereals, oilseeds and protein crops. In addition to biofuels, various chemical, plastic and paper products are eligible under certain conditions.

In 1993, total non-food use set-aside area in France reached 73,000 ha on which rapeseed occupied over than 50%. In 1994, the total non-food use set-aside area was about 280,000 ha, the main part being sown with rapeseed. Planting of rapeseed was 408,000 ha in 1995 but decreased by 30% in 1996 according to SIDO. This development was accompanied by a development of the processing plants. Their capacity for bio-ethanol is 450,000 hl in 1996.

Catchment area protection

There is a specific provision in French Law for Public Health dealing with the protection of catchment areas. This requires designing and implementing a zone around every catchment in order to protect the quality of the resource by banning or regulating various uses. Moreover public ownership is prescribed for the central portion of the protected area. The Water Act of 1992 emphasizes these provisions and prescribes that a protection programme has to be implemented within a 5-year period. Because a limited number of wells are protected, at the national level only 24% of the total were protected by 1993; there is some evidence that the 1997 deadline will not be met (Bonnieux and Rainelli, 1996).

Registered installations and intensive rearing (classified installations for environmental protection)

The Act of July 19, 1976 on registered installations for environmental protection, and its implementation orders replace the 1917 Act on hazardous establishments which did not cover intensive livestock rearing. An activity coming under this Act has to be registered. This applies to livestock rearing installations of a certain size. According to size, they are subject to either a reporting regime (i.e. 'déclaration') or a permission regime (i.e. 'autorization') (Table 7). The permission regime is more exacting, since the applicant has to provide an impact assessment giving details of the source, nature and magnitude of any disamenities liable to result from the installation concerned. Disamenities include noise, use and discharge of water, the protection of underground waters, and waste disposal.

Table 7   Legal regime for classified livestock rearing installations according to animal type and farm size

	Reporting regime	Permission regime
Pigs over 30 kg	50 - 450	over 450
Poultry	5,000 - 20,000	over 20,000
Rabbits over 30 days	2,000 - 6,000	over 6,000
Veals or fattening bovines	50 - 200	over 200
Milk cows	40 - 80	over 80
Suckler cows	over 40	

The Act on classified installations has an important impact on agriculture since it limits intensive livestock rearing in areas with intensive animal production.

Use of sewage sludge and municipal waste on agricultural land

Total amount of sludge which was 865,000 tonnes in 1993 is projected to reach a level of about 1.3 million tonnes in 2003. Nowadays some 58% is disposed on land by using it as fertilizer or by processing it into compost also for agricultural purposes. But the spreading of such waste on agricultural land has potential harmful effects on the environment, human health and agriculture itself since it contains heavy metals like zinc or copper and more toxic products like lead, cadmium and mercury.

The spreading of sewage sludge and municipal waste is limited in order to prevent any long-term harmful effects on the environment and human health (Directives of June 12, 1986). In France the AFNOR Norm U 44041 defines the Maximum Allowed Concentration for the main heavy metals. According to this norm the number of years to reach the permitted concentration is about 170 years for lead and nickel. In fact for intensive cropping the compost application can be 10 times greater than the authorized amount. To avoid any problem concerning the quality of the products, the food industry does not accept in some regions (North of France) vegetables grown on sewage sludge.

3.4   Conditions in environmental policies

Among the general laws is the Water Act of December 16, 1964, which deals with management and distribution of water and the control of water pollution. This act created the qualitative and quantitative management system for the six major French hydrographic networks through advisory basin committees and executive basin agencies (The 'Agences Financières de Bassin', i.e. Water Authorities). Under that Act, these agencies determine and define quality targets for watercourses. Any issues related to quality and flows are controlled through a system of economic instruments with fees, which has to be extended to the agricultural sector through the implementation of the Nitrate Directive.

The nitrate policy, which aims to reduce and prevent water pollution by nitrates from agriculture, plays an important role in influencing farming practices through command and control regulation but also by means of economic incentives.

Here we examine environmental regulations concerning the implementation of the Nitrate Directive. Three administrative circulars were elaborated by the French authorities, in March 1992, June 1992 and November 1992 to implement the Nitrate Directive in France. The latter explains how to define the vulnerable zones at the regional level. A task force was set up in each 'département' to improve co-ordination and information. The zoning is based on the average level of nitrate concentration during the past years. Waters for which the level of NO_3 exceeds 50 mg per litre are automatically considered as polluted. Consequently the area is classified as 'vulnerable'. Waters below 25 mg are supposed safe. Special attention is paid to intermediate levels of nitrate concentrations. Time-series are set up in order to define any possible trend in an attempt to foresee possible problem areas. If the trend indicates a potential level of NO_3 over 50 mg in 2005, the aquifer or the river is classified as polluted. In waters where the nitrate concentration reaches 40 mg per litre and where time-series data do not exist, the prospect of becoming fully polluted is assumed to be an imminent possibility.

The agricultural pollution control programme

Since 1975 there has been a clear legal basis for imposing fines on non-point sources. In 1982 the water authorities proposed a system of levies on pig units according to the number of places available and slurry spreading possibilities, according to slope and the proximity of water courses. However this system never entered into use because of farmers' unions' objections.

In June 1991, the French government decided to combat water pollution caused by nitrates from agricultural sources and to introduce a levy on nitric sources of nitrogen. The aim was a progressive introduction of the polluter-pays principle into agriculture as was already the case in other sectors. To achieve this purpose, a special agreement was concluded between the Ministry of Environment and the Ministry of Agriculture on March 11, 1992. This agreement includes various Directives of the European Union which are related to water quality and those concerning the protection of waters against pollution caused by nitrates from agricultural sources.

Even if priority is given to measures to prevent pollution, especially regarding the extension service (more efficient application of fertilizers by more accurate timing or rates of application and a better use of animal waste), the role of economic incentives is important as well. Subsidies are provided to encourage farmers to reduce their pollution by using methods more compatible with the environment, by preventing water pollution by run-off and the leaching of liquids containing livestock manure and effluents from silage, and by processing manure.

Water effluent charges

The extension of the fee system expands a process originated in the late sixties. The basic principle for determining the amount of the fees remains the same. At the moment, these fees cover four substances in livestock farming: suspended solids, oxidisable matters, reduced nitrogen and phosphates. In the future, other substances (e.g. heavy metals and some organic toxins) would be taken into account. There are three steps in calculating the levy.

(i)   Emissions in physical units are calculated for each pollutant and each category of livestock (dairy cows, pigs, poultry). This is based on industry averages since technical coefficients are used to transform number of animals into quantity of polluting substances. Monetary coefficients then are used to obtain a gross charge per farm. These coefficients are pollutant specific and do not vary according to the various industries.

(ii)   Farms are classified according to a number of parameters including their manure storage capacity, location of the buildings, run-off from buildings, manure spreading scheme and livestock density. This process defines 9 environmental classes which are based on individual performance. The abatement of polluting emissions is then estimated for each farm. This outcome is converted into monetary terms in order to obtain a premium per farm.

(iii)   Net charges are then calculated, i.e the difference between gross charge and premium. The charge system and specially the technical coefficients result from negotiations between parties and are therefore a compromise. It must be emphasized that net charges do not include either suspended solids or phosphates because a 100% abatement is assumed for these two substances. Farmers will have to pay the net fee if it is lower than 6,458 FF (in 1996), which corresponds to 200 population equivalent 1). This is the monetary equivalent of farmers' rights on the environment.

The charge system was planned to be gradually enforced, according to the following calendar:

*   for those rearing installations classified before January 1, 1991 charges will be paid from January 1, 1993;

*   for those rearing installations classified during 1992 charges will have to be paid from January 1, 1994;

*   for all animal husbandry and crops the charges will have to be paid from January 1, 1995.

For each category subsidies will be given during the year before the levies are charged to help the farmers to meet the required standards. This system was criticized from a theoretical point of view because the regulatory mechanisms to control non-point source pollutants are more difficult to implement than those for point source pollutants. There are different cost structures for the acquisition of information and informational differences between polluters and regulators can be found. From a political point of view, French farmers protest against the proposed schedule and also against the principle of taxation itself. The farmers' unions argue that agriculture is a special case because of the special circumstances of the farming economy, of agricultural pollution and agricultural policy. Concerning the levies on intensive animal rearing, pig producers argue that the tax is unfair because mineral fertilizers are excluded. From an economic point of view the large increase in European pig production has caused prices of pork to fall sharply from around 150 ECU/100 kg in the summer of 1992 to 125 ECU by mid-October before sinking to around 100 ECU in April 1993. So their financial position has deteriorated and they refuse to accept any extra cost.

---

1)   One person equivalent is a reference unit of pollution from one person (3.65 kg of nitrogen per year). Farmers are exempted from taxation if their pollution level does not exceed 200 person equivalents.

All these elements led the former government to delay the application of the taxation system. The largest farms entered the scheme in 1995. The process is expected to be completed within a five-year period. Moreover, in order to reduce the effect of the charge system on farm income, there is a transition period during which eligible farmers will only support a proportion of the total amount. For example, for 1995 eligible farmers had to pay 40% of the total, in 2000, 76%. It is expected that farmers will pay the total of the charge in 2003.

3.5 Concluding observations

Traditionally in France, policymakers, farmers and the food industry have oriented agricultural policy towards a more productive agriculture. The main agri-environmental concern was not the environmental impacts of farming practices, but agricultural land abandonment and the return of once cultivated or grazed lands to the natural state. It is revealing that agri-environmental measures have been slow to penetrate the Ministry of Agriculture and the farming community. In consequence, it is not surprising that the French agricultural interest groups did not try to influence the new CAP towards less intensive farming practices. The possibility to consider maize as a cereal or as a fodder is a good example; the definition of high historical yields for irrigated crops is also revealing.

More generally, the fact that the new CAP is not completely decoupled has two broad consequences. First of all, the arable crop regime leads to compensatory payments based on fertility. As a consequence, environmentally detrimental use of fertilizers and pesticides is not really discouraged in the more favoured areas. On the other hand, the semi-decoupled regime operating for beef, sheep and to a certain extent for the milk quotas, does contribute to more friendly environmental practices in the less favoured agricultural areas.

In a general way, the differentiated agricultural support makes sense in maintaining the private competitiveness of cereal growers and the French returns from the CAP budget.

Finally, most of the environmental regulations more or less oriented towards agriculture are still too weak to meet the objective of resource conservation and sustainable agricultural development. But a successful integration of agricultural and environmental policies supposes that both the amenities provided by agriculture and the negative externalities originating from agriculture are correctly assessed.

References

Allaire, G., P. Daucé (1995) *Etude du dispositif de préretraite en agriculture;* Fascicule 8, l'impact structurel de la préretraite. INRA-ESR Dijon

Baromètre Porc, 1996, n°233, septembre

Benoit, M., G. Laignel, G. Liénard (1997) *Elevages ovins de race rustique en Massif Central Nord RAVA, BMC,* Limousine. INRA-ESR Laboratoire d'Economie de l'Elevage, 38p.

Bignal, E., D. Mc Cracken (1996) *The ecological resources of European farmland;* In: M. Whitby (Ed.) The European Environment and CAP Reform: Policies and prospects for conservation. Wallingford, CAB International, pp. 26-42

Blogowski, A., Ph. Boyer (1994) *La réforme de la PAC: quels effets sur les revenus?* Agreste-Données chiffrées, n°55, March 1994

Boisson, J.-M., H. Buller (1996) *Agri-environmental schemes in France;* In: M. Whitby (Ed.) The European Environment and CAP Reform: Policies and prospects for conservation. Wallingford, CAB International, pp. 105-130

Bonnieux, F., P. Rainelli (1996) Public Policies for catchment protection. *International Conference on Multiple Land Use and Catchment Management,* Aberdeen, 11-13 september 1996, INRA-ESR Rennes, 17p.

Boussemart, J.-P., G. Flichman, F. Jacquet, H.-B. Lefer (1996) *Prévoir les effets de la Politique* Agricole Commune sur deux régions agricoles françaises: application d'un modèle bio-économique; Revue Canadienne d'Economie Rurale, 44, 121-138

CNASEA (1996) Rapport d'activités 1995

Cohen, J., P. Dupraz, D. Vermersch (1996) *Nouvelle PAC et nouveaux projets d'irrigation;* Cahiers d'Economie et Sociologie Rurales, n°39-40, 223-250

Dourmad, J.-Y., C. Le Mouël, P. Rainelli (1995) Réduction des rejets azotés des porcs par la voie alimentaire: évaluation économique et influence des changements de la Politique Agricole Commune. *INRA Productions Animales,* 8 (2), 135-144

Guyomard, H., X. Delache, X. Irz, L.-P. Mahé (1996) *A Micro-econometric Analysis of Milk Quota Transfer: application to French Producers;* Journal of Agricultural Economics, 47, 2, 206-223

Léon, Y. and M. Quinqu (1994) Réforme de la PAC: un faible impact sur la répartition régionale des valeurs ajoutées, *INRA, Actes et Communications* (12), 119 140

Litvan, D. (1996) *Des aides à la surface: pour un élevage plus extensif, une meilleure régulation du marché et une plus grande intégration au territoire;* Document de travail, Ministère de l'Economie et des Finances Direction de la Prévision, 63p.

Perrier-Cornet, P., (1995) *Les structures depuis 1990 et leur évolution. in Production Laitière: structures, coûts, rentabilité;* Les Cahiers de l'ONILAIT, n°14, May

Piot-Lepetit, I., D. Vermersch, R.-D. Weaver (1997) *Agriculture's environmental externalities: is there a free lunch?* DEA evidence for French agriculture. Applied Economics, 29, 331-338

Vermersch, D., F. Bonnieux, P. Rainelli (1993) *Abatement of agricultural pollution and economic incentives: the case of intensive livestock farming in France;* Environmental and Resource Economics, 3, 285-296

4. IRELAND

Brendan Flynn

4.1 Introduction

On the eve of long promised substantive reform of the CAP 1), this chapter aims to examine to what extent perceptions, policies and research has changed among the agriculture and environment policy community in the Republic of Ireland. It is argued that notwithstanding evidence for some positive change, the goal of serious policy integration still remains substantively far off. In particular perceptions are still led and dominated by a powerful network of state-farmer interests (Collins, 1993), who view the environment more as a tactical issue to be exploited for continuing transfers and subsidies at a time of impending liberalization of the CAP's price regime. Policy therefore centres typically around massive EU transfers (see Table 1). However, these may be quite ineffectual as they can merely end up displacing wastes and also lack a targeted approach towards specific problems. Additionally it is argued that the state of research in Ireland into agri-environment problems remains wanting. Much is conducted in an ad hoc or fragmented way. What studies are undertaken remain often very heavily dependent on EU funding and guidance, domestic interests in the area being limited.

Overall then a relatively conservative picture emerges from Ireland. Unlike some other European states, notably Austria (Hofer, 1997), there are few signs of large-scale support for domestic initiatives which could articulate and support new patterns of smarter, greener farming or even a more extensive, organic type of production.

4.2 Irish response to the CAP reform in general: A conservative consensus?

Traditionally, the perceptions of the CAP held by a powerful corporatist state-farmer axis in Ireland have been largely pragmatic and positive. Moreover in terms of general public opinion, Ireland has consistently shown a high level of support for European integration, which sustains a wider popular interpretation of the CAP as perhaps almost synonymous with what European inte-

---

1) See for example Commissioner Fischler's Agricultural strategy paper, Commission of the European Communities, 1995. 'Study on Alternative strategies for the development of relations in the field of agriculture between the EU and the associated countries with a view to future accession of these countries.' CSE(95)607. The strategic changes which this approach promises for the CAP may be thought of briefly as being: (1) Increasing demands for continuing liberalization, (2) fuller integration of environmental issues into the CAP, (3) continuing need for a deepening of the Integrated Rural development approach, (4) Simplification and reducing the bureaucracy of the CAP, including greater national descretion and control, and finally, (5) the question of enlargement and the imminent entry to the CAP of the Visegrad east-central European states.

gration means in practice: subsidies for farmers and thus a historic lifeline for rural Ireland 1). Indeed one very distinctive theme in the Irish perception of the CAP, has been to focus on the politics of redistribution during mooted reforms rather than the issues of economic efficiency, much less environmental concerns. For example, during the Mac Sharry Reform proposal of 1992, both of the main farming organizations argued vehemently that cuts in price supports would devastate rural Ireland as a whole. Indeed one might go as far as to say that the structure of Irish state-farmer interests are such that any reform of the CAP will be viewed very guardedly.

This is firstly because the Irish state and the two main farming groups in Ireland, the Irish Farmers Association (IFA) and the Irish Creamery Milk Suppliers Association (ICMSA), continue to dominate official responses to reform proposals. Secondly it should be noted that a huge investment in political capital and energy has been made by successive Irish governments into the CAP system as it has existed in the past (Kearney et al., 1995). This strategy is felt by the key Irish actors to have both modernised Irish agriculture and secured an effective power base for Irish interests at the heart of Europe by which to secure substantial transfers. Therefore it is very likely that the Irish consensus is relatively inimical to any further change of the CAP.

However, the associated reform demand for CAP liberalization in particular, is complex. In 1992 the GATT deal presented unique challenges to farmers, the state and agribusiness in the way it pitted free trade against farm subsidies 2). Should such a fulcrum effect of forces emerge again one cannot predict that the state and agribusiness sector will support the likely opposition to reform from the established farmers' interest groups. In such a context the arguments for further liberalization may be persuasive to state and agribusiness concerns, especially as key Irish actors

---

1) This assertion is based on Eurobarometer Trends data quoted in Laffan, Brigid. 1992. *Integration and Co-operation in Europe.* London: Routledge. pp.124. This shows that in 1975 for example some 50 per cent of Irish respondents thought that membership of the Common Market was generally a good thing, making Ireland about the sixth most enthusiastic supporter. However, by 1990 some 75 per cent of Irish respondents suggested that membership was a good thing, a huge increase putting Ireland in third position of a ranking of Member States whose citizens are most favourable to the Common Market. The percentage who suggested that membership was 'bad' for Ireland fell from 20 per cent to 8 per cent in the same period.

2) Whereas the Irish farmers remained essentially uncompromising and stalwart in their opposition to both the CAP reforms and some details of the GATT agreement, in contrast trade unions and industrial leaders were critical of this stance, being quick to point out that on a more complete national costs-benefit analysis, it would have appeared to be in the national interest to see the GATT deal signed even at the expense of some losses in the agricultural sector. In short, some of the Dublin based media, and the urban based industrial interests generally, were pointing out that Ireland was first and foremost an industrial trading nation with a very open economy, and not simply any longer an underdeveloped agricultural society. Equally there was a certain hostility on the part of the Irish trade union movement in particular to the established power and influence of the Irish farmers' lobby and the sense that they could 'get their own way-costing us urban industrial jobs if GATT fails.' All this after years of huge subsidy from the taxpayers of Europe but a very low share of national tax contributions from the farming sector? What was also significant was that some agribusiness concerns broke ranks with the farmers over the GATT related issues to the Mac Sharry reform. They simply feared the trade related consequences of failure to sign the GATT agreement, and also clearly it was in their interests to see the costs of many of their raw material inputs come down under Mac Sharry's proposed pricing regime. Thus in many ways, the 1992 reform-GATT crisis represents perhaps a key turning point in perceptions of the CAP in Ireland.

should be seen as not ideologically opposed to further liberalization, but are instead pragmatically ambiguous about where they stand on such issues.

In fact Irish views on liberalization may be more receptive than was previously thought 1). In private for example, some officials of the farmers' representative bodies will admit that there would be attractions and advantages for the more competitive farmers in a more liberalized regime. For environmentalists, liberalization might mean less waste and largesse from farmers than we have hitherto known, and perhaps even a re-deployment of funds towards environmental concerns. Indeed the two smaller farming groups, the Irish Organic Farmers and Growers Association (IOFGA) and the United Farmers Association (UFA), are in particular more open to the Mac Sharry Reforms. This is because they see in such measures a possible ending of the absurdities of oversupply, bureaucracy, and primarily what they see as unfair subsidies for large intensive farmers.

It therefore appears that questions of environmental side payments are merging with the tentative beginnings of a debate on the merits of liberalization. Currently some of the policy elite within the Department of Agriculture, Food and Forestry (DAFF), Teagasc and the IFA and ICMSA (as distinct from their rank and file) seem to be cautiously accepting the inevitability of some further liberalization reforms. This might involve some highly competitive producers operating at close to world market prices but within environmental constraints, but also a diversified sector of small scale 'niche' farmers would remain, involved in agri-tourism, alternative enterprises, environmental stewardship activities, and supported by direct income support where necessary.

Above all it appears that the key Irish actors will pragmatically accept environmental projects under the CAP reform in return for income lost through price support reduced, as was arguably done in the Mac Sharry reform with its attendant Regulation 2078/92. Therefore further moves towards integrating environmental and social criteria as Commissioner Fischler has identified as necessary (CSE (95)607), may actually from an Irish perspective make liberalization more palatable and acceptable. Certainly there is little of a view which is fundamentally opposed to liberalization on the grounds that it will lead to greater environmental externalities (ranching, biotechnology, American style agribusiness). Rather concerns when they are expressed seem focused on the social issues of rural decline (Maloney, 1994; Sheehy, 1992). From a comparative perspective the defensive use of environmental issues to fend off appeals for liberalization (suggesting that it leads to serious environmental problems) has not emerged as a strong tactic by the leading Irish actors as yet. Insofar as they articulate any coherent strategy, they prefer to see environment reforms as a key bargaining chip in the context of support for liberalization rounds so as to secure new incomes.

The impending integration of the Visegrad states 2) after the year 2000 also represents a major challenge to Irish perceptions of the CAP as a fund for income transfers when that fund

---

1)   This became apparent in a number of interviews conducted with DAFF, IFA, ICMSA officials, September 1996.

2)   Usually taken to include Poland, Czech Republic and Hungary. Slovakia has not been considered an acceptable applicant for EU membership by the EU Commission as of late 1997, although Estonia, and Cyprus and Malta provisionally have been included.

must of necessity become drastically more sparse for the Irish, considering how much has been won in the past. The real concern appears to be at the pace and character in which continued CAP funds will decline; will it be sudden and dramatic or will it be more gradual and planned? For the main farmers groups, the IFA and ICMSA, the prospect of yet more competition from the Visegrad states, together with continuing liberalization cannot be a happy one.

Conversely if the Irish share of transfers drop, a rationing principle for those who need the support most, in line with Commissioner Fischler's observation about the need for greater targeting of aid, may seem to offer more marginal rural groups and those seeking a clearer focus on rural development rather than agriculture, the chance relatively speaking to win the lion's share of a much smaller amount. In this context groups such as the UFA and IOGFA have been continually arguing for greater justice between farmers within the CAP, and may welcome a chance to become the focus of smaller but from their point of view, a more beneficial type of state support. In contrast the larger farmers' groups have a more direct view; targeting may be a code word for austerity and may simply mean less money for rural Ireland as a whole. Thus the environmental issue has perhaps some potential to open up the myth of unity between Irish farmers, the state, and agribusiness, something that was itself perhaps merely a fair weather system of the golden days of the unreformed CAP 1)?

Pressure for change in perceptions is also likely to arrive as a result of increasing spill-over effects (such as the BSE crisis in the UK) and a general trend towards more consumer and retailer emphasis of food quality and safety. There are already some indications from marketing sources in Ireland that no necessary 'clean and green' image is enjoyed in the wider perception of Ireland abroad (Mulrennan, 1995b). In this light there will be a continuing likely need to actually prove Ireland's good environmental credentials substantively.

4.3 Greening the CAP: Have Irish perceptions taken onboard the environment agenda?

In the above section the dominant consensus on the CAP reform in Ireland was explained, while some of the challenges to that were also outlined. In this section we consider to what extent environmental issues of the CAP have altered perceptions among the key actors. It is suggested that little real change has occurred.

In fact environmentalist critiques of the CAP in Ireland have been painstakingly slow to develop and perhaps this is more generally a function of the comparatively low levels of environ-

---

1)   State-farmer relations have historically been quite varied in Ireland. During the 1930s there were big political conflicts between the interests of larger farmers and smallholders which translated into electoral politics, while in the 1960s the National Farmers Association (which later became the IFA) embarked on a very conflictual pattern of relations, with street marches, etc.

mental awareness which the Irish public display compared with some other European states 1). For example there still does not exist any one single environmentalist group focused specifically on the CAP, much less any examining the CAP reform process from an environmental point of view. Interestingly neither does Ireland have any real comparable state organization to the cohesive UK countryside agencies which serve to integrate together most environmentalist concerns about the rural environment and landscape. Irish environmentalist NGOs remain small-scale and of basically three distinct types.

Either they are nationally organized, generalist organizations which occasionally run campaigns on environmental issues related to agriculture, such as Greenpeace Ireland 2), or else they are very specific, such as Birdwatch Ireland (the Irish equivalent of the RSPB but nowhere near as large or as effective). Alternatively they take the form of localized, short-lived individual campaigners on specific instances of water pollution related to agricultural development, for example groups like SOLD (Save Our Lough Derg) or the Lough Sheelin Anti-Pollution Group. By utilizing local clientelistic networks with the established political parties these groups are often the most effective in securing ad hoc responses from the state (McTeare, 1991, Nugent, 1995, Allen and Jones, 1990).

What is also perhaps worth nothing about many of these smaller local groups, is that they are as much concerned about decline in fish stocks from the point of view of local angling and amenity interests, as they are from any environmental perspective (Ibid.). Where such 'local pressure' is not successful in resolving disputes, a quite 'adversarial' pattern of local environmental politics can develop. The Environmental Action Alliance is perhaps a typical example in this regard, being loosely allied with the Trout Anglers' Federation of Ireland, and eschewing a more sedate professionalized role of political and policy lobbying at national level, for a more localized direct, legal, and confrontational style 3).

Indeed in contrast to such small groups, some of the leading notable national environmentalist groups, such as Earthwatch, have little or no interest or expertise on the agriculture issue,

---

1) From Commission of the European Communities: 1986. *The Europeans and their Environment.* Brussels: CEC, reported and discussed more fully in, Whiteman, David. 1990. 'The progress and potential of the Green party in Ireland', *Irish Political Studies*, Vol.5:45-58. Interestingly the magnitude of the gaps between some of the Irish positions and that of the EC norm was such that one must suggest that, even if this data is comparatively out of date, Irish environmental awareness had by European standards a lot of distance to cover by 1996 for it to be approaching the EU norm. For instance the Irish are some 17 points below the norm on whether the environment is considered *'An urgent and immediate problem'* and some 15 per percentage points above the norm in agreeing with the assertion that *'Development should take priority over environmental protection.'* Even bracketing Ireland with the less environmentally aware Southern EC states does little to diminish the stark lack of general environmental awareness exhibited. (Op.cit.)

2) Something of a sign of just how weak Irish environmentalist NGOs are can be seen from the fact that from the middle of 1997 Greenpeace Ireland was officially wound up as a lobby group, ostensibly for financial reasons. As a result its profile on the agriculture issue in the last few years has been minimal, although some of the personnel involve show signs of regrouping into a new national environmental lobby.

3) Taken from interviews.

takes some additional measures, such as switching to organic production, or stocking rare breeds. The REPS is advertised and administered by both the State agriculture advisory and development field service, Teagasc, and uniquely by a network of private farm consultants who account for about 55% of current entrants to the scheme. Circa 23,000 farmers have currently joined the scheme, which is over half its original target of 40,000 (Leavy, 1997).

Although it is clearly too early to comprehensively judge the success of the REPS, nonetheless a few observations can be made which would tend to suggest pessimism about whether REPS really represents the move required toward a more truly preventative policy. To begin with, there has been a collapse of a supporting EU finance scheme, the Control of Farmyard Pollution Scheme (CFYPS) for 1994-1999. This has made the scheme much less attractive. Additionally there has been evidence of some breech of the rules and a general unease about the quality of private contractors' advice to farmers. One issue which has arisen is the lack of employment of appropriately qualified scientific staff by these private consultants to give expert advice or the fact that the training received by these planners in administering the scheme from the DAFF was quite short and seemed rudimentary.

Indeed this novel form of 'privatised delivery' of an environmental policy does give grounds for concern, as some private REPS consultants, with a background perhaps in farm financial advice rather than ecology would not seem ideal to secure the detailed environmental goals required by the REPS. Several private REPS consultants have been censured by the Department of Agriculture, Food and Forestry and penalties have been introduced to ensure compliance and a general tightening up of the whole payments system - for example a 100% first year payment in advance has been ended.

Yet as it stands the REPS is not perhaps as attractive as it could be for the most marginal smallholders, as until recently income has been accounted for social welfare requirements and maximum income from the scheme is circa IR£4,800-£5,000. These sums are limited enough in comparison to an average farm income for 1994 of IR£15,700 in Ireland (Mulrennan 1994, p.3). While this social welfare co-ordination problem has since been ameliorated by a recent decision to discount 50% of REPS income from social welfare considerations, nonetheless, REPS remains essentially as but one 'scheme' which farmers may opt for led by obviously rational concerns about income.

Arguably then, REPS represents little change towards a more genuine integration of the environment and agriculture. One can argue that the established corporatist networks between the Irish state, the EU institutions and organized farming interests have managed to ring-fence the issue into a largely technical debate within the narrow confines of the CAP beneficiary community. As a result a potentially political issue about farm pollution has in some ways been de-politicised and turned into a question of broader farm support and modernization. If one wanted to be perhaps overly sceptical, one might say that organized farming interests have succeeded in turning the principle of the polluter pays on it head, so that it has become one of, 'pollution pays' 1).

---

1)   In contrast to the huge amounts which have been lavished on the various CAP-led farm wastes projects, it is instructive to compare the amounts spent in Ireland on the dedicated EU fund for the environment, (LIFE). This has involved a total of £3.28 million in 1993 and some £5.06 million in 1994, which compares poorly to circa £500 million lavished on farmers for pollution control (EPA, 1996, p.213).

The REPS as it currently stands exudes the influence of income support rather than a rationale of ecological risk abatement. For instance payments are set at a threshold of 40 ha with voluntary participation, which means comparatively small to medium farmers will the join the scheme in a randomly distributed scattering across counties a pattern consistent with a fair share out of income but not one necessarily with tackling environmental problems?

Perhaps, were the scheme to be designed by an environmentalist with a hierarchy of ecological risks in mind, its structure would be different. To begin with the budget would be perhaps targeted at trying to entice high risk categories: sheep hill farmers in the west, intensive pig and poultry operators, small dairy farmers with poor waste infrastructure, those who are close to riparian zones, and those on sensitive, porous or acid soils, etc. But the REPS appears not to do this properly 1). It does not target financial aid in accordance with a hierarchy of risk, instead leaving it up to farmers to decide to join on the basis of whether the scheme might look financially attractive to them.

4.5 Have Irish research agendas changed towards integration of agriculture and environment?

As one might expect from the dominant perceptions of the CAP reform which continue to stress the productivist concerns of price liberalization or indeed the emphasis which policy has had on income transfers and subsidies for farmers, there is not a very developed level of research into agri-environmental problems in Ireland. Indeed one can say in general that there appears to be a grave paucity of serious scientific research with regard to the long term impacts which the individual CAP commodity regimes have had on the environment. Instead there is all too frequently an assumption that Irish conditions do not show malign impacts. Consequently there are only a series of somewhat ad hoc papers and reports that suggest, rather than authoritatively prove, what the environmental impacts of the CAP are.

These included an early paper by a Teagasc expert Harte (1992), whose conclusions reflect the still widely held view by Irish actors, that the impacts of the CAP on the environment have been essentially limited. A more recent paper, by Lee (1995), attempts to assess the overall agriculture situation in terms of the notion of 'sustainability' and reaches a more critical conclusion. Another key turning point perhaps in terms of levels of knowledge and awareness was the organization of a conference by the Royal Irish Academy on Agriculture and the Environment, which provided a welter of papers of mixed quality (Maloney et al., 1994). The fragmented and limited nature of such studies must be emphasised: there exists no seriously quantified long range study of the CAP's environment effects, either from a social science or natural sciences background. In fact one could say that the knowledge base appears fragmented into a series of small-scale studies, and is dominated by the natural sciences at the expenses of social sciences.

The level of domestic Irish support for research into the environmental impacts of the CAP appears very small, and it is significant that many research actors appear to have undertaken pro-

---

1)   This is not just my view. See for instance: Murphy, John. 1996. 'REPS not targeting Wildlife'. *Wings-Journal of the Irish Wildbird Conservancy-Birdwatch Ireland.* Autumn:21.

jects with EU funding, for example several projects have benefited from the Research Stimulus Fund. Comparatively, the amount of Irish government spending of R&D is generally less than other states, and agriculture despite its position within the Irish economy has not attracted a strong indigenous research and development community or share of that spending,

One can suggest that without the significant levels of EU funding what scientific and expert knowledge as does exist on the environment in Ireland would be even worse. In terms of actors who actually do research, by far the most important and authoritative actor is Teagasc, the Irish state agriculture and food advisory service. Yet while they indeed spend a significant amount of their budget on R&D (some 30% on agriculture and rural development R&D, with a further 11% on food research 1), we do not know precisely how much of this is devoted to agri-environmental studies. Nor does the £6m it is estimated to have spent on agri-environment research between 1985 and 1994 seem anything but ineffectual when compared with the budget for the EU managed fifth framework programme: £260m between 1994-1999 (See Table 1, Department of Employment and Enterprise, 1996).

Equally although Teagasc have developed a specific centre for the study of environmental problems relating to agriculture, no specific studies have been undertaken assessing the impact of the CAP on the environment 2). Instead research has taken a pragmatic slant with the emphasis on producing scientific findings which could be used as a basis for guidelines. In that regard a recent official code of practice was recently produced jointly by the Departments of Environment and Agriculture, Food and Forestry (DAFF/DOE, 1996). It remains however of only limited use as it is merely advisory. More worrying has been the financial cutbacks which Teagasc's organic programme has suffered from (Hickey, 1996).

Notwithstanding this environment, Teagasc have though produced some interesting research in a few areas of the commodity regime which make the case for more research to be carried out so that policymakers can be better informed. For example they have conducted a number of ad hoc experiments and small-scale own initiative studies on the impact of sheep grazing on upland soils in the west of Ireland, an issue which is becoming very sensitive within the REPS program-

---

1) This is based on 1995 figures provided in: Teagasc. 1995. *'Research-Advisory Training Services 1995-1999.'* pp. 12.

2) Teagasc has completed some attempts at assessing the impact of the Control of Farmyard Pollution Scheme, but this has not be written up in serious report format, while according to the Head of Centre at Johnstown Castle. He indicates they are however, contemplating embarking on a large-scale evaluation of the effectiveness of the Rural Environmental Protection Scheme, Ireland's programme to implement Regulation 2078/92. One stumbling block for this is to what extent the impact of the scheme on riparian zones can be feasibly tested, as REPS recipients are widely dispersed across counties. My thanks to Joe Lee, Head of Centre at Johnstown, for this information.

me under Regulation 2078/92 1). A tentative interpretation of some of the results from these studies suggest that some of the upland peat soil erosion which has occurred in the last few years has
been exaggerated by climatic conditions as well as high stocking densities.

Another recent programme, completed under the former agricultural research programme
of DG VI (CAMAR), looked at the reduction of fertilizer inputs in grassland management and the
reduction in use of processed feeds with a view to determining whether carcass weight and quality
were adversely affected by reducing such. This would tend to suggest that reducing 'artificial inputs' seemed to be neutral. That is, farmers could: (a) reduce chemical inputs and not suffer a decline in weight, but also (b), in physio-chemical terms the animals would not be obviously any
healthier or 'natural' for the consumer, as trace elements were minimal anyway 2). What is perhaps
intriguing here is that the even limited results thrown up by both these small-scale studies suggests implications for policymakers. It would seem obvious that there is a need to fine tune investment and the direction of policy even further if research were expanded and integrated within an
overall agri-environment policy.

In that regard evaluations of the measures taken to date appear limited to once off studies
rather than any systematic long-term approaches. For example, evaluations of both the Control
of Farmyard Pollution Scheme (1989-1993) and the Environmental Sensitive Areas Programme
(1993), have been produced by private consultants (Fitzpatrick Associates, 1992). These studies
have taken a rather matter of fact approach and conclude that such schemes were of benefit, without perhaps raising a debate about alternative policy options and the use of cost-benefit analysis.
Additional research initiatives currently underway or about to start in Ireland are described briefly
in Table 2.

Other state bodies with competence on research into environmental affairs have been slow
to act. The relatively new Environmental Protection Agency (EPA), set up in 1992, is expressing
a growing interest in the area, for example recently funding a study of phosphate losses in Irish
agricultural conditions. The Heritage Council, a body only recently established in particular to
develop expertise and co-ordinate information with regard to land designations and heritage areas,
has expressed an interest in moving into funding studies which assess the impact of productive
sectors on the environment. Currently they have funded a UK consultancy to assess the impact
of forestry policy on the environment and expect eventually to undertake studies with regard to
agriculture proper. The Universities and third level sector generally does engage in some broadly
related research, however, again most of this is focused on essentially limited narrow scope ques-

---

1) In this regard a key facility in such studies has been the use of a special Hill Sheep Farm at Leenaun, Co.
Galway. The broad objective of this research station is to 'demonstrate in a scientific and practical way how
sheep farming in hill areas can be conducted in a cost effective and sustainable manner.' Some of the Leenaun
centre's work appears to be own initiative but other parts were clearly directed by EU funding. In particular
under the Research Stimulus Fund a project was funded in 1995, which evaluated the impact of livestock on
the hill-environment-soil, water and vegetation in western Ireland. This project was co-ordinated between
Teagasc and UCD, with Michael Walsh, Teagasc Resource Centre, Athenry, Co.Galway heading up the
project. See: Eagan, N. J.F. Collins and M.Walsh. 'Teagasc Hill Sheep Farm, Leenaun, Co. Mayo,etc.' in
Farm & Food., Spring, 1996.

2) Many thanks to Gerry Keane for this information. Contact: Grange Research Centre, Dunsany, Co.Meath
Ireland. Tel: +353 (0)46 25214, Fax: 353 (0)46 26154.

tions. In terms of social science, the dominant discipline is that of economics with regard to a study of the CAP. Unfortunately this has not produced much by way of an environmental economics literature with perhaps an exception being some work on the possible introduction of a regime of fertilizer taxes by Convery and Scott (1997). Private bodies and environmental lobbies have produced some publications on the impact of the CAP on the environment. Yet again mostly this has been only in a fragmented way and not fully scientific in terms of proving a link between various commodity regimes and actual problems (Murphy, 1996; Hickie et al., 1995; Hickie, 1996; IPCC, 1994; Mealdon et al., 1992; Meadon and Skehan, 1996). For instance from the environmental NGO sector, Earthwatch, Greenpeace and Birdwatch Ireland all have produced from time to time, some small scale reports which suggest an association between incidences of environmental decline and farm activities (Ibid.).

Table 2   Recent Irish research on agri-environment related issues

---

- Teagasc together with the Natural Resources Development Centre at Trinity College Dublin is undertaking a large scale study into Eutrophication of Fresh waters due to phosphorous losses from agricultural sources. This will run from 1994-1999. This research is funded by the Environmental Protection Agency (EPA) a).
- The Department of Agriculture, Food and Forestry's Pesticide Control Unit recently completed a five year programme (1990-1995) of detailed scientific sampling of raw agricultural produce (both domestic and imported) which included both fruits, vegetables, grains, and meat products. The results of this were that only 42 samples out of 8,590 were found to exceed the Maximum Residue Level (MRL) b). This represents about 0.48% of the total sample. The sensitivity of testing techniques has been increased ten fold since 1994 and later in 1997 the results will be published in a user friendly format.
- Partially related to meat quality and the beef sector, and indicative of Ireland's response to the whole BSE saga, the Irish Food Research Committee of the DAFF allocated some IR£5 million into topics broadly related to Transmissible Spongiform Encephalopthy (TSE). The main focus of this body of research is more consumer confidence related than substantively environmental, with for instance a focus on diagnostic tests, traceability of animals etc. Contact: DAFF (As above)
- The National Food Biotechnology Centre, University College Cork, does some work which is broadly related to food safety concerns, however, it appears no specific environmental work is being done for instance through bio-engineered new low input crops, etc. The National Agricultural and Veterinary Biotechnology Centre appears to engage in environment related research only tangentially to its main works which engages in technical support for intensive mushroom producers, the development of a vaccine to fight a particular disease of poultry, and a broad programme of crop quality c).
- Cork Regional Technical College has a Clean Technology Centre which although largely focused on the needs of industry, nonetheless does involve the dairy-food processing sector in its projects. Most of these take the form of environmental audit and administration projects and run for about 2 years on average. In particular it has targeted partnership with a number of small-to-medium enterprises for the development of training methods and documents on Integrated Environmental management d).

---

a) Contact: Gearoid O Riain Natural Resources development Centre, Trinity College, University of Dublin, Dublin 2, Tel: +353-1-6081244, Fax: =353-1-6718047; b) In fact of these 42 cases only 23 were domestic in origin meaning that Irish produce which exceeded the MRL represented about 0.26% of the total sample. For further details contact: Department of Agriculture, Food and Forestry (DAFF). TEL: +353 (0)1 607279; c) Contact: Forbairt: http://www.forbairt.ie.reserach; d) Contact: Dr.Dermot Quirke: http://www.rtc-cork.ie/rd/cleant/Clean tech.

4.6   Concluding observations

A number of core themes are evident from the above discussion of Irish responses to mooted the CAP reform. Firstly it appears there has been relatively little substantive change in perceptions. Most of the key actors still retain a focus on environmental reforms of the CAP not for their intrinsic environmental merit, but for whether or not they can be used as side payments to compensate for the likely continuing liberalization of the price regime.

Equally this means that the content of policy is constrained in how much it can change. Politics dictates that policy will remain fixated on subsidies and transfers, rather than any broader debate about wider instruments and other approaches (such as taxes, and clean production technologies). It is also fairly clear that insofar as the environment has been accepted onto the agenda of dominant Irish actors, this has not been thanks to domestic political or social pressure. Irish environmentalists remain largely marginal in influence. Indeed one can argue, that were it not for the pressure of various EU water quality objectives and even traditional countryside resource lobbies, such as anglers, domestic pressure might in fact be threadbare.

Note however, that would not seem to preclude the development of policy. For there certainly has been policy change in the sense that a lot of programmes, schemes and initiatives, involving very large amounts of money have emerged. Yet it remains to be seen how effective such measures will prove, dominated as they are by a logic of income support for marginal farmers, displacement of wastes and an almost exclusive reliance on subsidies. Above all policy has perhaps not changed in the sense that it is beginning to target ecological risks - that has not happened in any scale as of yet.

This is not to say that the agri-environment situation in Ireland is stagnant. If anything the environmental awareness deficit of Irish agriculture is quickly reducing and the future is unpredictable in gauging whether environmentalism as a political pressure will emerge in a more systematic way. In this regard Ireland may be unusual in leading off with the animals rights issue as its first major public controversy which has challenged modern farming methods.

Even if environmentalist sentiment does grow however, no one should underestimate the power of the state-farmer relationship to dominate responses to the CAP reform. Should serious environmental reforms be proposed, it must be remembered that key Irish actors are basically very conservative and cautious on the issue of the CAP reform generally, baring in mind how favourable the status quo has proven. In the event of such a reform line being taken, it is predicted that the same pragmatic and productivist-oriented approaches will be to the fore in determining the official Irish response: that is the environment will be treated as a potential future resource for maintaining direct and indirect transfers.

However, this does not preclude the possibility that such an approach is not without its contradictions and political difficulties for policymakers. For the linked issues of the CAP reform and environmental side payments has the potential to make even more transparent divisions between farmers in Ireland. In particular such measures appear more attractive to the smaller more marginalized cohort, who are already more favourable to REPS like measures. It is worth recalling that the whole issue of the CAP reform has in the past proven problematic for the state-farmer axis, as was seen during the related GATT crisis of 1992. In such settings agribusiness and the

State may sometimes part company with organized farmers and endorse a more radical package of measures.

As regards the state of research it was suggested there is something of a discrepancy between the scale of funds being invested and our knowledge about how effective policies to date have been, and how serious problems actually are. The picture that emerges is one of some small-scale research, but it remains largely uncoordinated in very limited applied projects. While Teagasc has built up a good knowledge base and struggled to develop more expertise with limited funding, there is as yet no large-scale authoritative study of the impacts from the CAP, nor the effectiveness of what limited measures it has mandated to abate these. Until such time as a genuine domestic interest develops, it is likely that such research can only feed into the policy process in an ad hoc if not haphazard way.

It is certainly possible to end on the note, that Ireland has come a long way in just a decade regarding the sophistication of its agri-environment measures, particularly when one considers the flagship REPS policy. Yet for all these trends and the huge sums involved, measures of environmental decline continue in a slow but steady way (Murphy, 1996; Murphy and Maclochlainn, 1997; Maclochlainn, 1996; EPA, 1996). Ireland may only be just beginning to turn a corner where complacency has become unseated. Therefore the future suggests if anything, at least a continuing vociferous demand for change in perceptions, policies and research agendas, even if that demand is likely to be frustrated by the essentially conservative approach to green reforms of the CAP which the dominant policymakers still display.

References

Allen, Robert and Tara Jones (1990) *Guests of the Nation. the people of Ireland versus the multinationals;* London, Earthscan

Collins, Neil (1993) *Still recognisably Pluralist? State-Farmer relations in Ireland;* In: Hill, R.J. and M. Marsh (eds.). Modern Irish Democracy-Essays in honour of Basil Chubb; Dublin, Irish Academic Press., pp. 104-122

Commission of the European Communities (CEC)/Teagasc (1989) *Intensive Farming and the Impact on the Environment and the Rural Economy of Restrictions on the use of chemical and Animal Fertilisers.* Study prepared for Directorate General for Agriculture, prepared by Teagasc

Commission of the European Communities (1995) *Study on Alternative strategies for the development of relations in the field of agriculture between the EU and the associated countries with a view to future accession of these countries;* CSE(95) 607

Convery, F.J. and Sue Scott (1997) *Giving substance to the Polluter Pays Principle; Paper given at Conference in Environment and EU Treaty Revisions: The IGC review;* Irish Institute of European Affiars, Dublin, Monday 10th March 1997

Crowe, D.P., A. Markey and P.J. Phelan (1993) *A study of the Pilot Environmentally Sensitive Area Scheme in the Slieve Bloom Mountains: A report prepared for the Slieve Bloom Rural Development Committee;* Dublin, Department of Agribusiness, Extension and Rural Development. University College Dublin; UCD

Department of Agriculture, Food and Forestry (1994a) *The Rural Environmental Protection Scheme*

Department of Agriculture, Food and Forestry (1994b) *The Operational Programme for Agriculture, rural development and forestry 1994-1999*

Department of Agriculture, Food and Forestry/Department of the Environment (1996) *Code of Good Agricultural Practice to Protect Waters from Pollution by Nitrates*

Department of Employment and Enterprise (1996) *White paper on Science, Technology and Innovation.* Dublin: Government Publications Office

Department of Environment (1992) *Agriculture;* Part 1-Environmental Impacts and Trends-Sectoral Analysis, pp. 18-20 and 'Agriculture and the Environment,' Part2-Policy Responses, pp. 37-39; In: Ireland- National Report to the United Nations Conference on Environment and Development. Dublin: Department of Environment

Department of the Environment (1991) *Agriculture and the Environment;* pp. 12; In: An Environment Action Programme-1st progress report. Dublin: Department of the Environment

Department of the Environment (1995) *Agriculture and Farm development-The Planning Issues;* Planning Leaflet 6

Department of the Environment 1996-1993 inclusive. *Environment Bulletin- Developments in the area of Environmental Protection;* Issues 32, 28, 25, 22, 20

Environment Research Unit (ERU) (1989) *Farm Wastes and Water Pollution-The present Position;* Dublin: ERU

Environmental Research Unit (ERU) (1993) *Irish Environmental Statistics*-2nd edition. Dublin: ERU

Environmental Protection Agency/Stapleton, L (Ed.) (1996) *State of the Environment in Ireland;* Ireland: EPA

Eurostat (1993) *Research and development: Annual Statistics;* Luxembourg, Official Publications of EC

Harte, L. (1992) *Impacts of the CAP for Activity at Farm Level;* pp. 271-284; In: Feehan, John (ed). Environment and Development in Ireland; Dublin, Environment Institute, UCD

Hickey, Michael (1996) *Johnston Castle Organic Farm: whither to now?;* Organic Matters-Bi-monthly Magazine of the Irish Organic Farmers and Growers Association. No.30. Feb-Mar. 15

Hickie, David, Roger Turner, Clive Mellon and John Coveney (1995) *Ireland's Forested Future-A Plan for forestry and the Environment;* Discussion paper prepared for Royal Society for the Protection of Birds, An Taisce and the Irish Wildbird Conservancy; Dublin, RSPB

Hickie, David (1996) *Evaluation of Environmental Designations in Ireland;* Dublin, Heritage Council

Hofer, Karin (1997) *Shared responsibility- a viable concept? Dimensions of voluntary instruments in environmental policy. Food-labelling as a case study;* Paper presented at the Summer Symposium at the University of Bolonga, 'The Innovation of EU environmental Policy.' 21-25th July

Irish Peatland Conservation Council (IPCC) (1994) *Irish Peatland Conservation Plan 2000;* IPCC

grassland and previously uncultivated land were ploughed up, thereby extending the extensive margin of cultivation. Simultaneously, through a system of input grants and subsidies, farmers were encouraged to adopt 'modern' management practices utilizing agrochemicals and more powerful machinery. This increased the intensive margin of cultivation, raising yields per hectare. Although these changes were in motion prior to the UK's accession to the EC, adoption of the Common Agricultural Policy (CAP) reinforced the pattern of change.

CAP reform

Within the arable sector, the 1992 reform sought to abate production by reducing both the area of arable crops and the intensity of cultivation. The latter was achieved by sharp reductions in support prices (30% for cereals, 50% for oilseeds) which were anticipated, *ceteris paribus*, to encourage farmers to reduce their usage particularly of agrochemicals, thereby lowering yields. To compensate farmers for the expected effects on their income they were offered payments for each hectare of a particular crop that they grew, under the so-called arable area payments scheme (AAPS). This was an attempt partly to decouple incomes from agricultural output levels. The level of area payments varies by crop and by region, reflecting yield differences. For example, the 1995 AAPS offered £270/ha for cereals and £475 for oilseeds in England, whilst the same crops were worth £240 and £485/ha respectively in Wales (MAFF, 1996a).

In order to qualify for the AAPS, however, farmers 1) had to set-aside a proportion of their arable land. That is, a reduction in the arable area was achieved by obliging farmers to take some of their land out of agricultural production, for which they were also compensated. In 1995, such payments ranged from £205 to £340/ha of set-aside, depending on the UK region and the type of set-aside (MAFF, 1996a). Originally, two forms of set-aside were offered, flexible and guaranteed. In the former case, a specified amount of land (initially, at least 15% of the farmer's arable area) was set-aside on an annually rotating basis. In the guaranteed case, the same piece or pieces of land (amounting to at least 18% of the farmer's arable area) was set-aside for five years. This distinction has now been dropped and currently the proportion of a farmer's arable area which has to be set-aside to qualify for the AAPS is only 5%. This reduction reflects changes in the world grain markets where agricultural production and stocks have declined and prices have risen sufficiently to take some of the pressure off the CAP. This is in marked contrast to earlier expectations that the set-aside requirements would have to be increased year-on-year to peg production back in the face of technological advances and slippage (i.e. the tendency for a given area reduction to achieve a less than proportionate decrease in output as farmers reallocate inputs, particularly that of land of differing quality).

Environmental impacts

Expansion of the arable area, its regional concentration, and the changes in both the mix of crops and their associated management practices have all had environmental impacts. The rural land-

---

1) Small farmers producing less than 92 tonnes of grain were exempted from this set-aside requirement.

scape has been greatly affected, with a loss of much of its variety and intricacy, through the enlargement of fields and holdings and the loss of boundary features, pasture and uncultivated land. There has also been a reduction in biodiversity. In the past the focus of the conservation interest has been on land-use change - the dramatic shift from grazing to arable - whereas in more recent years there has been greater emphasis on processes of intensification and changing farming systems.

The conversion of grassland to arable land has an immediate impact on the abundance of grassland plant species. This represents a loss of plant diversity, but also has habitat and food chain knock-on effects for associated invertebrates and higher animals. The loss of hedgerows, field margins and other small farm features has a similar effect. In addition, increasing land use homogeneity in a given area, whether it be via displacement of non-arable land or arable land, adversely affects highly mobile animals, such as birds of prey or badgers, which require an extensive habitat composed of a mosaic of land uses including grassland, arable and woodland. Ploughing of grassland also releases 'pulses' of nitrates from stored organic matter, thereby contributing to current and future pollution levels.

Increasing intensity of arable cultivations has had less visible, but equally significant environmental impacts. Changes in rotational practices and the shift to autumn sown cereals have led to habitat reduction, for example for hares and ground nesting birds, while pesticides have diminished their traditional food sources. A number of bird species characteristic of arable and mixed farmland have shown sharp declines over the past 25 years, e.g. grey partridge 82%, corn bunting 80%, turtle dove 77%, lapwing 62%, skylark 58% and linnet 52% (English Nature, the Game Conservancy, RSPB, 1997). As a result of changes in cultivation and agrochemical use, many formerly common arable flowering plants have also declined in recent decades, some (such as the corn buttercup) to the point where they are endangered as British species (a few are already extinct). Unlike many other groups, these species generally have no alternative habitat as they are dependent on regular cultivation and so can only be conserved on arable land, but not under intensive treatment. There have been pollution problems too. Inorganic fertilizer applications are considered to be contributing factors to increased nitrate and phosphate pollution levels in both surface and ground water. Soil erosion from intensively cultivated arable fields is another potential problem. Finally, the use of pre-emergent cereal herbicides has tended to leave persistent agrochemicals on soils at the time of year when the most erosive rain is likely to fall which has increased the chances of pesticide run-off (Evans, 1990). In 1992, the Drinking Water Inspectorate found 33 different pesticides present in British drinking water supplies, and it was estimated that 14.5 million people in England and Wales lived in areas where pesticides were present in drinking water at levels higher than those stipulated in the Drinking Water Directive (Ward, 1995). However, recent years have seen a continual decline in the amounts of pesticide applied. Between 1982 and 1992, the tonnage of active ingredient applied to cereal crops dropped by 28% in Britain (Department of the Environment, 1996, p.17). Much of this decline has been due to the use of more potent products at much lower dose rates.

Although the primary reason for the 1992 CAP reform was to reduce arable production levels, it was anticipated that the means of doing so - by diverting land out of production and lowering yields - would also have beneficial environmental impacts. Thus, for example, lower product prices would be anticipated to reduce agrochemical application rates whilst obligatory set-aside would be expected to reduce total usage. In the event, though, applications of agrochemicals, particularly fertilizer, appear to have increased in the period since 1992 (Asby and Sturgess, 1997; Winter and Gaskell, 1997). It was also hoped that set-aside would contribute to habitat regeneration. Whereas set-aside managed for conservation objectives can deliver a wide range of benefits, its management for agronomic or economic expediency can be environmentally damaging (Firbank, 1997). Any potential benefit, moreover, has been reduced by the lowering of set-aside obligations 1) and revisions to the regulations that have increased the scope for non-food production on set-aside land, leading to substantial areas of, for example, industrial oilseeds 2). The overall picture regarding changes in the environmental impact of the arable regime remains unclear for a number of reasons.

First, the reform package has not been in place for long and, moreover, was phased in gradually. Little survey information is yet available relating to changes in farming practice and environmental impacts. Consequently, there has been a reliance on modelling exercizes, whose predictions depend crucially on their underlying assumptions. In general, these models predict reductions in the areas of arable crops (primarily through diversion into set-aside) coupled with reductions in intensity of nitrogen fertilizer usage (Wallace and Kirke, 1992; Skea, 1993; Moxey et al., 1995). It has also been calculated that a 10% reduction in expected output prices could lead to as much as a 30% reduction in pesticide use (Russell et al., 1997). Second, changes to the arable regimes have not occurred in isolation. For example, withdrawal from the Exchange Rate Mechanism (ERM) and buoyant world markets led to UK price rises rather than falls, the introduction of headage quotas in the beef and sheep sector has hindered moves towards mixed farming, and the BSE crisis has sent indirect shocks through other sectors. In addition, natural variation in farming conditions, notably the weather and pest or disease incidence, causes year-on-year fluctuations in agricultural practices. These masking factors make it difficult to identify direct impacts of the AAPS and associated set-aside schemes.

Such factors also mean that the environmental damage sustained under the CAP in recent decades will not necessarily be reversed simply by altering arable market and policy pricing signals. Confounding influences from other sectors, continuing technological advances, changes in rural infrastructure and farming styles, and changing climatic conditions all make an automatic return to 'traditional' landscape or habitat features unlikely. In addition, achieving desired environmental change often requires positive management, not simply the withdrawal of negative practices: reducing output prices *may* lead to less intensive agriculture, but this may not occur in the right spatial location nor be the 'right' sort of deintensification. This is already recognized in the guidelines for managing set-aside, for example with the requirements for farmers to maintain a 'green-cover' to avoid bare ground and to encourage ground-nesting birds by not cutting this cover during certain periods (MAFF, 1995; Farmers Weekly, 4/4/97). Specifying appropriate prescriptions depends crucially, however, on an adequate appreciation of the heterogeneity of both the business structure and environmental circumstances of the farm population (Moxey et al., 1995;

---

1)   708,000 ha were set-aside in 1993/4, only 633,000 ha in 1994/5 (MAFF, 1996b).
2)   Approximately 16% of set-aside land in England was under industrial, non-food crops in 1994/5 (MAFF News Release 262/96 of 22nd July 1996).

Weaver et al., 1996). Such identification and analysis requires rigorous farm-level survey work, which is only now being reported (Winter et al., 1997).

5.3 The dairy regime

Dairying in the United Kingdom

Dairy farming plays a key role in UK agriculture and occurs on about one in every four farms. Although dairy farming incomes are currently depressed because of the strong exchange rate for the pound, in recent years dairy farmers have enjoyed high levels of profitability due to:
(I)   falling feed prices, following the CAP reform in the arable sector;
(ii)   the reorganization of the marketing arrangements for milk, with the demise of the UK Marketing Boards; and
(iii)   the quota restriction on milk output which has relieved the pressure to reduce support prices, which otherwise would have had to fall significantly.

The number of dairy cows in the UK has declined markedly since the introduction of milk quotas in 1984. However, aside from an initial shock response, structural change in the dairy sector has continued in much the same fashion as prior to 1984. That is, production is increasingly concentrated on fewer units with larger herds and higher milk yields per cow.

The milk regime under the CAP

The milk regime under the CAP dates from 1964 and centres on a 'target' price for milk which is supported through import protection, intervention buying and export subsidies. In the face of rapidly rising budgetary costs, associated with steadily increasing surplus production, the EC introduced milk quotas in 1984. These were originally set at a level equal to 1981 deliveries plus 1%. Subsequently there have been reductions in the overall quota (for which farmers have been financially compensated) but production of milk in the EC still exceeds domestic demand by around 20%. It is this surplus which has to be bought into intervention and/or sold in non-EC markets with the aid of export subsidies.

The transfer of quotas between farmers, through annual leasing and outright purchase, is possible in some Member States, including the UK, though inter-country transfers are as yet outlawed. Transfers ease the process of structural adjustment, allowing farmers to expand, contract or cease milk production (Hubbard, 1992). This has facilitated the continued concentration of production. Small farms and disadvantaged regions used to be sheltered from these pressures by the state control of milk purchase and distribution. The deregulation of the UK milk market in 1994, in part to conform with EC rules, has therefore exposed smaller and more remote dairy farmers to greater economic pressures.

The CAP reforms of 1992 have had little direct impact in the dairy sector, with the level of quota remaining unaltered. It was originally envisaged that quotas would be cut by 1% in 1993/94 and by a further 1% in 1994/95, but in the event these cuts were not implemented.

Environmental impacts

Dairy farming's most pressing environmental problem in Britain in recent years has been the pollution of water courses with farm wastes. The number of recorded water pollution incidents from farms more than doubled during the 1980s from around 1,500 in 1979 to over 4,000 in 1988. Approximately 80% of farm pollution incidents in the 1980s were associated with livestock farms, with intensive dairy farms being the most implicated. Such pollution was a key contributory factor in the aggregate decline in river water quality in England and Wales during the 1980s (Lowe et al., 1997).

The pollution problems arose from the intensification of dairy farming that the CAP fostered. Technical changes in the industry were compounded by the concentration of production they facilitated, towards fewer farms and larger herds. Average dairy herd size in England and Wales trebled to 63 cows between 1960 and 1987, while average stocking rates increased over the same period such that there was, on average, 43% less land per cow. The main technical changes have been a shift to housing cattle in cubicles which results in a more liquid slurry than the traditional, straw-based farm yard manure; a greater use of water to clean out farm buildings and yards which results in a much greater volume of effluents; and the shift from hay as the main source of winter feed to silage, which produces a potent effluent. An average sized dairy farm generates the equivalent pollution potential to a town of around 11,000 people.

In addition, there has been a considerable amount of investment in pollution control technologies to store farm effluents, helped by a 50% grant available between 1989 and 1993. However, investment has concentrated on 'engineering solutions' to the problem, including the construction of tin tanks or concrete pits to store slurry, and systems of settlement tanks and pumps to collect farm yard effluents and spread them on farmland through sprinkler systems. The technologies are a form of 'technical fix' which takes as given the intensive production system, and the nature and volume of the effluents it produces. Pollution control technologies that seek to improve the storage and disposal of these effluents are simply 'bolted on' in an effort to minimize the pollution of water courses. But two problems remain. First, these pollution control technologies are not 'fool-proof' and can be subject to catastrophic failure, particularly if regular and thorough maintenance procedures are not adequately followed. Secondly, a strategy of 'store and spread safely' may be shifting the nature of the pollution problem from one of periodic and catastrophic point source pollution incidents to a more insipid and chronic problem of diffuse pollution as more and more effluent comes to be spread on less and less land (Lowe et al., 1997).

Milk quotas are holding national production of milk more or less static. This said, structural change in the sector is continuing to result in a greater concentration of production, in terms of fewer but larger units. At the same time, the deregulation in the marketing of milk in the UK, with the demise of the Marketing Boards, is likely to lead to some regional relocation of dairy farming closer to areas of high consumer demand at the expense of more remote areas. In consequence, on the one hand, there is a danger of more river catchments becoming 'over-stocked' with dairy cows, with insufficient land available for farm wastes to be safely spread. On farms that are expanding their production of milk (through purchase or renting of quota) there is also a tendency towards greater intensification of grassland management, particularly for winter feed (mainly silage), with increased fertilizer applications in some cases and pressures on any remaining semi-

natural habitats. Another recent trend in dairying has been the cultivation of forage maize which has aroused concern regarding soil erosion and phosphate leakage (Royal Commission on Environmental Pollution, 1996), but which, it is suggested, could reduce nitrogen loss if substituted for grass silage and appropriately managed (Jarvis et al., 1996; Wilkins, 1996). On those farms, on the other hand, that are reducing milk output (through selling or leasing quota), stocking rates may be falling with some concomitant environmental benefits, although the outcome here is dependent on the alternative enterprises to which these farmers are turning. The decline of dairying in remote areas, particularly the hills and uplands, is removing a component which used to add diversity and character and provide an additional source of labour for feature maintenance in marginal landscapes.

5.4   The beef regime

The CAP beef regime

Within the UK, beef production has exceeded home consumption since the early 1980s (MAFF, 1997). In 1995 the beef herd totalled some 1.8 million animals having grown from 1.3 million in 1985. The UK was 112% self-sufficient in beef and the sector accounted for 15% of agriculture's gross output.

Prior to the Mac Sharry reforms, the beef regime relied on price support and intervention buying measures together with the payment of the Beef Variable Premium (BVPS) on finished animals. The BVPS ended in April 1989 and was replaced by the Beef Premium Scheme (BPS). Under the BPS, an annual ceiling of 90 male animals per holding was introduced. Within the so-called Less Favoured Areas (LFAs) there were, and still are, additional payments for hill cows - the Hill Livestock Compensatory Allowances (HLCAs - see below).

Under the 1992 reform there has been a continuation of support through the operation of a guide price against which import levies are calculated, plus continued intervention buying. However the reform package sought to reduce both the need for import levies and intervention buying through the following four main policy measures:

(I)   Prices paid to farmers for transferring beef into intervention stores were cut by 15% over three years from 1993/94, with ceilings introduced progressively on intervention purchases from 1993/94 to 1997/98.

(ii)   A new scheme to replace the BPS was introduced to offset the costs incurred through falling intervention prices. The Beef Special Premium Scheme (BSPS), like the BPS, pays premia only on male animals (steers) to a limit of 90 eligible cattle per holding in each of two age categories. Unlike the BPS, payments are not confined to point of slaughter. Instead, premia can be claimed twice in the life of an animal - between 8 and 21 months and 21 months or more. Two further innovations were introduced in an effort to limit claims. First, a regional ceiling operates. For example, in England and Wales if claims exceed 940,380 head, amounts paid to all producers are reduced in proportion to the excess claims. Secondly, entitlement is limited by stocking density rules. The stocking density limit has fallen progressively from 3.5 livestock units (LUs) per hectare of forage area in 1993, to 3.0 LUs in 1994, 2.5 LUs in 1995 and 2.0 LUs in 1996 1).

(iii)   The Suckler Cow Premium Scheme (SCPS) entitlement, paid to farmers since 1980 for rearing animals from a beef breed for meat, was made conditional on possession of a producer quota, based on the number of animals receiving SCPS payments in 1992 (minus a 1% siphon to form a national reserve). As with BSPS, entitlement depends on compliance with stocking density rules. The amount of SCPS premium received varies between the following geographical areas: the English Less Favoured Area (LFA), the Welsh LFA, the Scottish Highlands and Islands LFA, the remaining Scottish LFA and Great Britain non-LFA.

(iv)   An Extensification Premium has been made available to producers with a stocking density of less than 1.4 LUs per hectare of forage area. This is payable on both the BSPS and SCPS. Once again, regional ceilings on premium claimed apply. In 1996, an additional payment was introduced for an even lower stocking rate of 1 LU per forage hectare.

In summary, these measures were designed to: safeguard farmers' incomes whilst reducing the budgetary costs of CAP; encourage extensification as a crude environmental concession; reduce beef production in dairy herds as a contribution towards reducing the beef surplus; and maintain seasonal equilibrium in the beef market.

BSE policies

BSE first emerged in the mid-1980s but it erupted as a public health crisis in early 1996. The impact of the BSE crisis cannot be ignored for not only has it blown off course the above policies of reducing surpluses and the cost of the beef regime, it has also led to the introduction of a raft of new policies to support the ailing beef sector and these are summarized below (see Table 1). They are of particular importance because the net effect is greatly to reduce the size of the beef sector in the UK with considerable environmental implications (Gaskell and Winter, 1996).

Environmental impacts of beef

The impact of beef production on the natural environment is highly complex with both negative and positive connotations. On the one hand, intensive beef production, especially when it takes place alongside other intensive forms of agriculture such as dairying or arable farming, may have a negative environmental impact. On the other hand, grazing is essential to the maintenance of

---

1)   The calculation of both stocking densities and forage area are complex and require a submission under IACS (unless a farmer is exempt from the stocking density rules because of claiming less than 15 LUs in total on the holding). The following stock have to be taken into consideration in calculating stocking rates: dairy cows, breeding ewes on which Sheep Annual Premium has been claimed, male cattle on which Beef Special Premium has been claimed, aged under two years on date of claim, male cattle on which Beef Special Premium has been claimed, aged over two years on date of claim, suckler cows on which Suckler Cow Premium has been claimed (including replacement in-calf heifers). Forage area refers to land actually available for feeding and grazing livestock.

important habitats and landscapes in both the lowlands and uplands. Few farm systems are solely reliant on beef enterprises and much depends on the stocking levels and the relationship with other enterprises. In lowland areas, beef herds are often established as a secondary enterprise in predominantly arable and dairy systems. In upland areas, beef cattle typically exist alongside sheep.

Table 1   BSE: Summary of selected key measures in the UK

30 months plus cattle slaughter scheme	£550 million (70% funded by the EU) to compensate for the removal of cattle from the food chain. By mid-October '96 nearly 630,000 animals had been slaughtered under the scheme in the UK.
Additional premia payments	An additional 23 ECU (£19.70) on BSP and 27 ECU (£23.13) on SCP payments to be paid 1996/97 only (budget = £81 million).
Beef marketing payment scheme	A one-off payment, flat rate headage payment of £66.76 for adult clean cattle marketed between 20 March and 30 June 1996 for slaughter for human consumption. More than 29,000 claims were made in the UK covering some 450,000 animals (budget = £29 million).
Beef Assurance Scheme	This allows animals of 30-42 months to be slaughtered for human consumption provided they are from specialist beef herds never exposed to the risk of BSE. Cattle under this scheme will have been reared mainly on grass.
Calf processing aid scheme	An EU scheme to destroy very young male calves. By mid-October 1996 almost 270,000 calves had been slaughtered under the scheme in the UK.
Intervention purchases	Opening up of intervention from the beginning of April 1996. From mid-April to mid-October, more than 350,000 tonnes of beef were purchased into intervention in the EU, nearly 30,000 tonnes of which was in GB.

In the 1980s and early 1990s there was a shift of beef production from the uplands to the lowlands. On the one hand, lowland dairy farmers, in particular, turned to beef production to supplement milk production pegged by quotas. On the other hand, the decline of cattle in the uplands was due to the relative economics of cattle and sheep production prevailing over a couple of decades but reinforced more recently by the operation of sheep and suckler cow quota rules.

In the lowlands the environmental consequences were largely negative. Some farmers did take the opportunity to sell their dairy quotas and switch production to beef, with reduced intensity of grassland management and lower pollution risks as a result. But the expansion of beef production amongst dairy farmers in the main exacerbated the trend towards more intensive grassland management with attendant effects of habitat loss and pollution. It might be thought that the stocking rate rules would serve to alleviate the problem of intensification in the lowlands. In fact the situation is far from clear cut. The rules are not designed to achieve environmental benefits as such but to place limits on levels of support payment. They only apply to specified categories of livestock and not to the actual number of animals on the farm. Farmers may keep animals for which no claim for premium is made and animals which are ineligible for premia, such as non-breeding female beef stock, calves under six months of age, lambs or alternative categories of livestock such as deer. Moreover, the stocking rates are set at such a level that few farmers have had to reduce stocking to qualify: fewer than 8% of farmers with beef enterprises were affected by changes to stocking density restrictions up until 1995 and just 20% anticipated a change in herd management to take account of the new 1996 stocking rates (Winter and Gaskell, 1997; Winter et al., 1997).

The consequences of fewer cattle in the uplands are more complex. Upland heather (*Calluna vulgaris*) is one of the most important and distinctive habitats in north-western Europe, with a considerable proportion of its distribution in Great Britain (Thompson et al., 1995). It is important for its distinctive plant communities and associated animal species, particularly birds, as well as from a landscape and amenity perspective. However, much heather has been lost or de-graded as a result of heavy grazing by sheep. The grazing problems are exacerbated by winter feeding on old heather, poor burning and competition from bracken (Johnson and Merrell, 1994). There is also concern about grass moors where course species, such as *Molinia* and *Nardus*, are spreading at the expense of finer grass species and a diverse ground flora. Cattle help maintain the important heather and grass moorland communities by trampling the bracken and grazing the coarse grasses. The loss of cattle from upland moors allows these invasive plants to take over. In many areas the cattle have been replaced by additional sheep. Being more selective grazers than cattle, the sheep do not control the coarse grasses that compete with heather, nor do they trample bracken as effectively as cattle. The sheep may also add to grazing pressures. Vegetation change is not the only issue of importance. Cattle, especially the traditional and distinctive hill breeds (such as Welsh Black, Highland and Galloway), play an important role in the cultural landscape of British upland areas (Evans and Yarwood, 1995).

The response to the crisis in the beef sector and in beef consumption brought about by BSE is leading to a significant reduction in the UK beef sector across the board. In the uplands this will accelerate the decline in cattle and exacerbate the problems described above. In the lowlands the BSE measures and the drop in demand may serve to encourage some farmers to withdraw from beef production altogether. The environmental consequences will depend on the alternative uses to which the land is put - there is anecdotal evidence of switches to non-IACS cropping and of land sales to non-farmers. There are sites of considerable conservation importance in the low-lands, such as fragment grasslands within arable areas and grazing marshes, which could suffer from neglect if extensive beef systems decline. In some cases, the conditions are such that cattle are even more essential to the grazing regime than in the uplands. For example, the alluvium or alkaline peats of the Somerset Levels support vigorous swards in wet conditions of considerable conservation importance but that are unsuitable for sheep (ENTEC, 1996).

Cross-compliance

An element of cross-compliance exists in both the BSPS and SCPS. Livestock must not be al-lowed to graze in numbers which damage growth, quality or species composition of vegetation on that land (MAFF, 1996c). MAFF will advise farmers on the number of animals it thinks is appropriate to graze on land where overgrazing is deemed to be a problem. MAFF also reserves the right to recommend changes in management practices. Scheme premia will then be paid according to the number of animals advised to the farmer or be withheld if the conditions are not met.

5.5   The sheepmeat regime

Sheep in the United Kingdom

In 1996, sheepmeat and wool production accounted for 14% of gross livestock output or 8% of total gross agricultural output (MAFF, 1997). This was achieved from a breeding flock of approximately 20.5 million ewes, or 43 million sheep and lambs. By comparison, during the 1940s there were only around 11.5 million ewes, or 20 million sheep and lambs (MAFF, 1947). This doubling in flock size, and the slightly more than doubling in the number of lambs, reflects various changes in both grassland productivity to support higher stocking rates and in ewe management to improve lambing rates and quality. These changes have been stimulated by prevailing agricultural policies (Bowers and Cheshire, 1983; Buckwell, 1990).

In the 1940s a system was set up to offer hill and upland farmers subsidies on their breeding stock, and grants on land improvement through drainage and liming. This led to areas of rough grazing being converted to more productive pasture, with altered grass species and soil nutrient status, and to shifts towards breeds with faster growth rates and higher lambing percentages. Although input subsidies were gradually phased out by the late 1970s, expansion in sheep numbers was also encouraged through price support. In the UK, this was done via a deficiency payment scheme. In 1980 a CAP sheepmeat regime was introduced (it also included provisions for goatmeat, but the UK government never chose to operate the goatmeat supports). Its price support system included an annual ewe premium (the Sheep Annual Premium or SAP) to which was added, in 1991, a supplementary payment for sheep farmers in the Less Favoured Areas. There was also the option of using private storage aid and intervention to support the market. The UK, though, received derogation to continue operating its deficiency payment scheme (which became known as the variable premium scheme) - an arrangement which only ended in 1992. Changes in sheep and grassland management have had various environmental effects.

Environmental impacts

Many managed grassland covers support native flora and fauna and are important semi-natural habitats. Prime examples include, in the uplands, heather moorland and blanket mires and, in the lowlands, heath land and grazing marshes. Changes in sheep and grassland management during the post-war period have contributed to a loss of plant species diversity, a reduction in semi-natural vegetation and the fragmentation of habitats in many of these areas. Since much of the intensification has been achieved through drainage and ploughing up of land, it is essentially irreversible (NCC, 1990; Cook and Moorby, 1993).

More subtle, within-cover-type changes have also occurred. Grassland and rough grazing swards are typically managed at low intensity and support a mix of vegetation types, but are susceptible to grazing pressure - both the direction and rate of succession between vegetation types can be influenced by grazing management (Oglethorpe et al., 1995). Perhaps the most obvious is overgrazing. This can lead to loss of species diversity, especially tall grasses and herbs, which in turn affects many invertebrate and higher animal populations. For example, bird species such as chough, corncrake, curlew and lapwing rely on low intensity semi-natural cover for nesting and feeding sites. Increased stocking rates on shrubby heaths can adversely affect species such as heather, leading to a rougher pasture cover type. High stocking rates on improved pasture can lead to dominance of tolerant species such as Italian Ryegrass and, in some cases, poaching and soil degradation. At the other extreme, under grazing can also cause habitat change. The vegetation cover of many semi-natural communities is maintained by low level grazing that prevents development of a climax community. If grazing pressure is reduced or removed, grassland may be invaded by tall grasses, scrub and even trees. Improved pastures (e.g. ryegrass) can degrade relatively rapidly to rough pastures (e.g. bent-fescue with rushes). Vestiges of drainage and fertilizers may, however, delay degradation quite considerably (Ball et al., 1982).

The use of agrochemicals on grassland has also increased over time leading to reduced species diversity. In particular, nitrogenous fertilizer usage on permanent pasture has increased from approximately 4kg/ha in the 1940s to over 100kg/ha in the 1990s, with even a small proportion of rough grazing land now receiving limited applications (Spedding, 1983; British Survey of Fertilizer Practice, 1994). Similarly, although herbicide and pesticide usage is relatively low on grassland compared to arable land, it too has increased with around 10% of pasture land receiving treatments (Davis et al., 1989).

CAP reform

Although sheepmeat is a comparatively recent addition to the CAP, it was revised alongside other regimes in the 1992 Mac Sharry reforms. The EU sought to control both the costs of supporting the sheep and the levels of production. Finite limits were set for the number of animals to be supported and these were translated into quotas for individual producers. Allowance was made for quota to be transferred between producers in certain circumstances but with restrictions on inter-regional transfers to protect regions dependent upon sheep production. Just before the Mac Sharry reforms, the UK had replaced its price support with the Sheep Annual Premium (SAP). In 1996, the annual SAP rates were £13.66 for a lowland ewe and £19.04 for a ewe in the Less Favoured Areas. The annual rate is divided into three payments made at specific times throughout the year to reflect the fact that some sheep are transferred between farms during the production cycle.

Eligibility for SAP payments depends upon entitlements which are measured by 'quotas', set initially to 1991 flock sizes. That is, *ceteris paribus*, farmers may only claim SAP on the same number of ewes as were on their farm in 1991. It is possible to buy or lease additional quota to increase entitlements, although there are restrictions on this, particularly between lowland and upland farms but also between UK regions. It was anticipated that quotas would effectively freeze the size and spatial distribution of the UK breeding flock (Saunders and Moxey, 1994), and therefore little immediate change in the environmental impacts of sheep production was expected.

There is some evidence, however, to suggest that decoupling income from production (i.e. the removal of price support) has led to a significant reduction in lamb numbers (Vipond, 1995). Lamb numbers declined by 10% between 1992/96. The effect may have been to reduce grazing pressures, but, if so, the environmental implications are uncertain since susceptibility to over or under grazing depends on various factors including the site's history, current state and proximity to other habitats, and detailed information on these factors is scarce. An added complication arises from implicit linkages to the beef regime, where eligibility for various premia is dependent on total livestock units, calculated on the basis of both cattle and sheep numbers. There has been a progressive tightening of stocking rate limits and, to continue to qualify, farmers may respond by reducing their sheep numbers. This has environmental consequences since sheep and cattle exert quite different grazing pressures by feeding on different sward species (Ellett, 1984).

Given the short time that the reforms have been in place it is difficult to decide whether apparent shifts in sheep management represent trends, or merely short-term responses to market fluctuations. Winter (1996) raises the possibility of long-term changes in the structure of the sheep sector. The current pattern of production is stratified into three altitude tiers: hill, upland and lowland production (Cooper and Thomas, 1991). This stratification ensures that grazing resources in each tier are exploited optimally, matching sheep characteristics and requirements to environmental constraints. The highest tier, hill production, is characterized by pure-bred flocks of traditional breeds such as Cheviot and Swaledale which are able to utilize the limited hill grazing resources at low stocking with low lambing rates. The middle, upland, tier is characterized by more intensive production. Stocking and lambing rates are higher, reflecting more favourable conditions and different sheep breeds. In particular, cross-breeding is common with more productive longwool or down rams used to sire higher quality lambs. Beef and dairy enterprises may also be present, together with limited arable (primarily fodder) crops. The lowland tier is characterized by the highest stocking and lambing rates in the sector. A wide variety of breeds, some specific to lowland areas, are kept. Lowland sheep enterprises compete for land with other enterprises and often have to fit into arable rotations.

The three tiers are linked through movements of lambs and ewes. For example, lambs born in the hill tier are typically sold to upland farms as store animals for fattening. Hill ewes are also sold to upland farms to produce crossbred lambs and ewes, which are then often sold as store or breeding animals to lowlands farms. Despite the quota system, it is possible that linkages between the tiers are being weakened by changes in market demand and revisions to the SAP scheme. Specifically, there is evidence that the traditional demand for Easter lamb is declining at the same time that the first instalment of the SAP is being rolled into the second instalment. Traditional production arrangements have also been affected by new export markets and the finishing of lightweight lambs on upland farms as a slaughter product for continental markets (Ashworth et al., 1997). Together, these factors may reduce the viability of early fattening of sheep on lowland farms, especially relative to other enterprises. Winter (1996) points to the price differential between the traded SAP quota for the lowlands (of £18) and that for the uplands (of £45) as evidence of this. If this is the case, sheep will become rarer in the lowlands, leading to a loss of grassland through either conversion to arable or degradation through under grazing. If hill and upland farms have the resources, sheep that would previously have moved to the lowlands may remain in the higher tiers, with implications for increased grazing pressure. This has been observed over more than a decade with hill and upland farmers finishing more of their own stock or adding more value to them before sale.

5.6 Agri-environment policies

United Kingdom tradition of agri-environment policy

The main instrument used for agri-environment policies in the UK is a contract between government agencies and farmers whereby the farmers are paid to undertake or refrain from specified land management practices for environmental benefit. Before its introduction by the EC, this mechanism had been applied in the UK to the provision of access in the countryside and the management of Sites of Special Scientific Interest and of land in the National Parks. Its use evolved in a context in which government sought to achieve the maintenance of landscape and wildlife through the cooperation of farmers and landowners. Given this substantial experience of contract-based countryside management, it is not surprising that, in 1984, the UK Government proposed an amendment to what became EC Structures Regulation 797/85, allowing Member States to make payments to farmers in environmentally sensitive areas. This provision was firmly based in British experience, and not surprisingly, it was enthusiastically taken up by UK officials and farmers.

Agri-environment policy in CAP reform

With the CAP reform, the scope of EC agri-environment policy was greatly expanded and put on a firmer basis. The aims of Regulation 2078/92, which was one of three measures 'Accompanying the CAP reform', are to contribute to the EC's objectives relating to the environment and to provide an appropriate income for farmers. The Regulation is financed out of the Guarantee Section of EAGGF (=FEOGA) and the rate of reimbursement for Member States is up to 75% of expenditure inside the Objective 1 regions and 50% elsewhere. The fact that these are upper limits underlines the flexibility for Member States as to the types of measure they may implement if they are prepared to support the cost. The Regulation lists the activities eligible for aid as: reducing the use of fertilizers, increasing extensive land uses, reducing stocking rates, promoting farming methods consistent with environmental protection and the continued existence of rare breeds of livestock in danger of extinction, ensuring the upkeep of abandoned farm and woodland, introducing 20-year set-aside for reserves or parks, managing land for access and leisure, and providing training for farmers in environmentally friendly practices.

In place of the requirement under Article 19 of 797/85 for designation of specific areas, Regulation 2078/92 calls for multi-annual zonal programmes and specifies that such programmes shall contain a definition of the relevant areas, their environmental characteristics, proposed objectives, conditions for grant aid, expected expenditure and the means of informing agricultural and rural operators. There is also the possibility for Member States to 'establish a general regulatory framework providing for the horizontal application throughout their territory' of the eligible aids.

The UK response to the regulation was to implement it selectively, ignoring the parts concerned with rare breeds in danger of extinction, upkeep of abandoned land and training in environmentally friendly practices. The House of Commons Agriculture Committee (1997) has recently suggested that training should be provided for farmers and that the Government should undertake a consultation exercise to determine the strength of support for a scheme to aid rare breeds. Much more than for any other member state there has been considerable continuity with earlier policy developed under Article 19 of 797/85.

Environmentally Sensitive Areas

The main instrument applied under these policies has been the ESA. The additional resources available under 2078/92 have allowed a considerable expansion of the programme geographically. Areas are designated as ESAs where traditional farming practices are vulnerable to change that would threaten the ecology, landscape or heritage. Farmers within ESAs are offered contracts whereby they are reimbursed for farming according to more or less elaborate management packages. The packages are specific to each ESA and may involve two or three 'tiers' of constraint which farmers may accept in return for standard payments geared to reimburse them for the resulting losses in farming income (the higher tiers involve greater constraints and higher payments) and to offer them a positive incentive to undertake conservation or landscape management under the scheme. Agreements in ESAs are for ten years with the possibility of termination after five years. Participation is entirely voluntary.

Table 2   Environmentally Sensitive Areas: Area (hectares) and Number of Agreements a)

	1992/3	1993/4		1994/5	1995/6	Total
England						
Area under agreement	129,358	137,100		79,933	63,571	409,062
Number of agreements	3,265	1,249		1,627	1,322	7,463
Wales						
Area under agreement	na	73,909	b)	12,635	26,999	113,543
Number of agreements	na	860	b)	23	510	1,393
Scotland						
Area under agreement	na	90,033	b)	131,424	141,344	362,801
Number of agreements	na	583	b)	134	360	1,077
Northern Ireland						
Area under agreement	na	na		na	na	83,000
Number of agreements	na	na		na	na	2,800
United Kingdom						
Area under agreement	129,358	301,042		223,992	314,914	968,406
Number of agreements	3,265	2,692		1,784	4,992	12,733

a) The individual year entries refer to the net increase in number of agreements held; b) These numbers represent cumulative totals since the inception of the schemes.
Source: House of Commons Agriculture Committee, 1997.

There are 43 ESAs in the UK (see Figure 1). They cover some three million ha and are mainly located in areas of upland grazing. There are, nevertheless, some important lowland ESAs, some of them in predominantly arable areas although the scheme mainly focuses on remnant semi-natural habitat within these areas. The distribution and area of land under ESA agreements are detailed in Table 2. There are currently nearly 13,000 agreements covering almost one million hectare.

Figure 1   ESAs in the UK

Other schemes

Other schemes pursued under Regulation 2078/92 are recorded in Table 3. Although most of them have only been on offer for a year or two, there are now just under 10,000 agreements covering more than 200,000 ha. It will be noticed that, whereas nearly all schemes are present in England, several do not appear in the other parts of the UK (a detailed account of the current state of these schemes can be found in House of Commons Agriculture Committee, 1997). These schemes are now described in more detail.

Countryside Stewardship: During 1991 an innovative scheme - Countryside Stewardship - was introduced designed to be much more flexible than ESAs and to deliver more specific public benefits. The elements in each ten-year contract are selected from a menu of standard payments for annual management and capital works, but the precise mix of measures and management details for each measure are individually agreed with participants. Not only is participation voluntary on the part of farmers but also payment is discretionary on the part of the MAFF. It is applied to specific broad habitat or landscape types - for example lowland heath, water fringes, grassland and historic landscapes - rather than to designated areas. The measure encourages practices which contribute to maintenance or enhancement of landscape, wildlife and historic features and the provision of access. After being piloted by the Countryside Commission (an official conservation agency), the administration of Countryside Stewardship was passed over to MAFF in 1996 which plans for its expanded application. At the same time as reassigning responsibility for this policy, it was brought under regulation 2078/92. It is notable that this scheme accounts for more than 5,000 agreements covering over 90,000 ha. In policy terms, Countryside Stewardship is identified as the Government's main incentive scheme for the wider countryside outside ESAs (Rural White Paper, 1995)

Tir Cymen: Sometimes described as the Welsh version of Countryside Stewardship, Tir Cymen applies in Wales where it provides for three levels of payment: first a general (horizontal) payment for the whole area of the farm; second, payment for management of specific habitats on the farm; and third, capital grants towards a schedule of agreed investments of an environmental nature. It is aimed at particular types of landscape and habitat and additionally seeks to promote new permissive paths. Like Countryside Stewardship the elements for inclusion in schemes are negotiated farm by farm. It is popular in the areas where it has been tested, where there are nearly 500 agreements extending over more than 60,000 ha.

Countryside Access Scheme: The least successful single instrument introduced under the Regulation is the Access Scheme which applies to land under the Arable Area Payments Scheme (AAPS). To be eligible farmers must enter land into the AAPS for five years and undertake to facilitate recreational access by keeping grass short, maintaining entry to the area and keeping it tidy and safe. Payments additional to the AAPS are designed to cover the cost of maintaining the access area. However, this has secured only 77 contracts in England covering less than 1,000 ha. A similar scheme in Scotland has elicited practically no response.

Habitat Schemes: There are different Habitat Schemes across the UK which aim to establish or improve habitats of significant conservation value. In some cases the aim is also to create buffer strips between intensive agricultural land and natural sites or water courses. The schemes apply to particular habitat types varying between the countries of the UK but including water

Table 3 Other UK Schemes under Regulation 2078/92: Area (hectares) and number of agreements a)

	1992/3	1993/4	1994/5	1995/6	Total
Countryside Stewardship (E,W)					
Area under agreement	56,248	19,427	11,644	5,266	92,585
Number of agreements	2,358	1,289	1,105	532	5,284
Countryside Access (E)					
Area under agreement			993		993
Number of agreements			77		77
Tir Cymen (W)					
Area under agreement	24,897	21,027	15,133		61,057
Number of agreements	201	187	168		556
Habitat Schemes (E,W,S,NI)					
Area under agreement			3,700	3,164	6,864
Number of agreements			192	259	451
Moorland Scheme (E,W,S,NI)					
Area under agreement				6,694	6,694
Number of agreements				28	28
Organic Aid Scheme (E, S, W, NI)					
Area under agreement				21,151	21,151
Number of agreements				129	129
Nitrate Sensitive Areas (E)					
Area under agreement					
- Premium Arable Scheme			2,349	1,273	3,622
- Premium Grass Scheme			398	62	460
- Basic Scheme			10,446	5,083	15,529
Total Area			13,194	6,417	19,611
Number of agreements			241	118	359
Total: (UK)					
Area under agreement	81,145	40,454	44,664	22,692	208,955
Number of agreements	2,559	1,476	1,783	1,066	6,884

E= England
W= Wales
S= Scotland
NI= Northern Ireland
a) The individual year entries refer to the net increase in number of agreements.
Source: House of Commons Agriculture Committee, 1997.

fringe areas, former five-year set-aside land, coastal salt marsh and heath, grazed semi-natural woodland and specified types of permanent grassland. The schemes have brought nearly 7,000 ha under 451 agreements. Most of them are for 20 years and most involve the managed reduction or withdrawal of grazing. They generally apply to only small parts of farms (on average, to 15 ha per farm).

Moorland Schemes: Moorland Schemes, which again vary between the different countries of the UK, have brought 6,000 ha under some 28 contracts, each of 20 years' duration. Moorland areas are large and are extensively grazed, usually by sheep. The schemes seek to reduce grazing pressure thus contributing to the ecological re-instatement of heather moorland.

Organic Aids: Organic Aids Schemes are offered in England, Scotland and Northern Ireland although there has been virtually no uptake in the last of these. The Schemes only assist with the conversion of land to organic farming; they do not support existing organic production. The agreements are for five years, with an upper ceiling of eligible area per agricultural business of 300 ha. There are somewhat over 100 agreements within the schemes covering more than 20,000 ha. Three-quarters of this is rough or unimproved grassland in Scotland.

Nitrate Sensitive Areas: In Nitrate Sensitive Areas (NSAs) farmers are offered grants for constraining their use of fertilizer. The NSAs, which are mainly arable, are all designated in catchments of selected ground water sources of public water supply and all are within Nitrate Vulnerable Zones (NVZs) designated under the EC Nitrates Directive (91/676). NVZs impose compulsory restrictions on farming based largely on good agricultural practice. The NSA scheme provides a voluntary supplement to induce more demanding changes in agricultural practice. Most of the area under contract is in the Basic Scheme (which allows for a continuation of arable cropping with restrictions designed to reduce nitrate leaching), but there are small areas amounting to a quarter of the total which receive higher payments for conversion of arable land to extensive grazing and for conversion of intensive grassland systems to more extensive grazing.

Policy monitoring and evaluation

A considerable amount of effort has been devoted to the monitoring and evaluation of agri-environment policies in the UK. Some measure of the official effort can be gauged from the expenditure details in Table 4. The table shows that administrative and monitoring costs amounted to at least 46% of expenditure on agri-environment measures in 1992/93 but fell steadily as a propor-

Table 4   *Public expenditure on 2078/92 in the UK: 1992/3 - 1996/7 (£ million, current prices).*

	1992/3	1993/4	1994/5	1995/6
Payments to farmers	17,904	29,781	37,723	54,768
Monitoring and running costs	15,351	17,340	19,984	20,552
Total	33,255	47,121	57,707	75,320
Monitoring and running costs as % of total	46.2	36.8	34.6	27.3

Source: Agriculture Committee, House of Commons (1997).
Note: Monitoring and Running costs are not reported for Scotland and these data therefore underestimate the importance of those costs in the percentages reported.

123

tion (although rising in absolute terms) down to about 27% in 1995/96, as the policy took off. It is not known whether other EC Member States devote similar resources to monitoring and administration of these arrangements. It is also notable that this aspect of agri-environment policy is entirely funded by Member States. Some individual monitoring and evaluation efforts are reviewed below.

Evaluation of ESAs

The Agriculture Act of 1986, which provided for the establishment of ESAs (under Section 18), also provided for regular assessment of their impact on land use (sub-section 8). Two types of monitoring study have been undertaken. First, a substantial series of studies have been made of the ESAs in England, ten of which were published in 1992 (see Whitby, 1994 for further information). These were based on the period from designation in 1987 or 1988 to 1991 and, not surprisingly, found few major changes in land use to report. However, the changes that were found provided a basis for estimating the small reductions in output resulting from these policies (Whitby et al., 1997).

More recently, MAFF has published a further set of studies relating to the same ESAs. The studies do not distinguish between agreements which have been running for less than five years and those which have been in existence longer, which is unfortunate given that some ESAs were substantially increased in size at redesignation (in 1992 and 1993), and the possibility of relating the changes found per farm to the duration of agreements has thus been missed (House of Commons Agriculture Committee, 1997).

Despite these problems it is possible to detect some significant changes in ESAs which follow more or less automatically from the compliance with management prescriptions. These would include the repair of stone walls and barns, for example. Also, Whitby et al. (1997) estimate that the first three years of designation of the first ten ESAs in England saw the reversion of some 5,500 ha of arable land to grassland with benefits to the landscape but also providing for an increase of nearly 3,000 livestock units and an aggregate reduction in fertilizer use of more than one million tonnes.

Socio-economic studies have also been commissioned and completed for the first round of ESAs (Whitby, 1994). They have found that management contracts have been successful in supporting farm incomes but have not produced major changes in farm management practices. Since they were based on only three years of the application of contracts, this is no surprise. A further set of socio-economic studies has now been completed and their publication is awaited.

Evaluation of Countryside Stewardship

The Countryside Stewardship scheme was the subject of a four-year monitoring and evaluation programme (Land Use Consultants, 1996; Ecoscope Applied Ecologists, 1997). This found that the scheme was achieving a high degree of success. It had targeted its resources to landscape types and geographical areas that offered great potential for environmental improvement and public benefit. Some 91% of sites under the scheme that were sampled were meeting the scheme's environmental objectives and were providing environmental benefits whether for landscape, wildlife, history/archaeology or access, or a combination of these. However this did leave a small number of sites where no benefit had been achieved, including some cases where harm had occurred. Also, not all of the sites where benefit had occurred delivered the range of benefits potentially covered by the scheme. It was concluded that the continued success of the scheme depended on regular compliance monitoring and would be improved by expanding the technical skills of agreement holders and introducing targets for easy-to-measure indicators of environmental change. The socio-economic effects of Countryside Stewardship have been evaluated by CEAS Consultants (Wye) Limited and the University of Reading (1996). They concluded that in aggregate there were some 50 full-time equivalent jobs generated on farms through the scheme and another 220 generated with agricultural contractors directly from capital works funded under the scheme. They report both positive and negative effects on farm inputs with a small positive aggregate balance across the whole scheme which disaggregates into positive changes in machinery inventories and offsetting negative changes in fertilizer and pesticide use. Aggregate output was estimated to have fallen by nearly £5 million across the scheme. Farm incomes rose on some farms and fell on others. Small 'knock-on' effects of the expenditure through the scheme were reported.

Evaluation of Nitrate Sensitive Areas

Monitoring of pilot NSAs began for the winter of 1990/91 and of NSAs designated under the agri-environment regulation (2078/92) for the winter of 1994/95. Ultimately, the success of the scheme will be assessed by the measurement of nitrate concentrations in abstracted water, which is undertaken by the Environment Agency. However, because it will take some years for the effect of land-use changes to be demonstrated in this way, the effect of the scheme at the soil, or 'input', level is being monitored. This includes collecting data each year, from both participating and non-participating farmers, on cropping and husbandry practices for each field within the NSAs. Under a computer model, these data are then used to estimate nitrate leaching losses from the catchments. In addition, actual nitrate leaching from the soil zone is measured each year from a representative sample of fields, and the results used to validate the modelled nitrate losses.

Although it will take some years before the effects of NSA land use changes are reflected in average nitrate concentrations at the borehole (because of the time taken for water to reach and mix with the groundwater reserve), the monitoring results demonstrate that NSA measures are reducing nitrate leaching from the soil zone. In particular, the Premium Scheme options are extremely effective in reducing nitrate leaching losses to very low levels, and cover crops (which are required on Basic Scheme land which would otherwise have been bare over the vulnerable autumn/winter period) are proving effective at mopping-up residual nitrate in the soil before it can be leached.

5.7 Forestry measures in agriculture

The Farm Woodland Premium Scheme (FWPS) is administered by MAFF and the Agriculture Departments of the Scottish, Welsh and Northern Ireland Offices. It was introduced on 1 April

1992 to replace the pilot Farm Woodland Scheme. The scheme encourages farmers to convert productive agricultural land to woodland by providing annual payments for 10 or 15 years to off-set the income foregone. These payments are in addition to the full range of establishment grants available under the Forestry Commission's Woodland Grant Scheme (WGS). The FWPS and the WGS comprise the UK's national programme under Council Regulation 2080/92 on forestry measures in agriculture. The objectives are:

(I)   to encourage the planting of woodlands by farmers who will remain in farming, thereby enhancing the farmed landscape and environment;
(ii) to encourage a productive alternative land use to agriculture.

For the UK as a whole, in the period 1 April 1992 (when the Scheme opened) to 30 September 1996, some 5,360 applications were approved to convert some 34,000 ha, an average of 6.3 ha per application. This comprises 31% arable land, 39% improved grassland and 30% unimproved grassland. Nearly three quarters of planting is of broadleaves. The majority of planting has been small areas for amenity or sporting purposes.

Economic and environmental evaluations of the scheme were undertaken in all four countries of the UK. These were followed by a policy review of the FWPS and a public consultation on the proposed changes. Parliamentary and European Commission approval was sought for the following changes: revised objectives for the scheme, placing more emphasis on securing environmental benefits; increased rates of payment (except on unimproved land); separate higher rates of payment for land eligible for the Arable Area Payments Scheme; creating a single application system to cover both the FWPS and the Forestry Commission's Woodland Grant Scheme in Great Britain; broadening the definition of agricultural businesses that can be accepted into the scheme; removing the requirement for participants, once accepted into the scheme, to continue to run an agricultural business; removing the current rule limiting applications to 50% of the agricultural area of a farm; and introducing a maximum limit of 200 ha on the area per farm business that can be aided.

5.8   Objective 5a

The Guidance Section of the EAGGF complements the market support actions of the Guarantee section and is one of the so-called structural funds which are the main financial instruments used by the European Community to help reduce disparities and support social and economic cohesion across Europe. The application of the funds is targeted according to six objectives each addressing different types and levels of need. Objective 5 is promoting rural development and this is subdivided into: 5a, for speeding up the adjustment of agricultural structures in the framework of the reform of the CAP; and 5b, for promoting development of designated rural areas (see next section).

Objective 5a includes a series of measures for agricultural produce marketing/processing, increasing farm efficiency, help for cooperatives, farm diversification and income support for hill farming. UK co-funding for 5a, however, was wound down to release domestic funds for 5b. This

126

entailed the phasing out of the Farm and Conservation Grant Scheme and the Processing and Marketing Grant Scheme. The bulk of the support that remains goes on Less Favoured Areas.

LFA policy

The origins of LFA-type supports in Britain can be traced back to 1940 when a subsidy was given on every hill ewe; in 1943 this was extended to each hill cow. The 1946 Hill Farming Act added improvement grants. The hill farming area was taken roughly to be the land between the 600 and 800 feet (200 - 260 metres) contours. The subsidies paid on rearing hill livestock were intermittent and depended on the economic circumstances of the sector: when prices or lambing rates were low, subsidies would be paid, but when lambing rates and prices were high they were withheld. These arrangements were replaced with a system of permanent annual headage payments in the 1960s. In 1967 the range of financial aids for improvements was extended. When the UK negotiated entry to the EC, one of the conditions was that the UK would be able to continue to give special help to hill farming. In response, Directive 75/268 was introduced which provides for the payment of selective incentives to farmers in certain agriculturally disadvantaged areas in order to achieve the continuation of farming thereby maintaining a minimum population level or conserving the countryside.

Other directives specified the eligible areas in each Member State. There are two main categories of disadvantaged areas: mountain areas and less favoured areas. No areas in the UK have been listed under the former; and all but one of the eligible areas have been listed under the latter 1). Less Favoured Areas are defined as areas, regional in character, that are 'in danger of depopulation and where the conservation of the countryside is necessary' (Article 3 (4)). They must satisfy criteria concerning infertility, poor economic situation and a low or dwindling population dependent on agriculture. In Britain these requirements were interpreted as covering the hill regions. Some 42% of the UK agricultural land area was initially listed, and this was extended to 51% in 1984 and to 53% in 1991. The original areas are referred to by MAFF as severely disadvantaged and qualify for somewhat higher rates of support than do the areas subsequently added which are referred to as disadvantaged (Thomson et al., 1990).

The LFA Directives have always been linked to other Community farm structure policies which specify the type of aids that are eligible. At first the link was with Directive 72/159 on the moderation of farms, but subsequently with Regulation 797/85 which was itself replaced by Regulation 2328/91 in 1991. The types of aid permitted are:
(i)   Hill Livestock Compensatory Allowances (HLCAs). These are annual allowances in the form of headage payments, paid in the UK on breeding sheep and beef cattle. Since 1991 an upper limit of 1.4 livestock units/ha of forage has applied to the allowances eligible for reimbursement. This was advocated by the UK to discourage overstocking (Member States can set more restrictive limits and payment conditions in order to safeguard the environment).

---

1)   The exception being the Isles of Scilly which fall within a third category, for other small areas affected by specific handicaps in which farming must be continued to conserve the environment.

(ii)   Investment Aids to Farms. For approved investment plans farms may receive higher rates of capital grants in LFAs than elsewhere. Investment aids may also be given for joint schemes to improve pastures or fodder production.

Reimbursement of up to 25% may be paid from the Guidance Section of the EAGGF.

The payment rates for HLCAs in the UK for 1997 are shown in Table 5. Compared with 1996, rates for breeding cows were doubled in the severely disadvantaged areas and tripled in the disadvantaged areas in order to maintain the incomes of hill cattle farmers. This was done in response to the BSE crisis.

Table 5   1997 rates of HLCAs (£ per head)

Animal	Severely Disadvantaged Area	Disadvantaged Area
Breeding cows	97.50	69.75
Hardy breeding ewes	5.75	2.65
Other ewes	3.00	2.65

Source: MAFF and IB, 1997.

Table 6 shows the recent trend in public expenditure on the UK LFAs. The HLCAs make up by far the largest share of this expenditure. The steadily diminishing significance of these payments will have been reversed for hill cattle farming by the 1997 increase in the relevant rates.

Table 6   Public expenditure on LFA policy instruments in the UK (million £)

Policy instrument	1992/3	1993/4	1994/5	1995/6
Hill Livestock Compensatory				
Allowances - Cattle	83.2	62.0	57.8	62.3
- Sheep	56.2	62.5	59.1	49.6
Additional Capital Grants	8.8	8.2	7.4	3.0
Other Expenditure	13.5	6.8	3.9	6.7
Total	161.7	139.5	128.2	111.6

The effects of LFA policy in the UK

The main agricultural systems in the LFAs are based on sheep and cattle rearing. The supports farmers receive under the commodity regimes for beef and sheepmeat far outweigh the HLCAs. There are also other sources of support for which farms in LFAs may be eligible. For example, the bulk of payments under the agri-environment regulation goes to LFAs, and most of the areas

128

designated under Objective 5b are within LFAs. Separating out the specific effects of LFA supports therefore is not really possible.

There is evidence that the LFA policy may have helped to slow the rate of farming depopulation. Although agricultural employment in the LFAs has fallen - by about 7,000, or 4% in the past decade - it has actually risen as a proportion of national agricultural employment by 2% to about 27% over that period (MAFF and IB, 1996, 1997).

A justification of maintaining farming in the LFAs is to conserve the countryside. The LFA boundaries include most of the National Parks in England and Wales and a major share of other land designated for its conservation value, including Environmentally Sensitive Areas, Sites of Special Scientific Interest and Areas of Outstanding Natural Beauty. It is generally acknowledged that the continuation of extensive livestock farming is crucial to maintaining the conservation importance of the LFAs. However, there is also concern that agricultural policy has encouraged overstocking in certain areas that has caused damage particularly to heather moorland (RSPB, 1984). Although it is likely that the sheepmeat regime has been of greater influence in encouraging overstocking, it is not apparent, until very recently, that the LFA policy has been operated in such a way as to ensure specifically that conservation values were maintained (Felton and Marsden, 1990).

Following EC Regulation 2328/91, the UK government introduced measures in 1992 to reduce overgrazing of moorland, including a working definition of overgrazing from an ecological perspective and a Code of Good Upland Management to assist farmers to maintain the character of the countryside. Since then, a few farmers have actually had their HLCA payments withheld for refusing to reduce stocking densities but the assessment of overgrazing and application of the relevant controls are complicated and resource-intensive, not least because of the reaction they provoke from farmers. With enforcement pursued in only the most blatant cases these environmental cross-compliance measures are not considered to be that effective. The 1997 increases in HLCA for cattle may help to redress the balance of incentives which in the past, by encouraging the expansion of sheep grazing at the expense of cattle grazing, has led to undesirable habitat changes.

5.9   Rural Structural Funds: Objective 5b

EU Objective 5b policy

The purpose of the structural funds is to reduce disparities in development between the regions of the European Community. The distribution of regional structural funds is shown in Figure 2. Objective 5b of the funds is for promoting the development of rural areas. It combines the following three funds in geographically targeted areas:
(I)   the European Regional Development Fund (ERDF) which is mainly for infrastructure, business development and employment generation;
(ii)   the European Social Fund (ESF) which is for training and skills development;
(iii)   the European Agricultural Guarantee and Guidance Fund (EAGGF) which supports agricultural diversification.

Objective 5b areas are designated on the basis of low levels of GDP per inhabitant and the following main criteria: a high proportion of agricultural employment; a low level of agricultural income; and a low population density or depopulation trend. There are a number of secondary criteria including: sensitivity of the area to developments in agriculture; pressures exerted on the environment and countryside; and parts of the area having Less Favoured Area status. In most of the UK 5b areas these secondary criteria are more important than the main ones of which only low population density generally applies.

The structural funds must be matched with domestic sources of finance and be allocated to projects that meet the aims of agreed regional programmes. Both government and the private sector are eligible to put forward projects. Economic regeneration and job creation are the rationale for the structural funds and in rural areas they are pursued through measures for agricultural and rural diversification. Previously there was no significant environmental component, but this has changed under new rules for the current round of structural funds for the period 1994-1999 which incorporate the principles and goals of sustainable development (Regulation 2081/93). Structural fund programmes must be consistent with Community policy which in the field of the environment is designed to ensure a high level of environmental protection. Also Member States must supply an appraisal of the state of the environment and the environmental impact of the operations envisaged ... as well as the steps they have taken to associate their environmental authorities with the preparation and implementation of the plans.

The new structural funds have the potential to deliver considerable environmental benefits which can be integrated with the economic development of rural areas. The Commission's preamble to the regulation on EAGGF expenditure within the structural funds (2085/93) states that it should be able to finance measures for sustainable development of the rural environment including developing and strengthening agricultural and forestry structures which use methods and techniques that respect the environment. The Regulation puts forward the following indicative actions:

(i)   renovation and development of villages and the protection and conservation of the countryside (Article 5c);
(ii)   encouragement for tourist and craft investment (Article 5g);
(iii)   to develop and optimally utilise woodlands in rural areas (Article 5l).

Objective 5b in the UK

This approach to regional development is not that familiar in the UK. Britain's traditional approach to regional policy has been urban and industrially oriented. Development agencies have had responsibilities for marginal rural areas but they have tended to concentrate their efforts on attracting new industrial development rather than encouraging diversification based on the primary sector. The UK even has some difficulties in simply meeting the criteria for 5b designation. The areas designated have had to rely on the secondary criteria (see above), and to meet just one of the main criteria - low population density - quite artificial regions have had to be devised that in a number of cases omit key centres that service them (Ward and McNicholas 1997).

In the previous round of structural funds (1988/93) only a few rural regions qualified - Cornwall, Wales, Dumfries and Galloway, and the Highlands and Islands. With the expansion of

130

the UK's allocation, 11 rural regions have been included in the current round. They have a combined EC and UK budget of £1.3 billion, include 5% of the population and cover 27% of the land surface, including much of the UK's most naturally diverse areas. In addition, Northern Ireland and the Highlands and Islands enjoy Objective 1 status (the structural fund objective to support development in poorer regions).

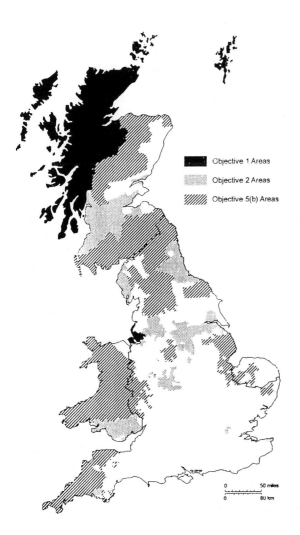

Objective 1 Areas

Objective 2 Areas

Objective 5(b) Areas

0   50 miles

0   80 km

Figure 2   Regional Structural Funds in the UK

The allocation of funds to projects in each 5b region is on the basis of a Single Programming Document (SPD) which is intended to deliver resources to the region in an integrated way. The SPD represents a review of the region analysing the salient strengths, weaknesses, opportunities and threats. Objectives are defined for a range of sectoral headings, typically business development, tourism, agriculture, environment and the community, and for each sector a number of measures are prescribed which detail targets and indicative actions.

The priorities set out in the SPDs provide the basis for the negotiations with the Commission on the specific distribution of resources between the structural funds. The breakdown of the overall allocations to the UK 5b regions is 65.2% ERDF, 16.4% ESF and 18.5% EAGGF. The last of these is the one with the most explicit environmental component (ESF, though, can support training in environmental and conservation skills; and ERDF can support investment in environmental protection and sustainable tourism).

EAGGF support makes up between 15-20% of the individual 5b programmes. Table 7 shows the proportion of these funds allocated within each SPD under the environment heading. However, this under represents the full financial commitments to the environment even under EAGGF, because other headings may also include some environmental measures. For example, the Midland Upland 5b SPD has no environment heading but the agriculture heading contains a number of on-farm conservation targets; and under a number of the regional programmes, advisory services to farmers on diversification and conservation are supported under agriculture (Wildlife Trusts and WWF, 1996).

Even so, just the EAGGF component directly under the environment headings amounts to £34.5 million (Table 7) which is equivalent to a quarter of the EU contribution to schemes under the agri-environment Regulation (2078/92) in the UK on an annualized basis. For regions without ESA designations (e.g. Lincolnshire), the environmental resources under 5b far exceed that under 2078/92.

The final column in Table 7 shows the percentage of the allocated funds that had been committed to approved projects by November 1996. Progress in establishing partnerships, developing projects, obtaining co-funding and winning approval for EAGGF projects has been rather slow. A number of criticisms have been levelled at the procedures. MAFF and the agriculture departments of the territorial ministries (the Welsh, Scottish and Northern Ireland Offices) lack the devolved regional structures which could give a lead in developing and promoting the programmes or take responsibility for project funding decisions. MAFF and the agriculture departments have also sometimes taken a narrow economic view of what it is appropriate to support. There have been problems in finding match funding. Project development as well as funding requirements may necessitate a partnership approach bringing together different organizations, public, private and voluntary. Establishing such partnerships can be very time consuming. Examples of successful projects include environmental improvement and management of ecologically important salt marshes and grazing meadows in a Devon estuary; moorland regeneration in the North York Moors; habitat and fisheries improvement in the headwaters of the river Tweed; hedgerow renovation in Wales; and native woodland conservation in Grampian.

132

Table 7   EAGGF allocations to Environment Measures in UK Objective 5b regions

Region	Allocation to the environment (x 1,000 £)	% of total EAGGF	% take up November 1996
Borders	1,833	61.1	22.4
Dumfries and Galloway	1,500	28.6	39.3
East Anglia	2,000	22.8	0.0
Lincolnshire	1,408	18.0	9.8
Marches	1,083	17.8	21.1
Midland Uplands	-	-	-
North and West Grampian	1,292	33.0	12.8
Northern Uplands	8,450	37.7	9.1
Rural Stirling and Upland Tayside	1,771	68.0	32.0
Rural Wales	5,583	18.8	80.8
South West Region	9,583	28.0	11.1
TOTAL	34,504	27.5	24.7

Source: Wildlife Trusts and WWF, 1996.
Conversion £1 = 1.2 ECU.

5.10 Concluding observations

The relationship between the reformed CAP and the environment is very complicated and is still quite uncertain. From the above review, however, it is possible to make some interim observations.

The provisional nature of any judgements

The implementation of the 1992 CAP reforms was phased in over a three to four year period. Certain measures, say under the agri-environment regulation, only became operational in 1996. The stocking density rule for the Beef Special Premium was progressively reduced each year between 1993 and 1996 and only began to bite for many farmers latterly. It is thus early days to be judging the impact of the reforms. More generally, we must accept that the full environmental consequences of many of the measures taken in recent years to counter the adverse effects of contemporary agriculture will not become clear for several years to come. For a start, while it is possible to refer to the take-up rates of various schemes and emerging changes in farming practices, the consequent second order effects in such areas as changes in enterprises, farming systems or technology may happen over an extended period. Secondly, the response time of the natural environment may be particularly long. Increasing the biodiversity of agriculturally improved grassland (an aim of some ESAs) can take up to a hundred years, for example. Likewise, the recovery period for ground water sources with excess nitrates may be several decades. The environmental reform of agriculture needs to adopt a long-term perspective.

The confounding effects of other contemporary events

The introduction of the CAP reforms has coincided with other major events which have greatly affected the economic climate for agriculture. These have swamped the impact of the reforms whose consequences are now that more difficult to discern. The reforms are also operating in a somewhat altered agricultural situation from that prevailing when they were formulated and are having to address problems not then envisaged.

One of the major events that has greatly affected British agriculture since the introduction of the Mac Sharry reforms is Britain's withdrawal from the Exchange Rate Mechanism (ERM). That event, coupled with a tightening of world grain supplies, confounded the expectation of reduced profitability in the cereal sector following the reforms. The implementation of the reforms over the first three years coincided with high world grain prices which boosted farmers' incomes significantly. In addition, British farmers were shielded from the effects of cuts in cereal support prices by successive devaluations of the green pound. For UK cereal producers, the years immediately following 1992 turned out in fact to be highly profitable ones which, for a period, gave a significant boost to intensification and specialization (Asby and Sturgess, 1997).

Another major event with more profound and longer term consequences is the BSE crisis. The reforms in the beef regime were intended to bring the supply and demand for beef back into equilibrium. The drop in demand following the crisis has frustrated that objective. At the same time the efforts to reduce cattle numbers have been greatly overtaken by the slaughtering programmes introduced in reaction to the BSE crisis. The result is a large reduction in the UK beef sector with serious consequences for the agricultural economy and the rural environment.

The Agri-environment Regulation as a minor component of the CAP

Agri-environmental expenditure, particularly when narrowed down to the sub-category of actual payments to farmers, remains very low in comparison with the overall size of the UK's agricultural economy and in comparison with total CAP expenditure in the UK. In 1995/96, for example, agri-environmental payments to farmers in the UK as a whole amounted to nearly £43 million, or 1.6% of total CAP expenditure in the UK of about £2.75 billion (MAFF, 1996d). Although agri-environment expenditure has expanded progressively over the past ten years, and is set to expand further in the next three years (for 1996/97 total expenditure will have amounted to about £80 million including £60 million in payments to farmers), there is little prospect of a sizeable shift towards environmentally-friendly farming practices until there is a more equal balance between such payments and the production incentives under the CAP's commodity regimes. As it is, commodity supports actively discourage take up of agri-environmental measures.

The bulk of expenditure under the agri-environment Regulation is on ESAs, followed by Countryside Stewardship (accounting respectively for 61% and 26% of expenditure in England in 1996/97). This means that the UK implementation of the Regulation has been very particular. In essence, Britain is still largely implementing on an expanded scale a policy devised under Article 19 of 797/85. The lack of a general extensification scheme (such as the French 'prime à l'herbe'), the restricted interpretation of some measures such as that for organic farming (which is used to support organic conversion but not production) and the limited take-up of some schemes (such as the ones for countryside access and moorland conservation) all have had the effect of constraining the growth of agri-environment policy in the UK. Arguably, this reflects a wariness in the UK towards an open-ended financial commitment under the Regulation (because of the UK's budgetary rebate under the Fontainebleau agreement, the marginal cost to the UK Treasury of match funded schemes such as this is actually 85% of the total cost). The approach adopted also reflects a UK preoccupation, in the development of agri-environment policy, to solve discrete and specific environmental problems through circumscribed and targeted measures rather than address the generic causes of agriculture's pressures on the environment or see agri-environment policy as playing a more strategic role in reorienting the CAP.

There are certain perverse consequences. The most important is that any beneficial effects of the agri-environment Regulation are likely to be swamped by the environmental impact of the rest of the CAP. There is a risk also that the intricacies of implementing the Regulation will distract attention from the bigger picture. Thus considerable resources have been devoted to monitoring and evaluating agri-environmental measures, but little to the environmental impact of the CAP commodity regimes.

Other direct environmental supports

The Agri-environment Regulation, though, is not the only source of direct environmental supports under the CAP. The Farm Woodland Premium Scheme (under Regulation 2080/92) is another source, but its impact so far has been rather modest. Funding for environmental schemes is also available under Objective 5b. The amount available is equivalent to a quarter of the EU contribution to schemes under 2078/92 in the UK on an annualized basis. In some regions, though, the environmental resources under 5b actually exceed those under 2078/92. The amount spent on LFA supports is larger than that for the Agri-environment Regulation. However, although LFAs are meant to be of conservation benefit, more could be done to ensure this was achieved.

De-intensification in the beef and sheep sectors

The impact of the reforms coupled with BSE is leading to a deintensification of livestock production. However, the environmental consequences are mixed and uncertain. Reduced grazing pressures are not necessarily leading to the revitalization of degraded moor and grassland. Likewise, semi-natural habitats, including species rich pastures, need active management, and the withdrawal of livestock may lead to unwanted environmental changes. In parts of the UK, livestock numbers are falling below desirable levels from a land management/conservation point of view. The interaction between the different commodity regimes may actually be exacerbating matters, for example, by encouraging farmers to special in sheep or beef to maxim their returns from the different schemes on offer. Stocking rate rules have been introduced to help correlate the separate livestock regimes geographically. However, the rules are not designed to achieve environmental benefits but to constrain support costs and the available evidence suggests that the extensification mechanisms are not working.

The plethora of overlapping schemes

The strong impression is of a confusing medley of overlapping and conflicting schemes and programmes. Individual schemes are complicated and it is difficult to grasp how they interrelate and what their overall effects are. Not only is this confusing for the analyst but also for officials and farmers that must operate the system. The way these schemes interact on the ground is difficult to judge and may be quite haphazard from an environmental point of view.

To give just one example: moorland use is affected by the beef and sheep regimes, and HLCAs. Moorland management may be eligible for a number of the schemes under the agri-environment Regulation. Moorland restoration projects are being supported under Objective 5b. Each of these schemes has somewhat different rules and objectives. Each supports different measures. Objective 5b covers capital costs; the others cover revenue costs. Only those under 2078/92 have moorland conservation as a specific objective. There is no effort to combine them at the European or national level into environmentally coherent programmes. Indeed, the funding rules may actually frustrate this (several Objective 5b SPDs, for example, rule out combining 5b and agri-environment funding) (Wildlife Trusts and WWF 1996).

Environmental cross-compliance

The UK is the only country so far to have taken up the option to apply environmental conditions to livestock supports. However the enforcement process has proved cumbersome and only a small number of farmers have had payments withheld for overgrazing. There is some scepticism now over the potential scope and effectiveness of this type of environmental conditionality attached retrospectively to farm payments.

Farm investment and environmental pressures

Arable area payments and other commodity supports are making a vastly greater financial injection into farm businesses than are agri-environment measures. Previously, the general downward trend in farm incomes since the mid-1970s had not only encouraged farmers to look for alternative income sources but had diminished the rate of investment in agriculture. In fact depreciation exceeded gross fixed capital formation over several years. In the years immediately after 1992, though, the trend was reversed as shown in Table 8 (However, 1997 saw a sharp downturn in

Table 8   Impact of the CAP reform on income and investment in UK agriculture (£ million)

	1993	1994	1995	1996	1997 (provisional)
Net Farm Income	3,384	3,407	4,276	4,101	2,276
Depreciation	1,680	1,730	1,838	1,939	1,968
Gross Capital Formation	1,652	1,893	1,980	2,048	n.a.

Source: MAFF, 1997.

farm incomes). By encouraging additional investment in mechanization, amalgamation and specialization, the commodity supports may have thus stimulated yet more environmental damage.

References

Allanson, P.A. and A. Moxey (1996) *Agricultural land use change in England and Wales, 1892-1992;* Journal of Environmental Planning and Management 39, 243-254

Asby, C. and I. Sturgess (1997) *Economics of Wheat and Barley Production in Great Britain, 1995/6;* Special Studies in Agricultural Economics Report No. 34; Cambridge, Department of Land Economy

Ashworth, S.W., A. Waterhouse, T. Treacher and C.E.F. Topp (1997) *The EU Sheepmeat and Goatmeat Regime and its impact on the environment;* A report prepared for the Land Use Policy Group of the Countryside Agencies of Great Britain; Scottish Agricultural College, Auchincruive

Baldock, D. and P. Lowe (1996) *The development of European agri-environment Policy;* In: Whitby, M. (ed.), The European Environment and CAP Reform: Policies and Prospects for Conservation; CAB International; Wallingford, pp. 8-25

Ball, D.F., J. Dale, J. Sheail and O.W. Heal (1982) *Vegetation Change in Upland Landscapes;* NERC/ITE

Bowers, J. and P. Cheshire (1983) *Agriculture, the Countryside and Land Use;* Methuen, London

British Survey of Fertilizer Practice (1994) *Fertilizer Use on Farm Crops 1993;* HMSO, London

Buckwell, A. (1990) *Economic signals, farmers' response and environmental change;* Journal of Rural Studies 5(2), 149-160

CEAS Consultants, (Wye) Limited and the University of Reading (1996) *Socio-economic Effects of the Countryside Stewardship Scheme;* Unpublished Report to the Countryside Commission, Cheltenham

Cook, H. and H. Moorby (1993) *English marshlands reclaimed for grazing: a review of the physical environment;* Journal of Environmental Management 38, 55-72

Cooper, M. and R. Thomas (1991) *Profitable Sheep Farming;* London, Farming Press Books

Countryside Commission (1974) *New Agricultural Landscapes;* Countryside Commission, Cheltenham

Cox, G., P. Lowe and M. Winter (1990) *The Voluntary Principle in Conservation;* Packard, Chichester

Davis, R., M. Thomas and D.G. Garthwaite (1989) *Pesticide Usage Survey Report 79: Grassland and Fodder Crops;* MAFF, Harpenden

Department of the Environment (1996) *Pesticides in Water;* London, HMSO

Ecoscope Applied Ecologists (1997) *Countryside Stewardship: Monitoring and Evaluation of the Pilot Scheme 1991/96;* Countryside Commission, Cheltenham

Ellett, J.S. (1984) *Sheep Stocking Densities and Vegetation Change;* Unpublished Dissertation, Conservation Course, London University

ENTEC (1996) *Options for Change in the CAP Beef Regime;* Report to Countryside Commission, Countryside Council for Wales, English Nature and Scottish Natural Heritage

Evans, N.J. and R. Yarwood (1995) *Livestock and landscape;* Landscape Research 20, 141-146

Evans, R. (1990) *Water erosion in British farmers' fields: Some causes, impacts, predictions;* Progress in Physical Geography 14, 199-219

Felton, M. and J. Marsden (1990) *Heather Regeneration in England and Wales;* Nature Conservancy Council, Peterborough

Firbank, L.G. (Ed.) *Agronomic and Environmental Monitoring of Set-aside under the EC Arable Payments Scheme;* Unpublished report to MAFF by ITE, ADAS and BTO

Gaskell, P. and M. Winter (1996) *Beef Farming in Great Britain: Farmer Responses to the 1992 CAP Reforms and Implications of the 1996 BSE Crisis;* Report to the Countryside Commission, Scottish Natural Heritage and English Nature

Grigg, D. (1995) *An Introduction to Agricultural Geography 2nd edition;* London, Routledge

House of Commons Agriculture Committee (1997) *Environmentally Sensitive Areas and Other Schemes under the Agri-environment Regulation, Volume I;* House of Commons

Hubbard, L.J. (1992) *Two-tier pricing for milk: A re-examination;* Journal of Agricultural Economics, 43, 343-354

Jarvis, S.C., R.J. Wilkins and B.F. Pain (1996) *Opportunities for reducing the environmental impact of dairy farming management: a systems approach;* Grass and Forage Science, 51, 21-31

Johnson, J. and B.G.Merrell (1994) *Practical pasture management in hill and upland systems;* pp. 31-41; In: Livestock Production and Land Use in Hills and Uplands Occasional publication No.18 of the British Society of Animal Production

Land Use Consultants (1996) *Countryside Stewardship Monitoring and Evaluation;* unpublished report to the Countryside Commission, Cheltenham

Lowe, P., J. Clark, S. Seymour and N. Ward (eds) (1997) *Moralising the Environment: The Politics and Regulation of Farm Pollution;* London, UCL Press.

Lowe, P., G. Cox, M. MacEwen, T. O'Riordan and P. Winter (1986) *Countryside Conflicts: the Politics of Farming, Forestry and Conservation;* Gower, Aldershot, UK

MAFF (1947) *Agricultural Statistics in the UK, 1939-1944;* London, HMSO

MAFF (1995) *How to Manage Your Set-aside Land for Specific Environmental Objectives;* London, HMSO

MAFF (1996a) *Agriculture in the United Kingdom: 1995;* London, HMSO

MAFF (1996b) *Digest of Agricultural Census Statistics;* London, HMSO

MAFF (1996c) *Beef Special Premium Scheme 1996;* notes for guidance; London, HMSO

MAFF (1996d) *Intervention Board Departmental Report;* Cmnd. 3204, London, HMSO

MAFF (1997) *Agriculture in the United Kingdom 1996;* London, HMSO

MAFF/IB (1996) *The Government's Expenditure Plans 1996-7 to 1998-99;* London, Ministry of Agriculture, Fisheries and Food and Intervention Board; Cmnd. 1903; HMSO

MAFF/IB (1997) *The Government's Expenditure Plans 1997-8 to 1999-2000;* London, Ministry of Agriculture, Fisheries and Food and Intervention Board. Cmnd. 3604, HMSO

Moxey, A., B. White and J.R. O'Callaghan (1995) *CAP reform: an application of the NELUP economic model;* Journal of Environmental Planning and Management 38, 117-124

Nature Conservancy Council (1977) *Nature Conservation and Agriculture;* Nature Conservancy Council, London

Nature Conservancy Council (1984) *Nature Conservation in Great Britain;* Nature Conservancy Council, Peterborough

NCC (1990) *Nature Conservation and Agricultural Change;* Focus on nature conservation series No 25, NCC, Peterborough

Oglethorpe, D.R., R.A. Sanderson and J.R. O'Callaghan (1995) *The economic and ecological impact at the farm level of adopting Pennine Dales Environmentally Sensitive Area (ESA) grassland management prescriptions;* Journal of Environmental Planning and Management, 38

Royal Commission on Environmental Pollution (1996) *Sustainable Use of Soils;* 19th Report RCEP, London

Royal Society for the Prevention of Birds (1984) *Hill Farming and Birds;* RSPB, Sandy

Russell, N., V. Smith and B. Goodwin (1997) *The effects of CAP reform on the demand for crop protection in the UK;* pp. 397-413; In: Oskam, A. and Vijftigschild, R. (eds); Proceedings of the EU Concerted Action Workshop on Pesticides Department of Agricultural Economics and Policy, Wageningen Agricultural University, Netherlands

Saunders, C. and A. Moxey (1994) *Livestock Movement and Location in England;* Report prepared for English Nature, English Nature Research Reports No. 88, Peterborough: English Nature

Skea, A. (1993) *The effects of CAP reform on Scottish farming;* Scottish Agricultural Economics Review 7, 1-14

Spedding, C. (ed.) (1983) *Fream's Agriculture;* John Murray, London

Thompson, D.B.A., A.J. MacDonald, J.H. Marsden and C.A. Galbraith (1995) *Upland heather moorland in Great Britain: a review of international importance, vegetation change and some objectives for nature conservation;* Biological Conservation 71, 163-178

Thomson, K.J., B.J. Revell, A.K. Copus, N. Tzamarias, M.C. Whitby, C.M. Saunders, D.A. Parsisson, G.W. Furness and S.A.E. Magee (1990*) Farming and the Rural Economy of the Less Favoured Areas of the United Kingdom;* National Economic Development Office

Vipond, J. (1995) *One million fewer lambs - time for a rethink;* The Sheep Farmer November/December 1995, 32-33

Wallace, M. and A. Kirke (1992) *CAP Reform and Northern Ireland Agriculture: A Case Study Analysis;* Dept. of Agricultural and Food Economics, The Queen's University of Belfast

Ward, N. (1995) *Technological change and the regulation of pollution from agricultural pesticides;* Geoforum 26, 19 - 33

Ward, N. and K. McNicholas (1997) *Reconfiguring Rural Development in the UK: Objective 5b and the New Rural Governance;* University of Newcastle, Centre for Rural Economy Working Paper No. 24

Weaver, R.D., J.K. Harper and W.J. Gillmeister (1996) *Efficacy of standards versus incentives for managing the environmental impacts of agriculture;* Journal of Environmental Management, 46, 173-188

Whitby, M. and P. Lowe (1994) *The political and economic roots of environmental policy in agriculture;* In: Whitby, M. (Ed.); Incentives for Countryside Management: The Case of Environmentally Sensitive Areas CAB International, Wallingford, pp. 1-24

Whitby, M.C. (ed.) (1994) *Incentives for Countryside Management: the case of Environmentally Sensitive Areas;* Wallingford, CAB International

Whitby, M., C. Saunders and C. Ray (1997) *The full cost of Stewardship policies;* In: Dabbert, S., A. Dubgaard, L. Slangen and M. Whitby (eds); Landscape and Nature Conservation CAB International, Wallingford

Wildlife Trusts and World Wide Fund for Nature UK (1996) *Objective 5b Structural Funds: New Environmental Opportunities in the UK;* Unpublished report for WWF, Godalming

Wilkins, R.J. (1996) *Environmental constrains to grassland systems;* In: Grassland and Land Use Systems, 16th European Grassland Federation Meeting, pp. 695-703

Winter, M. (1996) *Effects of CAP Reform on the GB Countryside. Report of the Third Desk Study;* Countryside and Community Research Unit, Cheltenham and Gloucester College of Higher Education, Cheltenham

Winter, M. and P. Gaskell (1997) *The Effects of the 1992 Reform of the Common Agricultural Policy on the Countryside of Great Britain;* Report to Countryside Commission, Countryside Council of Wales, Department of the Environment and Scottish Natural Heritage

Winter, M., A. Rutherford and P. Gaskell (1997) *Beef Farming in the GB LFA -The Response of Farmers to the 1992 CAP Reform Measures and the Implications for Meeting World Trade Obligations;* Paper presented to International Conference on Livestock Systems in Rural Development in Disadvantaged Areas, Nafplio, Greece (forthcoming in conference proceedings)

PART 2

INTENSIVE FARMING SYSTEMS

6. BELGIUM

Jean-Marie Bouquiaux, Marielle Foguenne and Ludwig Lauwers

6.1   Introduction

Belgium, outside of the capital Brussels, consists of two regions: the Dutch speaking Flanders to the north and the French speaking Wallonia to the south. This sociogeographic context is important to the subject of CAP and the environment, for a number of reasons:
-   the north is more densely populated than the south;
-   agriculture is more intensive in the north, not only due to the occurrence of granivore production and horticulture, but also to a higher use of inputs in dairy and arable farming;
-   the regional, linguistic frontier almost coincides with the frontier between two major agricultural areas (the sandy loam and the loam area);
-   responsibility for environmental policy and, to an increasing extent, for agricultural policy is devolved to the regional level.

Some agricultural trends and environmental features will be presented in the next section, illustrating differences between the two regions. Next, a survey of market and price policy effects are given for the main agricultural sectors (arable crops, dairy farming, beef production and granivore production). The subsequent section is dedicated to accompanying and structural measures. In Section 5, some key environmental topics are further elaborated, where national policies and changes to the CAP have together brought about significant changes in agricultural practice and it is difficult to distinguish their separate effects, e.g. for deconcentration of granivore production or for the drop in fertilizer use. Finally, some conclusions for further research and policy making are drawn.

6.2   Interactions between agriculture and environment

The relative evolution since 1965 of some agricultural activities is shown in Figure 1. The agricultural area in use declined until 1993, but increased following that year. This might be due partly to the reform of the CAP, partly to manure policy, both of which induced a more efficient use (and reporting!) of agricultural land. The total agricultural area in 1965 was 1.6 million ha, including 691 thousand ha of arable crops and 870 thousand ha of grassland and forage crops. The total number of dairy cows in that year was 1 million. The evolution also indicates a shift towards more intensive beef and granivore production. Beef production doubled during the past couple of decades. Between 1965 and 1995, the number of pigs increased from 1.8 to 7.3 million, and of poultry from 33.5 to 47 million. In the meantime, the number of farms decreased by more than 60%, from 225,198 to 72,865.

The value of arable crop production increased by more than three quarters between 1965 and 1995 (Table 1), and livestock and horticultural production have more than doubled. In 1995,

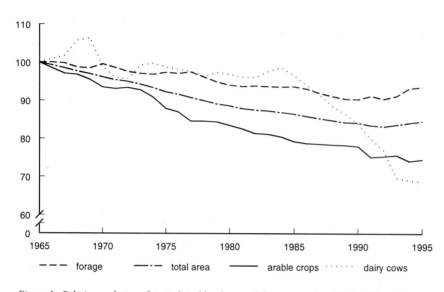

Figure 1 Relative evolution of agricultural land use and dairy cows since 1965 (1965=100)
Source: CLE-CEA.

the total value of Belgian agricultural production reached 243 billion BEF or 6 billion ECU
(1 ECU is about 40 BEF). Almost two thirds of total production value is generated by livestock.
The intensity of production is regionally differentiated. Nearly all indoor livestock production
(granivores and veal calves) is located in Flanders. Based on the May census of 1996, the number
of pigs, poultry and veal calves in Flanders amount to 96%, 95% and 99% of the national herds
respectively. The regional importance of horticulture is more difficult to illustrate through census
data because of its internal heterogeneity of intensity (glasshouse production, fruit, vegetables).
The total final production value of horticulture in Flanders amounts to 53.8 billion BEF, which

Table 1 Development of production value between 1965 and 1995 (1965=100) and production value in 1995 (million BEF)

Sector	Index of production value (1965 = 100)							Production value (million BEF)
	1965	1970	1975	1980	1985	1990	1995	1995
Arable crops	100	99	100	122	139	150	177	33,959
Livestock	100	136	147	156	163	173	214	152,298
Horticulture	100	129	126	108	125	143	216	56,635

Source: LEI, 1995.

144

is about 10 times the value produced in Wallonia. The distribution of the final value of agricultural production among the main sectors for both regions is shown in Figure 2. The total value of final production in 1996 was 191 billion BEF for Flanders (or 76% of national total) and 60.3 billion BEF for Wallonia. Sectors which are directly influenced by the CAP account for only 28% of Flemish agricultural production, but as much as 84% of the production in Wallonia. However, constraints in the CAP influenced sectors will undoubtedly induce shifts in product mix towards non-CAP influenced sectors.

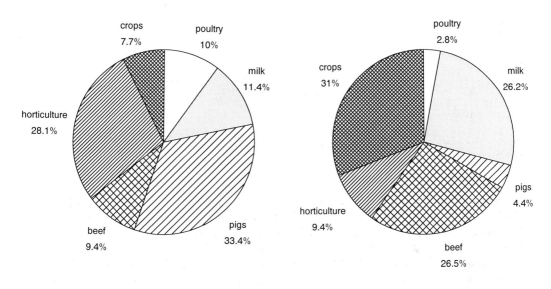

Figure 2   Final value of agricultural production by crops and livestock in Flanders (left) and Wallonia (right) in 1996

The regional difference in production intensity does not only apply to horticulture and granivore production, but also to the intensity of the remaining sectors. For example, mineral fertilizer use varies between the regions. The high nitrogen use in Flanders is due to the high intensity of dairy farming. An analysis of the 1991-92 records indicates that mineral nitrogen use on grassland was more than twice as high as in Wallonia: 250 kg N/ha against 110 kg N/ha (Lauwers, 1997).

The difference in agricultural orientation is strongly related to differences in urbanization. The population density in Flanders is approximately 436 inhabitants per km² which is more than twice that of Wallonia (197 inhabitants per km²). The pattern of land use is also quite different (Table 2). The percentage of urbanized area in Flanders is almost twice as high as in Wallonia. On the other hand, forests and nature constitute more than one-third of land use in Wallonia.

Table 2   Regional land use in Belgium according to the land registry data of 1990 (relative percentage of land use category in total cadastral area)

Region	Agriculture	Forests and nature	Urban area
Flanders	63	14	23
Wallonia	54	34	12
Brussels	7	22	70
Belgium	58	25	17

Source: Own calculations based on NIS-INS data.

Only 77% of the registered agricultural area is actually used for commercial agriculture. Regional differences in the degree of actual use - 61% for Brussels, 71% for Flanders and 82% for Wallonia - are linked to the degree of urbanization. In practice, strong urban pressures on agricultural land arise not only from the enlargement of urban areas itself but also, and to an increasing extent, from hobby farming. As a result, land prices in Flanders are almost twice as much as in Wallonia. Based on 1995 data on the agricultural land market (NIS-INS), the price per hectare of grassland in Flanders is on average 535,000 BEF, against 286,000 BEF in Wallonia. For arable land, these prices are 607,000 BEF/ha and 358,000 BEF/ha. High prices lead to an acceleration of the intensification process, which in its turn (through marginal productivity increase) stimulates prices.

With regard to environmental problems, the co-evolution of the land use pattern and agricultural development introduces an extra dimension to problem recognition and analysis. Historically, urban development has occurred near main water courses and on sandy soils which are environmentally vulnerable. The subsequent effects on externalities are manyfold, including:
- higher emission levels per unit of agricultural land;
- less self cleaning capacity (more leaching and the additional effects of urban wastes);
- aggregated utility functions.

Originally, therefore, the recognition of environmental problems due to agriculture in Belgium mainly started with those sectors located near urban areas: i.e. emissions from glasshouses, and the surplus manure problem associated with indoor livestock production.

The major environmental issues with regard to agriculture were reviewed in a joint publication of the Ministry of Agriculture and the Ministry of Public Health (Smet and De Keersmaeker, 1990). These include problems related to mineral emissions (eutrophication and acidification), pesticide residues and their emission to the environment, heavy metals and the dissemination of genetically modified organisms. This federal document gives a state of the art account of the damages, possible links with agriculture and actions to tackle the problems.

Janssen et al. (1991) present a larger list of environmental problems and a survey of their importance in relation to, inter alia, their geographic dimension, social relevance, abatement possibilities, costs and damages. The details of social relevance and agriculture's contribution to the problem are summarized in Table 3.

Table 3   Main environmental problems linked to agriculture: social relevance and agriculture's contribution to the problem

Issue		Social relevance	Contribution from agriculture (%)
Greenhouse gas emissions		low	15
Acidification (SO_2, NO_x)		high	5
Acidification (NH_3)		moderate	90
Heavy metals		low	10
Water reserves (stocks)		low	10
Nitrate	- in surface water	high	25
	- in groundwater	high	80
Phosphate	- in surface water	high	20
	- in groundwater	high	70
Biocides		very high	50-70
Erosion		low	high
Ozone layer depletion		low	0.1

Source: Janssen et al. (1990).

The public debate, political action and research efforts on these problems have increased considerably during the past decade. Owing to the transfer of government responsibilities in Belgium, most actions happen at the regional level. State of the environment reports are published both in Flanders and Wallonia on a regular basis (*Milieu- en Natuurrapport Vlaanderen; l'état de l'environnement en Wallonnie*).

After an initial period when professional agricultural organizations tended to refuse to acknowledge these environmental and social problems, they now feel the need for a constructive collaboration with government, society or environmental groups. An increasing number of cooperative actions and voluntary agreements attest to this shift.

6.3   Impacts of the CAP

6.3.1   The arable regime

The compensatory payments in the 1992 CAP reform are based on the cereal production observed between 1988 and 1992 for each agricultural region (in Belgium, 13 agricultural regions have been defined in terms of soil fertility). These payments are further differentiated according to small and large producers, with 92 tonnes per year as the threshold. The set aside requirement does not apply to farmers below this production level, but the maximum acreage eligible for payments is constrained by the quotient of 92 tonnes divided by the average regional production. Above this limit, compensation applies to the entire acreage, provided the set aside rule is met. In Belgium, many producers are within the 'small producers' regime, so the total set-aside area is rather small (Table 4).

Table 4   Trends in arable cropping since 1993

Feature	1993	1994	1995	1996
Cereals (ha)	312,261	309,412	308,824	294,823
Oil seeds (ha)	4,703	6,304	6,158	5,238
Protein crops (ha)	7,044	5,983	3,093	2,984
Total cereals, oil seeds and protein crops (ha)	324,008	321,699	318,075	303,045
Set-aside area (ha)	19,697	25,292	21,867	17,532
% set aside to cereals, oil seeds and protein crops	6	8	7	6
Set-aside requirement (%)	15	15	12	10
Maize (ha)	143,088	142,914	144,264	166,257
Sugar beets (ha)	99,087	95,178	98,810	97,990
Potatoes (ha)	48,120	51,591	55,846	61,403
Other arable crops (ha)	34,088	37,791	35,499	40,169

Source: NIS-INS.

As a result of lowered intervention prices, cereal prices decreased by about 27% from 1993 to 1996. Although prices fell considerably, the compensatory payments appeared to be sufficient to maintain farm profitability. Nevertheless, a shift occurred in the cropping pattern away from cereals and towards other arable crops, in particular potatoes (Table 4) (Thonon, 1994; Bernaerts, 1996). Conversion to sugar beet is hardly possible with the existing production rights

With regard to the indirect effects of the arable crop regime on field-based livestock production, the decrease of 27,000 ha in the green fodder acreage between 1983 and 1992 turned into an increase of 40,000 ha between 1992 and 1996. This is mainly due to fodder maize, which is not only a result of the arable crop regime itself, but also reflects an intensification of dairy farming and the necessity for manure disposal (see Section 3.2).

Compound feed mix produces indirect effects on granivore production. The content of soft wheat, barley and maize in compound feed dropped heavily towards the end of the eighties: from 23% in 1984 to 13% in 1989. Since the CAP reform, the cereal content (mainly soft wheat) has increased again and reached a 23% level in 1996. For 1997, a further increase to about one third of the feed mix is expected (Cornelissen, 1997). Compound feed prices, however, continue to decrease. For example, in 1995-96 the compound feed for finishing pigs cost about two thirds of the 1983-84 price (LEI, 1997). The impact of this evolution on granivore production is considered further in Section 3.4.

The nitrogen balances for two main arable crop regions - the sandy loam and the loam region - are compared in Figure 3. Due to a lack of data, the balance is only based on inorganic fertilizer use, atmospheric deposition and marketable products output. Between 1980 and 1986, the nitrogen balance is comparable for the two regions. The year 1987 was rather exceptional, with uncommon weather conditions. From 1988 on, the difference between the regions increases. As the sandy loam region is mainly situated in Flanders, a substitution of inorganic fertilizers for or-

148

ganic ones probably started at this time in response to the 1991 manure decree. The real balance will then actually be higher than the one shown in Figure 3.

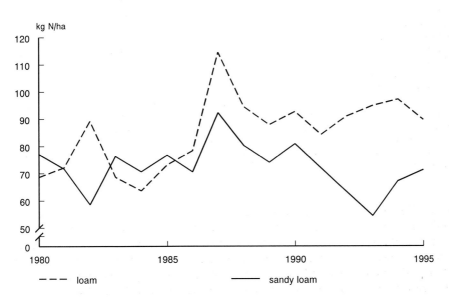

Figure 3 Nitrogen balance (kg N/ha) on arable farms in the sandy loam and loam region of Belgium, 1980-1995

Based on the same farm accountancy data, an analysis of the evolution of the pesticide use, expressed in constant prices, was made. There was a 10% decrease between 1980 and 1990, after which year the use of pesticides showed a slight increase. In the first CAP reform year a significant decrease appeared (10% as compared to the 1992 situation).

6.3.2   The dairy regime

Table 5 summarizes some important features of dairy farming in Belgium. Between 1983-1996, the number of dairy cows decreased by 0.3 million (or 34%). At the same time the number of farms with dairy cows dropped to less than half the 1983 number. In 1996, 20,320 farms with dairy cows remain, of which 7,461 also hold suckler cows. Dairy production has become more concentrated: overall production fell during that period by only 12%. Average milk production per farm increased by 120 tonnes, or 171%. The number of farms with more than 50 dairy cows increased by 33%. On specialized dairy farms, the average number of dairy cows went up by 16 head or 62% and the average fodder crop acreage increased by 48%, from 21 ha to 31 ha per farm. The stocking density increased from 1.52 dairy cows per ha in 1983 to 1.64 dairy cows per ha in 1996 (an 8% increase). As cow productivity also increased, productivity per ha went up at a faster rate: in 1996 almost 860 tonnes of milk were produced per 100 ha of fodder crops which is a 44% rise with respect to the 1983 productivity. The national quota is redistributed among the farmers (milk producers) and not among the dairy factories. As a consequence, the purchase of quota highly influenced the production cost of milk (Bouquiaux and Hellemans, 1998).

Table 5   Trends in dairy production between 1983 and 1996

Feature	1983	1996
Dairy cows (heads)	983,900	651,850
Milk production (tonnes)	3,871,646	3,415,705
Number of farms with dairy cows	48,967	19,768
Milk production (tonnes, with 3.6% fat matter) per farm (*)	70	190
Number of dairy cows per 100 ha of fodder crops (*)	152	164
Milk production (tonnes) per 100 ha of fodder crops (*)	598	859

(*) Only for professional farms (standard gross margin exceeds 0.5 million BEF).
Source: NIS-INS and CLE-CEA.

Indirect effects of the milk quota regime are mainly observed in beef production. Conversion from dairy to beef production (see also Section 3.3) is a typical course. Against the above mentioned reduction in holdings with dairy cows (29,000), the number of holdings with suckler cows has increased by well over 10,000.

The effect of the evolution of dairy production on nitrogen emissions on specialist dairy farms is analysed for two agricultural regions which differ in production intensity (Table 6): 'Kempen' (entirely situated in Flanders, with about 16% of the Belgian dairy herd) where dairy production is intensive and 'Liège - Haute Ardenne' (entirely situated in Wallonia, with about 13% of the Belgian dairy herd) with a more extensive production method. Based on farm accountancy data and on assumptions derived from mineral accountancy data published by Carlier et al. (1997), a nitrogen balance is simulated for the period 1980 to 1995. In 'Kempen' region, nitrogen balances first increased from 349 kg N/ha to 408 kg N/ha (in 1988), then showed a significant decrease of about one third to 275 kg N/ha. The emission per dairy cow follows a similar but less pronounced pattern. Finally, when expressed per tonne of milk, the balance shows a net improvement from 1980 on, in particular from 1990 on.

Improvements in nitrogen balances result from different factors. When looking at the latest five years, inorganic fertilizer use has decreased by 30%, whereas the concentrated feed contribution only increased by 4.5%. The main output (sold milk per ha) increased by 9%. The contribution of roughage maize in the dairy feed mix has strongly increased in 'Kempen' region (by more than 20%) which enables a higher amount of milk production per ha with more and more productive cows. The animal's basal metabolism becomes less and less important compared to the output.

150

Table 6   Nitrogen balance of dairy farms in 'Kempen' and 'Liège-Haute Ardenne' (kg N /ha)

	Kempen			Liège-Haute Ardenne		
	1980	1990	1995	1980	1990	1995
Input						
Inorganic fertilizer	226.5	222.3	171.3	142.3	155.9	137.5
Organic fertilizer	25.0	25.0	25.0	-	-	-
Concentrated feed	120.9	105.9	110.7	89.4	58.3	65.3
Roughage	4.9	7.2	6.4	2.3	1.6	1.9
Atmospheric deposition	40.0	40.0	40.0	40.0	40.0	40.0
Total input	417.3	400.4	353.4	274.0	255.8	244.7
Output						
Sold milk	42.8	49.4	53.9	32.4	32.3	33.8
Meat	25.5	25.4	24.0	21.5	17.9	17.6
Crops	0.4	0.2	0.4	0.2	0.3	0.1
Total output	68.7	75.0	78.3	54.1	50.5	51.5
Surplus						
(kgN/ha)	348.6	325.4	275.1	219.8	205.3	193.2
(kgN/cow)	188.6	175.2	154.8	139.2	158.0	152.1
(kgN/tonne milk)	40.7	33.0	27.2	34.1	32.0	29.8

Source: Own calculations based on the farm accountancy data of CLE-CEA.

In the 'Liège-Haute Ardenne' region, the milk production per cow and the cow density per hectare are both lower, and roughage maize is practically absent from the forage acreage. Similar trends with the nitrogen balance are observed, but the rate of improvement has been slower than in the 'Kempen' and the values are different. Expressed per ha or per cow, the balance is lower, respectively 30% and 2% in 1995, but in terms of per kg of milk produced, it is 9% higher.

Thus, although the nitrogen balances per ha or per cow are better in 'Liège - Haute Ardenne', if the national milk quotum (3 million tonnes) were be entirely produced under the conditions in the 'Kempen' (with its more productive cows and land) this would allow nitrogen emissions to be reduced by 6.7 tonnes and land use to be by 165,700 ha. At a first glance, then, the observed trend of intensification is not unequivocally harmful to the environment. More production per cow or per ha means a more efficient use of inputs: less basal metabolism and less land occupation. This corresponds with earlier findings that economic and ecological efficiency might coincide (Goffinet, 1985; Verbruggen et al., 1992). However, scale enlargement and the growing importance of maize have negative effects on the landscape (see also Section 5.3).

6.3.3   The beef regime

The beef sector in Belgium has faced an enormous growth since the mid-1980s (Table 7). Between 1983 and 1996, the total number of suckler cows increased by some 0.33 million or 220%

and the number of farms holding suckler cows by more than 75%. The occurrence of the beef breed 'Bleu Blanc Belge' ('Belgian Blue') on dairy farms has facilitated the conversion from dairy to beef production. In the CAP reform year 1993, the number of suckler cows increased by one third and the number of farms with suckler cows by 16%. The production of lean cattle is typical of the south of Belgium (extensive production) whereas further finishing is done in the North (intensive production). The average stocking density of suckler cows increased from 1.04 to 1.55 cows per ha of fodder crops. In the loam region, however, this density amounts to almost three cows per ha of fodder crops.

Table 7   Trends in suckler cow production between 1983 and 1996

Feature	1983	1996
Suckler cows (heads)	149,900	479,200
Farms with suckler cows	14,119	24,196
Suckler cows per farm	10.6	19.8
Stocking density (suckler cows per 100 ha of fodder crops)	104	155

Source: Own calculations based on NIS-INS data.

The profitability of specialized beef farms was heavily reduced following the price reduction resulting from the 1992 CAP reform (Bouquiaux and Hellemans, 1996). Compensatory payments proved to be insufficient for a number of reasons. The soil fertility allows a stocking density of 2 to 3 LU/ha, compared with the eligibility ceiling of 2 LU/ha.

The payments, moreover, are based on the Community average quality carcass and are not regionally differentiated, which discriminates against the Belgian Blue which has a high quality and high weight carcass.

If the situation of low profitability continues (or even becomes worse as a result of the Agenda 2000 proposals), many beef farms will have to stop their production in the near future. In regions with limited alternatives (this, for example, applies to regions where permanent grassland predominates, and it is difficult to switch to arable crops), large scale afforestation may be the outcome (e.g. the production of Christmas trees), still encouraged by compensatory payments, but resulting in a loss of open landscapes.

6.3.4   Granivore production

Belgium has experienced a major increase in intensive livestock production, mainly pigs and poultry. The main driving force initially was to achieve a reasonable income on small farms with a surplus of labour, but during the nineties there has been a restructuring of these sectors.

Pig production in Belgium changed from an insignificant but ubiquitously present agricultural activity to a highly specialized and regionally concentrated sector. In 1950, pig production was present on three quarters of all farms, but its scale never exceeded 50 animals per farm

(Bouquiaux, 1995). The number of pigs started to increase from 1965 on, which is closely linked to the GATT trade agreement of 1962 allowing soya and cereal substituting products to enter the EU freely. Poultry production is more related to the cereal price and showed a first important increase in the first half of the sixties, continuing to rise at a slower rate during the second half. From the beginning of the seventies, with the start of the increase in Community wheat prices, the slow increase of production was rather the result of improved feed conversion, because the production capacity (number of chickens) tended to decrease. The falling wheat prices from 1984 on is almost the mirror image of the rising broiler production. The trend in the number of laying hens has a similar pattern but is less pronounced. The most recent rise only started in 1990.

The 1962 trade agreement not only caused a shift in external trade, but also led to a structural change in production within the EU (mainly towards indoor production and regional concentration). Although the CAP reform responds to the problem of external trade deformation, it does not solve the structural deformation caused by the cheap feed supply for livestock (Sneessens, 1996).

Regional specialization occurred where the concentrated feed industry emerged. Proximity to the feed industry in national and regional harbours (Antwerp, Roeselare) has to be seen as the initial location factor for regional concentration, but the process of regional concentration itself generated its own regional advantages (regional scale economies). Towards the end of the eighties some important regional diseconomies became apparent: disease spread, manure saturation. This might initiate tendencies to change the concentration of production, and it was expected that CAP reform (the decrease of cereal prices) would enhance this process.

Table 8   Regional concentration of granivore production between 1991 and 1996

Concentration level in 1991	Increase of livestock population between 1991 and 1996 (%)			
	sows	finishing pigs	broilers	laying hens
Very high	18.1	5.2	13.4	50.9
High	12.1	12.1	16.3	14.2
Moderate	10.1	19.1	44.5	42.4
Low	1.9	13.9	53.5	22.1

Source: Own calculations based on agricultural census data (NIS-INS).

Table 8 summarizes some main findings of location analysis. There, the evolution of the production of sows and finishing pigs is analysed in four areas from 1991 to 1996. Each of these regions accounts for 25% of the total pig herd and is derived from a ranking of communalities based on their pig density. A similar approach has been used for the analysis of the trend in poultry production. While the number of sows is still increasing most in areas with the highest pig density, meat production (finishing pigs and broilers) shows a clearly different trend, with the largest increase in the regions of low density, although production continues to increase in the areas of concentration. The evolution in the laying hens sector is so different from one region to another that it is impossible to detect the effect of a specific location factor.

Parallel to this regional development, the land dependency of granivore production is increasing. In 1991, an average farm with pigs had 15.7 ha. Five years later, it was 19.3 ha. The average farm size of holdings which started in pig production between 1991 and 1996 was 19.2 ha. The average poultry farm acreage increased from 7.3 ha to 11.4 ha over this same period and new poultry farms had on average 15.1 ha per farm, indicating a conversion of existing land dependent farms.

The results presented may indicate that concentration is a side effect of the CAP; it however remains difficult to distinguish these effects from those produced by the manure policy. Nevertheless, centre function and concentration economies remain important location factors. Since the share of home-grown cereals in compound feed is only around 10% (Thonon, 1994), the effect of proximity to cereal production areas is less pronounced and perhaps remains inferior to the regional advantages of concentration areas (among which is the proximity to the compound feed industry). In this context, trends in the concentration of production is more the result of manure policy than of the CAP.

The location inertia will probably not remain in the future: manure saturation will become an important cost factor in Flanders, whereas some organizations in Wallonia highlight the regional potential for expansion. Based on the limitations under the Nitrates Directive there is still room for 14 ktonne N production, which is the equivalent to about 1.2 million finishing pigs, or 32 million broilers (Decock, UPA, Le Sillon Belge, 2/1/98). Within this context, incentives from the CAP reform to convert to granivore production will be encouraged. The final environmental effect among others, will depend on the substitution of inorganic fertilizer by organic manure.

6.4   The accompanying measures and structural policies

6.4.1   The accompanying measures

Until recently, the experience in Belgium with respect to agro-environmental policies remained very limited. Moreover, farmers' attitude towards the environment and the subsidizing measures was rather negative, in particular in Flanders (Everaet and Lenders, 1996). With regard to the 2078/92 Regulation, it should not be surprising that only a few of the measures are applied in Belgium. Moreover, a delay in application is due to the legal procedure of concertation between the Federal State and the Regions.

The following apply at the federal level:
- demonstration projects;
- monitoring programmes;
- organic farming.

Demonstration projects are organized mainly to promote reduction in the use of nitrogen fertilizers and pesticides and the conversion to organic farming. Annual costs are about one million ECU.

Monitoring programmes include a network of weather stations and reference fields for the most important crops (potatoes, cereals, Belgian endive, sugar beet, hops and several vegetables). Annual costs are about 0.3 million ECU.

Payments for organic farming are differentiated according to crop and farm situations (Table 9). Payments for existing organic farms range from 57% (grassland) to 97% (vegetables) of the authorized maximum set by Regulation 2078/92. Farms that still are in the conversion process (2 years) receive a higher payment (authorized maximum) than existing organic farms because they do not yet benefit from the higher prices for their products.

Table 9   Support for organic farming in Belgium

Crop	Authorized maximum under Regulation EC2078/92 (ECU/ha)	Support (share of authorized maximum, in %)	
		converting farms (during 2 years)	existing organic farms
Cereals, oil seed and protein crops	181	100	61
Other arable crops	302	100	73
Vegetables	302	100	97
Grassland	302	100	57
Permanent crops	845	100	87

Source: Ministry of Agriculture, DG3.

Limiting factors for a successful introduction of organic farming are the distribution and commercialization system, the obligation to be a full time farmer to benefit from the aid and the low level of the vegetables subsidy compared to the extra costs (Ghesquière, 1993). By 1996, there were agreed plans for only 145 farms or 3,671 ha.

At the regional level, the Flemish Region has set up two ambitious programmes on management agreements (Ministry of Agriculture, 1993, 'Programme d'aides introduit par l'Autorité fédérale belge, les Régions et les Communautés en application du règlement CEE 2078/92', not published). One is based on a management plan per specific area (ecological impulse area, regional landscapes or water protection areas) and the other has to be seen within the framework of the manure policy. These projects are not yet translated into legislation due to divergent points of view between agricultural and environmental organizations and to the interference with the manure policy. In particular with regard to the latter, the farmer organization 'Boerenbond' has asked for a quick judicial implementation of management agreements in order to compensate for income decreases due to strict fertilization limits. Besides, Flanders also started (in 1997) a demonstration and educational project on organic farming and the reduction of the use of fertilizers and pesticides.

155

Table 10   Agro-environmental measures at the Walloon level

		Grassland	Arable land	Premium (ECU/ha)
1a.	Delay of mowing	x		122
1b.	Diversification of seed mix on temporary grassland	x		196
2a.	Replacement of arable land by a strip of grassland or grassed headland		x	245
2b.	Establishment of extensive headland		x	122
2c.	Replacement of intensive grassland or bush tree orchards by a strip of extensive grassland	x		245
3.	Preservation and maintenance of hedges and lines of trees	x	x	50
4.	Keeping the stocking density equal or below 1.4 LU/ha forage surface	x		50
5.	Endangered local species (sheep and horses)		17-177 per head	
6a.	Reduction of the seed density (cereals)		x	88
6b.	Reduction of the use of synthetic herbicides (cereals)		x	88
7a.	Reduction and spot spraying of herbicides (maize)		x	147
7b.	Under-sowing (maize)		x	147
8.	Green cover during winter		x	98
9.	Delay of mowing with limited inputs	x		245
10.	Preservative measures in humid zones	x		50
11a.	Preservation of traditional orchards	x		122
11b.	Planting of old varieties	x		245
12	Traditional crops and vegetables		x	98-293

Source: Ministry of Agriculture, 1993, 'Programme d'aides introduit par l'Autorité fédérale belge, les Régions et les Communautés en application du règlement CEE 2078/92', not published.

In the Walloon Region two types of measures can be distinguished: horizontal and vertical measures. Horizontal measures are applied in the whole of Wallonia and concern the measures 1 to 5 mentioned in Table 10. Vertical measures only apply for some specific areas (e.g. vulnerable zones as defined in the Nitrates Directive, natural park) and must be integrated in an approved management plan. During the period 1994-1998, vertical measures remained in an experimental phase and only concerned 50 farms.

On October 15, 1997, 1,641 farms had already applied for horizontal measures involving 15,558 ha and 302 animals. About 75% of the dossiers concern the maintenance of hedges (Ministry of the Walloon region). Participation rates remain limited: only 10% of the farms and 6% of the agricultural area in Wallonia are concerned. Factors limiting a successful introduction are mainly:
- the low level of premiums (the average premium per dossier is 410 ECU);
- the maximum restrictions for the cumulation of premiums;
- the administrative procedure.

Again, the obligation to be a full time farmer can be seen as an additional constraint. Regional differences in application can be explained by the variable promotional efforts of the extension services.

The afforestation measure under Regulation 2080/92 has been adopted by the Regions. Nevertheless, the Federal State compensates income losses of full time farmers with an annual payment of 612 ECU during a period of five years.

Since forest and nature account for about one third of the land use in Wallonia (Table 2), the attention is more focussed on a qualitative management approach (conversion to foliage trees, combination of non adjacent plots) than on quantitative expansion. The programme is integrated in the general forestry policy and involves two aspects:
- Payments to compensate for the afforestation costs (unique premium per hectare, depending on tree type)
  - oak and beech: 1,837 ECU/ha;
  - other foliage trees: 980 ECU/ha;
  - conifers and poplars: 245 ECU/ha.
- Payments for planting hedges: a unique premium of 1.22 to 5 ECU per metre according to the hedge type, whether the planting is realized by a farmer or not.

The share of forest and nature in total land use is only 14% in Flanders (Table 2). Social expectations on recreation and nature conservation induce an increased afforestation demand. The objective of the afforestation programme is to increase the afforested area with 10,000 ha after five years. The specific programme is also made up of two aspects:
- Payments to compensate for the afforestation costs (premium per ha, depending on tree type)
  - poplars: 857 ECU/ha;
  - foliage trees: 1,469 to 3,673 ECU/ha depending on species and afforestation duration;
  - conifers: 980 to 1,959 ECU/ha.

Supplements are possible when marginal land is involved (245 ECU/ha), when under cropping is foreseen (490 ECU/ha) and when isolated plots are joined (122 ECU/ha).
- Payments to compensate for maintenance costs (annual premium per ha during five years). This premium is again differentiated according to type, further on according to time after planting (first two years and the three following years):
  - poplars: 245 and 196 ECU/ha;
  - foliage trees: 490 and 245 ECU/ha;
  - conifers: 245 and 122 ECU/ha.

The 2080/92 programme has only recently been started, and by October 1997 only 220 ha was included in the programme. The main reasons for this limited success are:
- the fact that afforestation is not profitable when compared to agricultural activity, particularly in Flanders (Gorissen and Schepens, 1997). These authors conclude that compensatory

payments have to be increased, in particular those that compensate for the income losses and that should cover the whole afforestation period;
- the strict land use planning measures;
- tenancy, conditions which restrict the afforestation of land.

On average, the acreages are rather small (1 to 2 ha per scheme). This results in a difficult forest management, but on the other hand, the landscape effects might be relatively important.

6.4.2   Structural policies

Structural policy interferes with market and price policy and influences agricultural development and externalities. Due to complex interactions, the effects are hardly known. In the future they will probably be better known as a result of a compulsory evaluation. Only the measures directly oriented towards the protection of the environment will be mentioned hereafter. They concern the Objectives 5a and 5b programmes and the regional aid to investment.

Within the Objective 5a programme, the South East of Wallonia is recognized as a less favoured area (LFA). Beef production (suckler cows) mainly of the 'Belgian Blue' breed is the predominant agricultural activity in this region. Three types of support are given to agriculture:
- compensatory payments to maintain extensive agriculture: 86 ECU/LU (with a maximum of 120 ECU/ha and 1,714 ECU per farm);
- aid to collective investments in forage harvesting machines (25% of the purchase price). The aid combined with technological progress incites silage production and intensification of grassland which result in negative effects on grassland flora (loss of biodiversity);
- supplementary aid to setting-up in farming and investment.

During the period 1994-1999, the Objective 5b programme applies for three areas in Belgium: 'Meetjesland' and 'Westhoek' (Flanders) and the South East of Wallonia. The Flemish projects have a comparable approach due to a similar situation and objective; to maintain a profitable agriculture with respect to the environment (sustainable agriculture). The projects are oriented to the improvement of profitability (analysis of regional potentials, product mix diversification including non food, marketing, agro tourism, on-farm processing of farm products, new technologies, agro informatics), to limit the negative externalities (promotion of organic farming, waste management, optimal manure application, consulting services, warning systems) and to promote the positive externalities (management agreements).

The 5b area of South East Wallonia is included in the LFA area. This programme since 1985 has mainly been oriented towards integrated rural development. The agricultural part of the programme aims at the promotion of extensive production, at the improvement of farm profitability and at agricultural participation in the maintenance of the rural structure. The projects concern:
- feasibility studies of innovations;
- quality label for grassland beef production;
- development of labelled livestock production;
- development of agrotourism;

158

-   commercialization and promotion of regional products;
-   product diversification (also at agribusiness level);
-   the sanitary framework of livestock production;
-   planting and maintenance of 'bocages' (hedgerows);
-   maintenance of remarkable agricultural sites;
-   preservation of endangered livestock breeds.

Aid to investment is an application of Regulation 950/97. This Regulation follows up on the former Regulation 2328/91, which has been reoriented towards integration with environmental aspects. Supplementary regional aid can be provided for investments related to the environment, even if they do not fit in the (economic) farm improvement plan. In practise, interest subsidies are given for loans related to (Crédit Agricole, 1995; Landbouwkrediet; 1996):
-   the manure and ammonia problem;
-   plant protection;
-   heating or cooling in horticulture (Flanders);
-   waste and used water management (Flanders);
-   economizing energy and water use (Flanders);
-   animal welfare (Wallonia).

6.5   Interactions between agriculture and environment

6.5.1   Nutrients

The mineral problem is considered the main environmental problem associated with agriculture, at least in the northern part of Belgium (Janssen et al., 1991; Verbruggen, 1996). In Flanders, the public debate started in the eighties and produced a set of policy measures aiming at a strict regulation of fertilization and of further livestock development (Van Gijseghem, 1997). In Wallonia, the problems are less pronounced, but transfer of manure from Flanders, concentration of granivore production or structural change to restore farm income might be new developments (Thonon, 1994) which could cause adverse effects on the environment.

Early problem description was mainly oriented to the manure surplus problem, directly linked to the indoor and regionally concentrated granivore production (Vanacker et al., 1988). Later, the importance of the other sectors became more and more apparent. In Wallonia, nearly all nitrogen produced in animal manure originates from cattle and even in Flanders, with high granivore concentration, cattle account for 60% of the total N production (Table 11). Nitrogen use through mineral fertilizers in Flanders is almost twice as high as the N production through pig production. In previous sections details were given of the mineral efficiency in arable and dairy farming.

The core of the mineral policies in Belgium consists of fertilization limits and the calculation of mineral production using livestock numbers and generic production coefficients. The application of the calculated mineral production (or 'polluting units' as defined in the Walloon legis-

Table 11   Mineral balances (in kg per ha of agricultural land) at regional level (Flanders and Wallonia), based on agricultural census data and farm accountancy data for the year 1991

Component of the system	Nitrogen balance (kg N/ha)		Phosphate balance (kg P₂O₅/ha)	
	Flanders	Wallonia	Flanders	Wallonia
Input	502.8	302.4	167.4	105.6
- animal manure	280.4	135.9	123.9	49.5
pigs	93.4	3.3	48.6	1.7
other livestock	17.4	0.6	14.4	0.5
cattle	166.1	130.4	59.5	46.7
other herbivores	3.5	1.6	1.4	0.6
- mineral fertilizer	184.4	128.5	43.5	56.1
- atmospheric deposition	38.0	38.0	0.0	0.0
Output	248.0	236.8	84.8	80.6
Surplus	254.8	65.6	82.6	25.0

Source: Lauwers, 1997.

lation) is then limited to the amount prescribed through the fertilization limits. Whereas these limits in Wallonia do not cause many problems for agricultural development and even leave possibilities for the development of granivore production on farms that are endangered by the CAP reform measures (Thonon, 1994), in Flanders they have given rise to the socio-economic problem of regional manure disposal. Therefore, the Flemish regulation also contains socio-economic measures to adjust the disposal problem, to control further structural development and to account for social corrections (Lauwers and Van Huylenbroeck, 1997; Van Gijseghem, 1997).

The increased supply of manure enabled some substitution of inorganic for organic fertilizers. Based on bookkeeping data, Lauwers and Van Steertegem (1996) found decreasing inorganic fertilizer use in arable crops when local manure disposal pressure increases. Nitrogen use on grassland, however, does not depend on local variation in manure pressure. The intensity of inorganic N per ha has been decreasing since 1988, both in Flanders and Wallonia, mainly as a result of the decreased use on grassland (see also Section 3.2). In 1995, compared to the 1988 situation, the inorganic N use per ha was 9% lower in Wallonia and 24% lower in Flanders. Table 12 shows the annual changes. The decrease already started before mineral policies became operational (1991) and it is difficult to discern any additional effect from the 1991 enactment. On the other hand, since the CAP reform the decline has slowed down.

Table 12   Annual change of inorganic nitrogen use in Flanders and Wallonia from 1989 to 1995 (%)

Region	1989	1990	1991	1992	1993	1994	1995
Flanders	- 4.5	- 5.3	- 3.3	- 3.4	- 5.4	- 1.9	- 3.2
Wallonia	0	- 5.0	- 1.5	- 3.1	- 3.1	+ 4.1	- 0.8

Source: Own calculations based on bookkeeping data CLE-CEA.

6.5.2   Pesticides

Pesticides, mainly phytopharmaceutical or plant protection products, have become an essential but rapidly changing factor in agricultural production. The number of active compounds went up from 30 in 1980 to over 400 in 1989 (Smet and De Keersmaeker, 1990). During the same period, their sales increased with an annual growth rate of 11% (Bouquiaux, 1995). Between 1980 and 1992, the importance of non-active ingredients (additives, mineral oils) increased considerably (Thonon, 1994), which might be an indication of a technological progress to more optimal use.

Although insufficient to account for efficiency and ecotoxicity (e.g. atrazine leaching in ground water), common indicators to describe pesticide use and its evolution are monetary values or kg of active ingredients. Table 13 shows the evolution of the annual sales in Belgium, expressed in tonnes of active ingredients.

Table 13   Annual sales of pesticides for use in agriculture in Belgium by product group (in 1,000 kg of active ingredients)

Product group	1980	1985	1988	1989	1990	1991	1992	1993	1994	1995	1996
Herbicides a)	4,591	4,617	5,145	5,264	5,213	5,091	5,120	5,560	5,775	6,240	5,953
Fungicides	1,109	2,123	2,583	2,637	2,743	2,837	3,292	2,789	2,284	2,659	2,402
Insecticides	1,234	516	430	506	459	365	387	328	312	360	388
Nematicides	-	1,133	927	842	808	778	857	774	696	744	779
Growth regulators	103	170	267	394	503	373	276	287	359	445	367
Other	170	414	466	443	538	524	494	548	462	491	514
Total	7,207	8,973	9,818	10,086	10,264	9,969	10,426	10,286	9,888	10,939	10,403

a) Including sales of herbicides for use outside agriculture.
Source: Ministry of Agriculture. DG4.

Total pesticide use has steadily increased to reach a level of 10,000 tonnes of active ingredients during the nineties. At crop level, some figures from panel data are available (KINT, 1997). The overall tendency of pesticide use in cereals and sugar beets is a decreasing one (Table 14). Maize has a constant and fairly low pesticide use, whereas for potatoes it is high and variable (highly weather dependent).

Table 14   Pesticide use in some main crops (expressed in 1,000 kg of active ingredients)

Crop	1990	1991	1992	1993	1994	1995	1996
Cereals	-	1,266.4	981.6	1,058.9	898.2	879.6	884.1
Maize	-	356.6	-	346.4	-	356.4	347.3
Potatoes	792	-	1,588.0	-	1,194	-	1,361.0
Sugar beet	-	619.4	572.9	476.7	418.7	4.7.0	442.9

Source: KINT, 1997.

In 1996, a voluntary agreement was signed between the Belgian Association of Drinking Water Companies (Belgaqua) and the Belgian Association of Phytopharmaceuticals Producers (Fytofar) agreeing on jointly tackling the problem of phytopharmaceutical residues in drinking water and to undertake preventive actions (KINT, 1996). The application of good agricultural practices is promoted. Among them are optimizing doses and application times, and the finetuning of spreading techniques. An action on collecting empty containers after use has not only proved to be an effective measure but it also has an important educational aspect. From 1995 on the obligatory control of spray equipment entailed its optimal application (CLO, 1995).

6.5.3   Landscape

Nearly all Belgian landscapes are strongly human influenced, the majority of them being shaped by agricultural practices and characterized by extensive production and a rather constant management. Traditional agricultural landscapes are usually valued positively. However, agriculture is more and more criticized for its changed management function. Negative elements of change are:
-    the disappearance of hedges and small natural features;
-    scale enlargement;
-    poor integration of new agricultural buildings;
-    decreased biodiversity;
-    enclosure of open landscapes.

The first two aspects are linked to the amalgamation of fields in order to optimize machinery use. Table 15 shows the evolution of fields at farm level between 1950 and 1990. The number of fields per farm did not alter very much. However, the average acreage per field went up from 1 ha to almost 3 ha, following the increase in farm size. Field enlargement means elimination of boundaries and consequently of various natural and landscape features associated with these boundaries. Besides landscape externalities, the trend has also negative effects for agriculture itself through increased erosion risks.

Intensification of agriculture has caused the appearance of new landscape elements: livestock housing, greenhouses, silos, storage barns. Scale enlargement has led to new buildings which sometimes are disproportional to the existing site. Uniform construction materials and bad integration in the existing landscape meet with public disapproval.

162

Table 15   Evolution of plot characteristics in Belgium between 1950 and 1990

	1950	1970	1990
Number of farms with land	251,334	179,462	84,963
Total acreage (ha)	1,713,933	1,540,306	1,357,366
Total number of plots	1,589,670	919,067	477,482
Average size per plot (ha)	1.08	1.68	2.84
% of farms with:			
1 plot	12	31	22
2 to 4	36	30	33
5 to 9	32	23	26
10 to 19	15	13	15
> 19	5	3	4
Number of plots per farm	6.3	5.1	5.6

Source: Bouquiaux, 1995.

In arable crops regions, considerable acreages of permanent grassland have been converted. This process started after the Second World War and was enhanced by the milk quota regime. Remaining grassland is used in a more intensive way (drainage, increased use of fertilizer). This has caused an important decrease of biodiversity and landscape amenities, e.g. through the disappearance of the typical wet grassland flora.

Opposite to the disappearance of natural landscape entities and plot enlargement (which cause an open landscape) is the process of landscape enclosure. Enclosure may be temporary or fairly permanent and depend on changing or leaving the agricultural business. The enormous expansion of maize in some regions (mainly in Flanders) has led to the loss of open landscapes during the main part of summer. A more permanent loss of open landscape results when abandoned agricultural land is used for Christmas tree production (mainly in the South-East) or for rural settlement.

6.6   Concluding observations

An overview of the effects of CAP on the environment and landscape encounters the difficulty to disentangle the direct and indirect effects of a complex system. Most assumed effects are defined in a mainly qualitative way, sometimes scarcely illustrated with some descriptive figures 'to prove' the qualitative assumptions. Many measures interfere with other constraints and obscure the net effect. Even in the case where policies are assumed to be synergistic, observations do not always (or entirely) confirm the assumption. This applies, for example, to the case of the presumed force of the cereals regime and regional manure policy to reduce concentration of intensive livestock production.

The lack of adequate information on the agro-environmental system is perhaps the main reason why criticism emerges concerning the linkage of environmental constraints to agricultural policies (e.g. some critical remarks of professional organizations during the Belgian colloquium on Agenda 2000). This attitude is completely in contradiction with the OECD principle of integrated policies (and perhaps also in contradiction with the intrinsic wishes of those who put forward the objections). To rebuild confidence in integrated policy systems, new research is wanted:
- what kind of incentives do emerging policy measures generate (e.g. livestock density constraints, per head premium in dairy farming);
- what are the environmental effects of farm changes coping with the incentives, e.g. effects on mineral balance, greenhouse gases, pesticide use, land use, landscape;
- what might be the differential effects on incentives, farm adaptation possibilities and final externalities, taking into account the regional differentiation in Belgium?

Attention should be paid to some generic measures which might be defined in such a way that some types of farming are completely ignored. This is in particular true for the highly urbanized areas in the major part of Belgium. Translating the environmental constraints into a less intensive agriculture will certainly generate adverse effects in those highly urbanized regions, where the most apparent way to cope with high land prices is to intensify agricultural production. For example, a livestock density rule (LU per ha) might exclude the more intensive types of dairy farming which have, however, a more efficient milk production (basal metabolism redistributed over a larger amount of produced milk). The high livestock density is then not only to be regarded as a high emission pressure but also as an efficient land use, a resource based criteria which might prevail in highly urbanized areas.

References

Bernaerts, E. (1996) *Evaluatie van de recente aanpassingen van het graanbeleid;* Brussel, Landbouw-Economisch Instituut, LEI-Publikaties Nr 582, 100 p.

Bouquiaux, J.M. (1995) *Evolution de l'agriculture belge et son impact sur le milieu depuis 1950;* Bruxelles, Institut Economique Agricole, Publications de l'I.E.A. N° 576, 40 p.

Bouquiaux, J.M. and R. Hellemans (1996) *Problématique de viande bovine en Belgique;* Bruxelles, Institut Economique Agricole, Publications de l'I.E.A., N° 583, 53 p.

Bouquiaux, J.M. and R. Hellemans (1998) *Analyse économique de la production laitière en Belgique;* Bruxelles, Centre d'Economie Agricole, Publications du C.E.A., N° 596, 36 p.

Carlier, L., J. Michiels, M. Beke, I. Verbruggen and E. Van Bockstaele (1997) *Mineralenbalans, in- en output van mineralen op Vlaamse melkveebedrijven;* In: Mestproblematiek (publicatie studiedag), Antwerpen, Technologisch Instituut KVIV, 8 p.

CLO (1995) *Centrum voor Landbouwkundig Onderzoek;* Gent, activiteitsverslag 1994-95, 283 p.

Cornelissen, R. (1997) *Het standpunt van de toeleveringssector;* bijdrage op Colloquium "België en het landbouwluik van Agenda 2000", 4 november 1997, 7 p.

Crédit Agricole (1995) *Agriculture et environnement;* Bruxelles, Crédit Agricole, 55 p.

Everaet, H. and S. Lenders (1996) En wat denkt de boer erover?; Brussel, Landbouw-Economisch Instituut, LEI-Publikaties Nr 586, 257 p.

Ghesquière, P. (1993) *Aide à l'agriculture biologique;* Jodoigne, Centre d'Animation et de Recherche en Agriculture Biologique (CARAB), 14 p.

Goffinet, R. (1985) *Problématique de la production laitière suite à l'instauration des quotas;* Bruxelles, Institut Economique Agricole, Publications de l'I.E.A., N° 249, 47 p.

Gorissen, D. and D. Schepens (1997) *Bebossing van landbouwgronden in het kader van Verordening (EEG) 2080/92;* economische afweging op basis van een structurele analyse van de betrokken actoren; Brussel, Centrum voor Privébosbouw vzw, 216 p (+bijlage)

Janssen, G., R. Tijskens and J. Snaet (1991) *Streven naar duurzaamheid, landbouw-milieunatuur;* Leuven, Belgische Boerenbond, 79 p.

KINT (1996) Koninklijk Instituut voor het Duurzame Beheer van de Natuurlijke Rijkdommen en de bevordering van Schone Technologie; Brussel, *jaarverslag 1996,* 54 p.

KINT (1997) Koninklijk Instituut voor het Duurzame Beheer van de Natuurlijke Rijkdommen en de bevordering van Schone Technologie, Brussel, Groenboek Belgaqua-Fytofar 1997, 44 p.

Landbouwkrediet (1996) *Land- en tuinbouw en milieu;* Brussel, Landbouwkrediet, 64 p.

Lauwers, L.H. (1997) *Effects of policy measures regarding the manure problem and pig production structure in Flanders;* In: Romstad, E., J. Simonsen and A. Vatn (Eds), Controlling mineral emissions in European agriculture: economics, policies and the environment, Wallingford Oxon, CAB International, pp. 193-208

Lauwers, L. and M. van Steertegem (1996) *Simulatie van nutriëntenstromen in de Vlaamse landbouw;* Mechelen, Vlaamse Milieumaatschappij, Wetenschappelijk Rapport in het kader van MIRA-2, 53 p.

Lauwers, L. and G. Van Huylenbroeck (1997) *Cost effectiveness of social corrections in flemish manure regulation;* Tijdschrift voor Sociaal wetenschappelijk onderzoek van de Landbouw, 12 (1), 58-72

LEI (1995) *Landbouwstatistisch Jaarboek 1995;* Brussel, Landbouw-Economisch Instituut, 102 p.

LEI (1997) *Technische en economische resultaten van de varkenshouderij op bedrijven uit het LEI-boekhoudnet (boekjaar 1995-1996);* Brussel, Landbouw-Economisch Instituut, LEI-Publikaties nr 592, 30 p.

Smet, M. and P. De Keersmaeker (1990) *Landbouw and Leefmilieu, oriëntaties;* Brussel, Ministerie van Landbouw, Dienst Informatie, 100 p.

Sneessens, J.F. (1996) *Stratégie pour une agriculture rurale;* Louvain-la-Neuve, Université Catholique de Louvain, 159 p.

Thonon, A. (1994) *La réforme de la politique agricole commune et ses conséquences pour l'agriculture wallonne;* Louvain-la-Neuve, Université Catholique de Louvain, 221 p.

Vanacker, L., L. Debaene and M. Macharis (1988) *De mestproblematiek in Vlaanderen;* rapport nr.1: mestproduktie, mestbalansen, mestnormen. Mechelen, OVAM, Directie Planning, 70 p.

Van Gijseghem, D.E.L.J. (1997) *Implementation of nitrate policies in Flanders;* In: Brouwer, F. and W. Kleinhanss (Eds), The implementation of nitrate policies in Europe: processes of change in environmental policy and agriculture, Kiel, Wissenschaftverlag Vauk KG, pp. 271-282

Verbruggen, A. (1996) *Milieu- en Natuurrapport Vlaanderen 1996: Leren om te keren;* Leuven, Garant, 585 p.

Verbruggen, I., L. Carlier, D. Van Lierde and E. Van Bockstaele (1992) *Mineralenoverschotten in de rundveehouderij;* In: Graslandgebruik en ruwvoederwinning met beperkingen (publicatie studiedag), Antwerpen, Technologisch Instituut KVIV, 14 p.

7. THE NETHERLANDS

Floor Brouwer and Siemen van Berkum

7.1 Introduction

The Netherlands, as one of the most densely populated and highly industrialized countries, also has one of the most intensive farming systems with high output levels supported by a considerable use of agrochemicals (fertilizers and crop protection products). The country is the third largest net exporter of food products; its share of global food exports at 7.9% (in 1995) is only slightly below that of France (9.1%) and the U.S.A. (13.2%). Farming intensified considerably over the post-war period with large increases in the use of agricultural inputs (fertilizers, feed concentrates, crop protection products and energy). The subsequent deterioration of the environment is one of the main issues of concern to society. During the past ten years the use of agrochemicals reduced by more than 20% for fertilizers and 40% for crop protection products.

The objective of this chapter is to review existing knowledge regarding the impact of the CAP on the environment for the Netherlands. Section 2 reviews the present state of the environment in relation to agriculture. The effects of market and price policies are examined in Section 3. This section starts with an overview of current linkages between the CAP and the development of the agricultural sector. An analysis is also provided of the environmental impacts of market and price policies related to cereals, dairy, beef, sheep, pigs and tomatoes. The CAP also includes the measures which accompanied the 1992 reform. The existing knowledge regarding the environmental effects of these measures are explored in Section 4. The various environmental policy objectives related to the agricultural sector are then identified in Section 5, including policies to reduce nitrates, crop protection products and improve energy efficiency.

7.2 Interactions between agriculture and environment

Major issues of environmental concern in the Netherlands that result at least in part from agriculture include:
-   pollution of groundwater as well as surface water and coastal waters (eutrophication) from nitrates and phosphate;
-   emissions of ammonia and subsequent acidification of water and soils;
-   deterioration of soils and water from the use of crop protection products;
-   rising levels of carbon dioxide and other greenhouse gases in the atmosphere, and their possible effects on global warming. These emissions result from, among others, energy consumption in agriculture and horticulture (CO_2) as well as from livestock production (e.g. emissions of CH_4);
-   desiccation of nature conservation areas through high usage of water from groundwater resources, drainage and land consolidation programmes.

These concerns have largely driven environmental policy for the agricultural sector. Agriculture is a major source of pollution in the Netherlands, both in terms of the deterioration of water quality through the leaching of crop protection products and nutrients, and emissions to the air. Table 1 gives agriculture's share of various types of pollutant, as well as the trend in agricultural emissions since 1990. Emissions related to the use of crop protection products have reduced substan-

Table 1 Emissions from the agricultural sector in 1996 and changes during the period 1990-1996

Emission type	Contribution from agriculture (%)	Change 1990-1996 (%)
Nematicides (usage)	100	-80
Herbicides (usage)	100	-3
Other pesticides (usage)	28	-29
Heavy metals (soils)	68	-11
Nitrogen load (soils)	100	11
Phosphorus load (soils)	100	-16
Ammonia	92	-37
Nitrous oxide (N_2O)	38	24
Methane (CH_4)	39	-6

Source: RIVM (1997).

Table 2 Some indicators of the impact of agriculture on the environment in 1990, 1995 and 1996

	1990	1995	1996
Crop protection products (1,000 kg)			
- Nematicides	8,937	2,390	1,750
- Insecticides	731	550	670
- Fungicides	4,140	4,490	4,100
- Herbicides	3.468	3,980	3,960
- Other	1,559	1,200	350
Total	18,835	12,610	10,830
Nitrogen surplus (kg N/ha)			
- Arable farms	170	188	n.a.
- Dairy farms	395	406	n.a.
- Average of all agricultural holdings	353	347	n.a.
Phosphate surplus (kg P/ha)			
- Arable farms	28	22	n.a.
- Dairy farms	27	28	n.a.
- Average of all agricultural holdings	33	25	n.a.
Emissions of ammonia (mln kg NH_3)	219	139	128
Energy consumption (PJ)			
- Horticulture under glass	116.0	127.3	n.a.

tially since the early 1990s, as has the phosphorus load applied to soils which was mainly achieved by adjustments in the feeding regimes of livestock production units. The nitrogen load to soils however has continued to increase.

The greatest reduction in crop protection products has been in the use of nematicides (Table 2). Currently, specific rules apply to the use of chemical soil disinfection: it is forbidden in horticulture under glass and allowed only once every four years for arable crops.

Emissions of ammonia also reduced considerably during the 1990s, mainly because all manure had to be administered on a low-emissions basis.

Almost 90% of total energy consumption by the agricultural sector is related to horticulture under glass. This sector is very energy-intensive as it covers less than 1% of utilized agricultural area. Carbon dioxide emission from energy consumption in horticulture under glass increased by some 13% during the period 1990 to 1993, and subsequently reduced to a level which is about 6% above the emission level of 1990. In absolute values, the emissions of CO_2 from horticulture under glass increased from 6.9 million tonnes (1989) to a level of 7.5 million tonnes (1995).

7.3 Impacts of the CAP

7.3.1 CAP measures and features of agriculture

On a relatively small area, agricultural and food products are produced not only for the home market but also for foreign markets. About 60-70% of Dutch agricultural production is exported to markets inside and outside the European Union. This strong international position has been achieved by a highly productive, capital-intensive, rather small-scale agricultural sector.

Table 3   Index of production volume of Dutch agriculture (1950=100) and production value in 1995 (in bln Dfl)

	1960	1970	1980	1990	1994	(1995)
Crop production	128	123	141	172	168	(2,930)
Livestock	150	170	303	332	345	(21,265)
Horticulture	148	245	437	787	889	(13,145)
Total	145	197	290	359	373	(37,145)

Source: Indices are own calculations based on CBS/LEI-DLO production data; production value 1995: Silvis and Van Bruchem (1997).

Dutch agriculture has developed strongly since World War II (Table 3). A fourfold increase of production volume has been achieved with less labour and land. Labour input in agriculture has declined by about 3% a year on average, while in 1995 the area of land used for agricultural purposes was 16% less than in 1950. Conversely, the use of non-factor inputs has increased substantially. Rutten (1992) estimates their use increased between 1950 and 1990 at an average rate of

4.3% per annum. Stolwijk (1992) indicates that the annual increase in use of inorganic fertilizers was 3% on average between 1950 and 1970 but only 0.3% between 1970 and 1990. Nitrogen use showed in both periods the largest increase. Use of chemical phosphate actually declined somewhat between 1970 and 1990 but this decline was more than offset by the increase in phosphate applied through animal manure.

The CAP is criticized for contributing to a deterioration of the environment. The general argument in this is that by offering higher prices to the farmers than would be the case without such a support policy, the CAP has stimulated the level as well as the intensity of agricultural production. These higher prices made it generally worthwhile to use more (variable or non-factor) inputs, on the assumption of decreasing marginal returns of inputs. The question addressed in this section is to what extent has the CAP contributed to the intensification of agricultural practices in the Netherlands.

Importance of CAP

At first sight the Common Agricultural Policy seems to be of little importance to Dutch agriculture. After all the production increase has been highest in horticulture and intensive livestock production (mainly pigs) (Table 3). In 1995 horticultural products accounted for 37% of total agriculture value, some 10 percentage points higher than a decade earlier. In 1995 also the production value of pig and poultry was almost 8 billion Dfl, accounting for slightly more than one third of all livestock production value. Together with horticultural products, these products account for more than 50% of Dutch agricultural production. Price supports, though, are of limited significance for these sectors. This suggests only a weak link between the CAP and production developments. However, these products may benefit from the implementation of trade policies as part of the CAP market measures. Furthermore, some indirect effects from the CAP on agriculture can be identified which may act as an incentive to increase production of pigs and certain horticultural products (e.g. tomatoes) (see remainder of this section).

The products of the core CAP regimes are of importance to Dutch agriculture too. Dairy products and beef account for about 25% of total agricultural value. Together with cereals and sugar these products are highly affected by the Union's market and price policies. Therefore, one may suggest that the CAP has stimulated farmers to increase production of those products by its relatively high price levels.

Such a direct relation between the level of support, production and input use is, however, debatable, because the exact relationships between inputs used and outputs produced in agriculture remains unclear (Brouwer and van Berkum, 1996:43-51). In other words, the production function is unknown, partly because technical relations are not (entirely) known, and partly because economic research into the effects of price changes is complicated by the occurrence of inter-sector shifts and by technical developments that increase supplies without extra inputs.

Indirect effects of CAP

Apart from the (assumed) direct effect of price support on production and input use, several indirect effects of the CAP have been implicated in the increase and intensification of agricultural

170

production. It has been recognized that the development in agricultural production has been stimulated by the realization of the internal market. The creation of the EC meant a steady growth of an easily accessible and stable internal market. Keener competition in this larger market gave rise to regional specialization and increased production in areas with comparative advantages. Particularly for the Netherlands with a relatively high share in products under a light market regime, production could grow mainly because of the free internal trade and the common market.

Furthermore, the stable prices guaranteed by the CAP market regimes (irrespective of the specific level of prices) promote production and technical developments which until recently were mainly aimed at increasing production rather than at reducing costs and/or non-factor input use. Such developments often cause environmental problems. A CAP with high prices influences land values as these values depend on average production yields: high production yields lead to high land values which are seen as a cause of intensified land use because of the increased opportunity cost of holding agricultural land for nature purposes or managing it extensively. Land prices, however, are also affected by the general economic situation and outlook (such as expectations of interest rates and inflation) and by causes internal to the sector (e.g the scale of the farming structure). This implies that end product prices are not the only explanation of high land values in the EU (Vermuë, 1993).

It seems that the changing price relations of the different inputs is one of the major explanations for the intensified use of non-factor inputs in the Netherlands. The costs of labour have risen sharply, motivating the search for labour-saving techniques. The relatively small-scale agricultural structure was not optimal for these techniques, so that land at farm level became scarce resulting in soaring prices for land. Scarcity at farm level encouraged the development of yield-increasing techniques and consequently a strong intensification of production. In contrast to labour and land, the costs of inorganic fertilizers, pesticides and animal feeding stuffs dropped in real terms in the Netherlands. At the same time CAP support prices have fallen in real terms over a long period, most noticeably since 1980. Therefore, intensification of agricultural production was caused by labour and land becoming more expensive relative to cheap other inputs rather than by end-product prices (Rutten, 1992; Vermuë, 1993). Only to a small extent can agricultural policy be held responsible for these changes in price relationships as the CAP hardly influences prices of inputs.

7.3.2   The arable regime

In the Netherlands growing cereals is part of a narrow cropping plan with a considerable acreage devoted to sugarbeet and potatoes. The possibility to replace cereals with other crops are therefore limited. Since the early 1990s the price of soft wheat has fallen in the Netherlands from 153.0 ECU/tonne (1990) and 157.6 (1992) to 132.1 (1994) and 130.0 (1995). Prices of feed wheat also fell although less sharply.

Usage of crop protection products on arable crops reduced substantially over this period, by almost 50%. This happened not only with cereal crops but with other arable crops like sugarbeet and potatoes too (Table 4). The overall reduction in the use of crop protection products was achieved largely through much lower use of nematicides. This was mainly due to a regulation introduced in 1993 to limit soil fumigants whereby treatment of parcels of land with fumigants was allowed only once every four years. Fungicide use was also reduced substantially through substitution by compounds requiring lower dosages (Venema et al., 1996).

The area of set-aside in the Netherlands in 1994 was about 28,000 ha (CEC, 1995). Total market set-aside includes 14,000 ha, with 12,000 ha under rotational set-aside and 2,000 ha under the non-rotational scheme (in 1994).

Table 4   Use of crop protection products on key arable crops (kg of active ingredients per ha)

	Insecticides	Fungicides	Herbicides	Nematicides	Other	Total
Winter wheat						
1990	0.4	3.2	3.2	0.2	0.9	7.9
1992	0.3	2.1	2.8	0.0	0.8	6.0
1994	0.2	0.9	2.3	0.0	0.6	4.1
1995	0.2	0.8	2.4	0.0	0.6	4.0
Sugarbeet						
1990	0.5	0.0	3.6	5.3	1.4	10.8
1992	0.4	0.0	3.3	2.0	1.6	7.3
1994	0.3	0.1	3.5	2.2	1.6	7.6
1995	0.2	0.0	3.5	0.7	1.9	6.3
Ware potatoes - clay areas						
1990	0.6	14.7	4.1	13.1	0.8	33.3
1992	0.7	17.2	4.0	2.6	0.3	24.8
1994	0.4	11.8	3.8	0.5	0.3	16.8
1995	0.5	9.1	4.2	0.8	0.3	14.9

Source: LEI-DLO, Farm Accountancy Data Network.

7.3.3   The dairy regime

Since the introduction of the milk quota the number of cows has fallen drastically in the Netherlands. In 1996 the dairy herd was 1.67 million milking cows, compared with 2.37 million cows in 1985 (CBS/LEI-DLO). The present dairy herd is equal to the herd in 1939! The number of milk and calf cows per 100 ha of grassland and green fodder declined from 176 in 1985 to 130 in 1996. The structure of the sector has also changed significantly. The number of farms has been reduced from almost 58,000 in 1985 to 36,250 in 1996. The small dairy farms with fewer than 30 cows have especially disappeared, their production capacity (milk quota and land) being taken over by those who remain. The average number of cows per dairy farm has increased since the introduction of the milk quota from 41 to 46. Furthermore, dairy farms have changed to other activities like sheep or beef when they have had to dispose of dairy cows. Utilized agricultural area per dairy farm increased by some 10% between the late 1980s and the early 1990s. Stocking densities also increased (livestock units per hectare of forage crops), although the share of dairy in total livestock production reduced during that period.

Table 5   Nitrogen balances of specialist dairy farms according to the farm gate approach, 1986-1995 (kg N/ha)

Balance	1986	1989	1991	1993	1995
Input	565	523	483	536	524
- Fertilizers	346	307	263	272	261
- Feed	159	143	143	180	184
- Others	60	73	76	84	79
Output	88	110	87	121	118
- Milk	64	64	64	62	64
- Animals	16	21	20	33	30
- Others	8	18	3	26	24
Surplus	477	420	396	415	406

Source: LEI-DLO

Regarding nitrogen use, on the one hand, the consumption of nitrogen fertilizers has reduced substantially since the mid-1980s. On the other hand, dairy producers increasingly depend on external feed because of the favourable price relationship between externally produced feed compared to home-produced roughage. This might contribute to a further intensification of dairy production. On the whole, though, the nitrogen surplus at specialist dairy farms in the Netherlands has reduced by about 15% since 1986 (Table 5). Natural conditions of agricultural production strongly affect nutrient surpluses. Depending on the weather, for example, the uptake of nutrients by crops might be depressed by as much as 30% compared with periods of regular plant growth. The nitrogen surplus on dairy farms in the Netherlands, for example, increased - after several years of a gradual fall - during the accounting year 1993/94. This was mainly due to unfavourable climatic conditions when nitrogen utilization was rather low, especially during the late summer period.

7.3.4   The beef regime

Beef production by raising young fattening bulls is a rather small activity in Dutch agriculture. In the beginning of the 1990s there were 20,500 farms involved, of which slightly more than 70% were grazing holdings. The total number of animals appeared to be around 470,000. By 1995, however, 5,000 of these farms had left the business, equally spread across size and type of holding. The number of fattening bulls on the farms had also decreased by almost 30%. The drastic fall in numbers of animals and farms was caused by the inadequate compensation for price reductions induced by the Mac Sharry reforms.

Male bovine premiums are subject to an individual limit of (2 times) 90 bovine animals per holding, while the ceiling for suckler cows premiums is equal to the number of animals for which a premium was granted in the reference year (1990, 1991 or 1992). The intervention prices for beef were reduced by 15% as of July 1993. Since 1996, compensation is subject to a maximum stocking rate of 2 livestock units (LU) per hectare of forage crops. Rules on livestock density had major implications for holdings with fattening bulls, because of the relatively small proportion of the holdings with stocking densities below this limit (Table 6). Less than 40% of the total number of fattening bulls was eligible for compensation.

Table 6   Characteristics of holdings with fattening bulls according to stocking density by farm size in 1991 and 1995

Stocking density (LU/ha)	Livestock population (number of fattening bulls)				Holdings
	1-199	200-399	>400	total	
< 2, 1991	73,374	670	401	74.445	8,574
< 2, 1995	72,441	541	0	72,982	7,452
> 2, 1991	308,200	69,784	33,990	411,974	12,058
> 2, 1995	217,964	50,306	28,009	296,279	8,150

Source: IKC, 1997.

Stocking densities fell only slightly on holdings with fattening bulls between 1991 and 1995. The extensification effects of the reform of the beef regime on these holdings have therefore been limited, although the total production capacity has been reduced substantially.

7.3.5   The sheepmeat regime

Since 1984 the sheep population in the Netherlands has increased mainly in response to the milk quota system. It more than doubled between 1984 and 1993. The increase was also partly due to the discussions in the early 1990s on limits for ewe premium. Also important were discussions to bring sheep production under quota. This was achieved in the manure legislation of 1992, in advance of which farmers had an incentive to increase their production capacity.

7.3.6   Pig production

Pig production receives hardly any protection under the CAP, and direct effects of the CAP on the environment therefore are likely to be limited in that sector. The indirect effects of the CAP, however, are considered to be rather important in this sector, i.e. through the use of feed. Because of the CAP, internally produced feed grains were expensive compared with imported substitutes, like protein cake and meal, tapioca, maize gluten feed, and so on. At the Kennedy Round of the

GATT in the 1960s it was agreed that import tariffs for these products would not exceed 6 percentage, resulting in a price advantage for feed concentrates based on these products against EU produced feed grains. As a consequence, livestock farmers replaced feed grains with these substitutes and specialized intensive livestock farmers were stimulated to develop especially in the vicinity of ports where these substitutes entered the Common Market. One of these areas was the Netherlands where pig production benefitted in this respect from a very favourable location. Table 7 shows the increase of the number of pigs and specialized pig holdings in the Netherlands since 1970.

Table 7   Number of pigs and specialized pig holdings in the Netherlands, 1970-1995

Year	Number of pigs (x 1,000)	Specialized pig holdings
1970	5,533	7,510
1975	7,279	9,325
1980	10,138	12,323
1985	12,383	9,972
1990	13,915	9,216
1995	14,397	7,738

The biggest increase in the pig herd occurred during the 1970s. This was due to the availability of cheap feed concentrates from the USA, Thailand and Brazil and the increasing sales opportunities within the EU because of the Common Market, which allowed for free internal trade. After Denmark, the Netherlands is the most important EU Member State exporting pigs and pork products. Most Dutch pig and pork products go to Germany and Italy. More pigs have meant more manure, and in the areas where pig production is concentrated, the manure has become an environmentally damaging waste product, instead of being considered a valuable fertilizer. Moreover, to feed the increased pig and poultry populations an increasing volume of feed concentrates was imported. The Netherlands is the biggest importer of compound feed in the European Union. The total import of feed (exclusive of cereals) in 1994 amounted to the equivalent of around 380 thousand tonnes of nitrogen, with more than two thirds of it originating from outside the EU (Brouwer et al., 1998).

Under the CAP reform the use of cereals produced in the European Union became more attractive compared to imported feed concentrates, through the rebalancing of costs for feed grains and cereals substitutes. This may eventually have an impact on the location and intensity of pig production in the European Union, resulting in a more balanced use of minerals in the sector and across EU regions. So far, the consequences for the level and location of pig production are rather inconclusive. As yet there is no shift in the concentration of pig production from the most important producing regions (Brouwer and Van Berkum, 1996:94). The producers who heavily rely on imported cereal substitutes, like those in the Netherlands, also benefit from cheaper feed cereals as prices for substitutes move in line with those of cereals. Furthermore, pig production is rather fixed as part of an agribusiness complex with supply and processing industries nearby. Therefore it is not to be expected that, in the short-run, pig production will move because of relative changes in prices of feed components. On the contrary, since the introduction of the Mac Sharry reforms on the cereal regime, the Dutch pig population has increased still further, the sector's performance having improved due to higher output prices and lower feed costs (Silvis and Van Bruchem, 1997:132).

Even so, more cereals will be used in feed when internal EU cereal prices decline further (see also Folmer et al., 1995:206). Then, cereal substitutes may only remain attractive for farmers close to harbours where the substitutes enter the Union. This would be mainly farmers in the Netherlands, but also Brittany, Northern Germany and Flanders. Such developments will affect the competitiveness and the level of production in certain regions in the EU in favour of those who can rely on cereals for feed rather easily. Therefore, on the mid and longer term relative price changes in feed components should have an impact on the relative concentration of pig production in the EU. Pig producers in countries like Denmark and France will benefit from the relative local abundance of cereals, while pig holdings in Belgium and the Netherlands will have a disadvantage in this respect (see for instance De Groot et al., 1994).

7.3.7   Tomato production

The European Union has a specific market and price policy for a number of vegetables and fruits, including a regulation on the market of tomatoes. The two most important instruments to influence producer prices of vegetables and fruits - including tomato prices - in the internal EU market, are intervention and border measures. These policies protect tomato production in the European Union against cheap import from third countries.

Until early 1997 production could partly be withdrawn from the market if prices fell below a certain minimum level. This minimum price level was in the range between 30 and 65% of the price producers received under regular market conditions. In the case of tomatoes, this minimum price level was not at such a high level that it provided an incentive to increase production. The instrument merely prevented major fluctuations in prices, production and producers' incomes. In general the level of production followed the changes in the market.

The system has altered since the beginning of 1997. The quantity which can be taken out of the market is now limited to a certain maximum which will be reduced gradually in the next couple of years. After a period of 6 years a maximum level of 10% of production will be eligible for intervention. The compensation paid to producers for the amount of tomatoes withdrawn is fixed at a certain level; and this compensation will also be reduced gradually in the course of the next six years. Generally speaking, intervention has become less attractive for producers.

The import of tomatoes is mainly regulated through border measures within the framework of the EU policies for vegetables and fruits. A system of reference prices existed until 1995. Producers were protected against low-priced imports through imposition of a compensatory levy on imports if their prices were below the reference prices which was based on production costs in the European Union. Without such a policy the production of tomatoes in the EU would have been lower than its present level, not least in the Netherlands. In this respect it has to be pointed out that the level of the reference prices depends on the season. Production costs in the EU in the first part of the season are much higher than in the height of the season. So reference prices were fixed much higher in the spring than during the summer and autumn. Just in that period, spring and early summer, the Netherlands has an important share in market supply.

With the implementation in 1995 of the GATT agreement the system of reference prices is replaced by a system of entry prices. The entry prices can be seen as minimum import prices. They are derived from the reference prices for the period 1986-1988. A levy has to be paid if the import price is lower than the entry price. This levy is limited to a maximum (the maximum tariff equivalent, MTE), which will be reduced during the period 1995-2000 by about 20% (see Table 8). Although the level of protection is thus set to decrease over time, in the year 2000 it will still remain considerable, the maximum average tariff equivalent being nearly half the entry price. The regulation remains of particular importance for tomato growers in the Netherlands as the entry price is, because of higher production costs, much higher in spring and in winter than during the rest of the year (see Table 9).

The use of agrochemicals to grow tomatoes has been reduced substantially and the integrated control of insects is now practically standard.

Total energy consumption to grow tomatoes under glass in 1994, based on an acreage of 1,250 ha, is estimated to be 755 million m^3 of natural gas which is about 20 percentage of the total consumption of natural gas by horticulture under glass. Total emissions related to consumption of natural gas by horticulture under glass amount to 7.5 million tonnes of CO_2 (situation in 1995).

Table 8   Market protection of tomato production in the European Union during the period November 1 - December 20, between 1995 and 2000

	Border tariff (%)	Entry price (ECU/tonne)	MTE (ECU/tonne)
Base	11.0	700.0	372.0
1995	10.6	687.7	359.7
1996	10.3	675.3	347.3
1997	9.9	663.0	335.0
1998	9.5	650.7	322.7
1999	9.2	638.3	310.3
2000	8.8	626.0	298.0

Assessments have been made on the efficiency of using natural gas in the production of tomatoes. This is expressed in terms of usage of natural gas (in m^3) per kg of output produced, and is assessed to be 1.42 (Brouwer et al., 1996). This implies that the production of 1 kg of tomatoes on average required 1.42 m^3 of natural gas. Differences however exist among holdings. Modern holdings with dynamic farm management generally have higher efficiency rates, require lower amounts of natural gas to achieve their production. Efficiency among two groups of holdings was in the range between 1.34 and 1.66.

Table 9   Relative change of the entry price of tomatoes during the period to implement the GATT agreement (ECU/tonne)

	Base	2000	Relative change (%)
MTE	372	298	20.0
Entry price			
1 November - 20 December	700	626	10.6
21 December - 31 December	750	676	9.9
1 January - 31 March	920	846	8.0
1 April - 30 April	1,200	1,126	6.2
1 May - 31 May	800	726	9.3
1 June - 30 September	600	526	12.3
1 October - 31 October	700	626	10.6

Source: Swinbank and Ritson (1995).

7.4   The accompanying measures

The implementation of the agri-environmental measures under Council Regulation 2078/92 are arranged in the Netherlands by the Regulation on Management Agreements and Nature Development (RBON). This was agreed with the European Commission in 1995. It includes also the national implementation of Regulation 2328/91 regarding the Less Favoured Areas (a maximum of 140,000 ha can be designated as LFAs).

RBON replaces the former RBO from 1993 and is to follow-up on the 1975 government document on the relation between agriculture and the conservation of nature and landscape (the so-called Relation Paper) (Nota relatie tussen landbouw en natuur- en landschapbehoud, 1974-75). The 1975 paper provided for the designation of up to 200,000 hectares of agricultural land where special conservation measures would apply.

A management agreement between the farmer and government specifies measures to be taken and the compensatory payments to be received. Payments are based on the consideration to compensate farmers entering a management agreement for any income losses compared to holdings under similar conditions without a management agreement (reference situation). The number of participants in Regulation 2078/92 and the acreage covered are shown in Table 10. The proportion of 'heavy' management contracts, with strict requirements on land management, increased substantially in response to the additional support measures for holdings which take additional measures on nature management. Examples of management recommendations are: no grazing or mowing of grassland before June 15; no use of crop protection products; no ploughing up of grassland; no harvesting or rolling before June 15; no fertilizer application on a 3-metre wide margin along field boundaries.

Management agreements may help reduce the decline in plant species diversity and the conservation of valuable botanical grasslands (Heinen et al., 1997). Local investigations also indicate positive impacts of such agreements on breeding populations of birds.

Table 10 Participation of management agreements by December 31, 1996, based on Regulation 2078/92 (and change compared to the situation in 1995)

Number of participants	Coverage (ha)	'Heavy' management contracts
5,891 (+ 4%)	43,320 (+ 10%)	39% (+36%)

Source: MLNV, 1997

7.5   Conditions in environmental policies

7.5.1   Introduction

Nutrients, crop protection products and energy are major issues of environmental concern and legislation has been developed over the past couple of years. Such legislation affects agriculture to a large extent.

7.5.2   Nutrients

To target the standards which are part of the Nitrates Directive, the Dutch government decided to use mineral accounting at farm level as a central part of the legislation. The implementation of a maximum application level of nitrogen from livestock manure (as formulated in the Nitrates Directive) moved in the Netherlands to standards of maximum allowable losses.

The implementation in the Netherlands of the Nitrates Directive is part of the Integral Note on Manure and Ammonia Policy. As of 1998, a mineral declaration system is required for all intensive livestock holdings with animal density which exceeds 2.5 LU/ha. On the basis of the

Table 11 Standards on nutrient losses, levies and limits on livestock density for the period 1998-2008

	1998	2000	2002	2005	2008/2010
Phosphate losses a)	40	35	30	25	20
Nitrogen losses grassland b)	300	275	250	200	180
Nitrogen losses arable land b)	175	150	125	110	100
Mild incentive c)	40 - 50				
Mild incentive d)		35 - 45	30 - 40	25 - 30	n.a.
Strong incentive d)	50	45	40	30	n.a.
Phosphate application a)		85	80	80	80
- grassland	120				
- arable land	120				
Livestock density to make					
a return (in Livestock Units)	2.0	2.0	2.0	2.0	n.a.

a) kg P_2O_5/ha; b) kg N/ha; c) Dfl 2.50/kg P_2O_5 by exceeding phosphate loss; d) Dfl 20/kg P_2O_5 by exceeding phosphate loss.

The use of pesticides in the Netherlands is lower than in southern European countries because of a temperate climate, high level of management practice and a high production intensity per square metre (Verhaegh, 1996).

Table 15 Usage of pesticides by horticulture under glass during the reference period of 1984-1988 (in 1,000 kg of active ingredients) and trends during the 1990s

Product group	1984-1988 (x 1,000 kg)	Development (reference = 100)			Policy target (reference period = 100)	
		1990	1993	1994	1995	2000
Insecticides	80	154	164	172	63	29
Fungicides	195	132	125	109	60	30
Nematicides	800	22	4	9	43	29
Other	260	110	115	206	62	65
Total	1,335	63	52	72	52	36

Source: Brouwer et al., 1996.

7.5.4 Energy

Environmental policy in the Netherlands related to energy consumption aims at a reduction in carbon dioxide emissions of 3-5% during the period 1989/90-2000. Agriculture and horticulture should also contribute to this reduction, but no explicit policy goal is formulated in this respect. The main target in energy policy is to improve the efficiency of energy use, and consequently contribute to a reduction in carbon dioxide emissions. Energy efficiency is the amount of primary energy required to produce one unit of product. This indicator is therefore affected by a combination of energy consumption and production level.

A Multi-Annual Agreement on Energy in Horticulture under Glass is to contribute to energy policy. This agreement is aimed to achieve an improvement of energy efficiency by 40% in 1995 compared to the reference year 1980. The long-term agreement on energy consumption was signed in 1993 between the government and the agribusiness. By the year 2000, energy efficiency in horticulture under glass needs to be improved by 50% compared to 1980. In the framework of this agreement the horticultural sector also aims to reduce the emissions of CO_2 by some 3-5% during the period between 1989/1990 and 2000. Similar agreements on the improvement of energy efficiency are prepared for other crop sectors as well as livestock production.

Energy-saving programmes taken up by the horticultural sector are financed through revenues generated under the agreement. Holdings pay a levy on the amount of natural gas consumed. In 1995 this amounted to 0.6 cents per m^3 consumed. The annual revenues thereby generated by the horticultural sector are around Dfl 20 million.

7.6 Concluding observations

1. Environmental policies in the Netherlands are crucial in the achievement of environment-friendly production methods. This applies to energy consumption and its contribution to carbon dioxide emissions, as well as the use of crop protection products and nutrients. Environmental legislation increasingly puts constraints on the agricultural sector; the agricultural sector has responded to such constraints by improving its efficiency in using inputs and by changing farm management practices. This applies for example to the production of cereals and the reduction of crop protection products.
2. Extensification effects of livestock production in response to the reform of the beef and sheep regimes remains limited in the Netherlands. The reform of the beef regime contributed to a reduction in production of fattening bulls. Nitrogen surpluses at dairy farms have remained rather stable over the past couple of years. The reduction of nitrogen surplus at such holdings was mainly achieved during the late 1980s by lower consumption of mineral fertilizers.
3. Two important driving forces to increase the intensity of agricultural production have been the internal market as well as the relative price relations among factor inputs. The internal market contributed largely to the increase in production volume over the past couple of decades. The scarcity of the available land resource also contributed to enhance the intensification of agriculture, as shown by pig production as well as horticulture under glass.
4. Participation in the agri-environmental measures under Regulation 2078/92 allows a group of farmers to contribute to the achievement of less intensive production methods. Given the scarcity of the available land resource, the premium needs to be relatively high in order to compensate for any income losses.

References

Brouwer, F.M. and S. van Berkum (1996) *CAP and environment in the European Union; Analysis of the effects of the CAP on the environment and assessment of existing environmental conditions in policy;* Wageningen, Wageningen Pers.

Brouwer, F.M., C.H.G. Daatselaar, J.P.P.J. Welten and J.H.M. Wijnands (Eds.) (1996) *Landbouw, milieu en economie: Editie 1996;* The Hague, Agricultural Economics Research Institute (LEI-DLO), Periodieke Rapport 68-94

Brouwer, F.M., P.J.G.J. Hellegers, M. Hoogeveen and H. Luesink (1998) *Managing nitrogen pollution from intensive livestock production in the EU: Economic and environmental benefits of reducing nitrogen pollution by nutritional management in relation to the changing CAP regime and the Nitrates Directive;* The Hague, Agricultural Economics Research Institute (LEI-DLO) (forthcoming)

Commission of the European Communities, CEC (1995) *Report of the Commission to the Council and the European Parliament on the objectives and implementation methods of extraordinary set-aside;* Brussels, Commission of the European Communities, COM (95) 122 def.

Folmer, C., M.A. Keyzer, M.D. Merbis, H.J.J. Stolwijk and P.J.J Veenendaal (1995) *The Common Agricultural Policy beyond the Mac Sharry reform;* Amsterdam, North-Holland Publishing Company, Series Contributions to Economic Analysis, Volume 230

De Groot, N.S.P, C.P.C.M. van der Hamsvoort and H. Rutten (eds.) (1994) *Voorbij het verleden. Drie toekomstbeelden voor de Nederlandse agribusiness 1990-2015. (Beyond the past. Three scenarios for the future of the Dutch agribusiness 1990-2015);* The Hague, Agricultural Economics Research Institute (LEI-DLO); Onderzoekverslag 127

Heinen, J., G. van Dijk and J. Nieuwenhuize (1997) *Netherlands: Experiences with nature policies regarding farmland. Country case studies of the Helsinki Seminar on environmental benefits from agriculture;* Paris, Organization for Economic Co-operation and Development, COM/AGR/CA/ENV/EPOC(97) 29, pp. 108-123

IKC (1997) *Wat waren de gevolgen van de Mac Sharry-regelingen voor de dierlijke sectoren in Nederland: Evaluatie 1993-1995;* Ede, Informatie- en KennisCentrum Landbouw

MLNV (1991) *Meerjarenplan Gewasbescherming; regeringsbeslissing;* The Hague, Tweede Kamer; Vergaderjaar 1990-1991; 21 677, nrs. 3-4

MLNV (1997) *Toepassing van Verordening 2078/92 in Nederland;* The Hague, Ministry of Agriculture, Nature Management and Fisheries (in Dutch)

Rutten, H. (1992) *Productivity growth of Dutch agriculture, 1949-1989;* The Hague, Agricultural Economics Research Institute (LEI-DLO); Mededeling 470

Silvis, H.J. and C. van Bruchem (Eds.) (1997) *Landbouw-Economisch Bericht 1997 (Agricultural Economic Report 1997);* The Hague, Agricultural Economics Research Institute (LEI-DLO); PR 1-97

Stolwijk, H.J.J. (1992) *De Nederlandse landbouw op de drempel van de 21ste eeuw.* The Hague, CPB; Onderzoeksmemorandum 95

Swinbank, A. and C. Ritson (1995) *The impact of the GATT agreement on EU fruit and vegetable policy;* Food Policy, vol. 20 (4)

Venema, G.S., J.A. Boone, WH. van Everdingen, J.H. Jager and J.H. Wisman (1996) *Bedrijfsuitkomsten en financiële positie (BEF); Samenvattend overzicht van landbouwbedrijven tot en met 1994/95;* The Hague, Agricultural Economics Research Institute (LEI-DLO), PR13-94/95

Verhaegh, A.P. (1996) *Efficiëntie van energie en gewasbeschermingsmiddelen tomaten en rozen in kassen: Nederland, Israël, Spanje en Marokko;* The Hague, Agricultural Economics Research Institute (LEI-DLO); Publicatie 4.142

Vermuë, A.J. (1993) *The Common Agricultural Policy and the environment: the case of the Netherlands;* The Hague, Ministry of Agriculture, Nature Management and Fisheries

8. DENMARK

Michael Linddal

8.1 Introduction

Environmental objectives and policies have been introduced in Denmark as a result of pressures from the general public, NGOs and the media, and the process was stimulated primarily through national policy, and to a lesser extent by EU legislation. The effectiveness of the implementation of national environmental measures in agriculture, however, is subject to the incentives and disincentives created through the CAP on national agricultural policy.

This chapter reviews the present state of knowledge regarding the effects of the CAP on the environment and landscape in Denmark, and the environmental requirements for the agricultural sector in national agricultural or environmental policy.

When agricultural policy takes environmental objectives on board, the agricultural sector and environmental impacts can be compared either with the unregulated agricultural sector (non-existing) or the regulated agricultural sector with market interventions (existing). The agricultural sector without regulated markets is assumed to cause environmental impacts, and these are probably enhanced through distortions from agricultural policy directed towards market interventions. In economic terms the excessive environmental impact of agriculture is a composite of both a *market failure* from the unregulated and a *policy failure* from the regulated agricultural sector. The former creates externalities while the latter creates market distortions and dead-weight losses thus exacerbating the overall social welfare loss. In this chapter no clear distinction is made between these two causes of environmental impacts of agricultural policy, although it is very much needed in order to make precise accounts of the impacts of the CAP.

8.2 Agricultural and environmental policies

This section includes a brief overview of issues relevant to the CAP and the environment in Denmark as they emerge in particular from the different EU regimes and the relevant national agricultural and environmental policy. National requirements in the Agricultural Act and national support schemes are not included in this section, but anyhow the EU regimes have to be viewed as additional measures to national legislation controlling the agricultural sector.

The arable crop regime

The reform of the CAP resulted in a decrease in market prices compensated through acreage premiums. Although it is difficult to isolate the specific impact of the CAP on the agricultural sector; the arable crop regime since 1992 does appear to have resulted in a reduced application of input factors and thus possibly reduced environmental impacts. The question, however, remains whether this follows the decrease in the crop price or is a direct result of the mandatory set-aside.

Another, and perhaps overriding influence is national environmental policy, for example, in putting constraints on the application of fertilizers and animal manure. Besides, in the long run increased production and/or relaxation of the set-aside requirements may offset any gains from the 1992 changes. After all, the area under compulsory set-aside has followed the reduction in the mandatory set-aside requirement from 15% in 1993/94 to 5% in 1996/97.

The acreage premiums in Denmark are independent of soil classes, and on poor soils they have maintained an opportunity value on agricultural land above a level that would allow a transfer to other land uses, such as those included in the accompanying measures.

The dairy regime 1)

The dairy sector is one of the agricultural sectors with a particular regulation resulting from EU regimes. It occurs with regulation of milk production through quotas. There is also national environmental regulation which seeks to harmonize livestock density and the land available for spreading of manure through a ceiling on the number of livestock units per ha of grazing. In a report prepared for the Ministry of Food, Agriculture and Fisheries the structural development of the dairy sector was reviewed. A tendency towards larger production units was clearly depicted. In 1995, 60% of the dairy herd was in production units with more than 50 dairy cows, compared with 35% in 1983. The Danish membership of the EU in 1973 resulted in an increase in milk production of 18% from 1973 to 1983. Since the regulation of milk production through quotas production has dropped 14%. The total number of dairy cows, though, has dropped by 29% 2) due to an increase in the average yield per dairy cow from 5.6 tonnes of milk per year in 1983 to 6.6 tonnes in 1995.

Regarding dairy production and the land harmonization requirements, the report estimated the number of excess livestock units (LU) on non-complying dairy farms, i.e. farms with insufficient land, as being 20,000 in 1995 but projected to double to 40,000 by 2003. The figure would be even higher if the allowable amount of nitrogen per livestock unit were to be tightened up. The problem is regionalized to areas with dense dairy production mainly in the Western part of the country.

The beef and sheepmeat regimes

No particular environmental impacts have been identified from either the beef or the sheep regime. The former is mainly integrated in the dairy sector while the latter is of virtually no importance in Denmark. From a landscape point of view, grazing with cattle or sheep may be of some relevance but the scale and role of EU regimes in this respect have not been identified.

---

1) Based on: Ministry of Food, Agriculture and Fisheries (1997). One livestock unit (LU) equals 100 kg nitrogen in forthcoming Danish policy.

2) The number of cattle in thousands (dairy cows in brackets) was 2,775 in 1972. It increased to a total of 2,852 (1,003) in 1983 and declined to 2,052 (712) in 1995.

Accompanying measures and compulsory set-aside

Beneficial environmental considerations are mainly related to the CAP through the accompanying measures and the compulsory set-aside. The agri-environmental measures (Reg. 2078/92) are, however, only applied to a limited extent in Danish agriculture. By 1997, out of 350,000 ha made eligible in designated environmentally sensitive areas (ESA), only about 58,500 ha (2% of the agricultural area) had been included in the scheme, spread across five main activities: grass in crop rotation (3.8%), grass outside crop rotation (83.7%), 20 years set-aside (1.7%), reduced application of fertilizer (10.6%) and 12-meter-wide zones without pesticide application (0.2%). An important administrative change has been the transfer of the administration of the agri-environmental scheme from central government to the 14 county authorities.

Another accompanying measure is afforestation of agricultural land (Reg. 2080/92). This has been well incorporated with a Danish policy objective established in 1989 to double the forest coverage from its current level of 11% over the coming 80 to 100 years. This is equivalent to an annual increase in afforested land of some 5,000 ha. Despite the grants made available the afforestation rate has so far been substantially below the anticipated rate. The national contribution and the incentives have been increased in a revised grant scheme implemented in 1997, which for some areas includes income compensation over 20 years. This has boosted the interest in afforestation.

A number of competitive payment schemes in agriculture result in distortions of the land rent, thus making payments for agri-environmental objectives less effective. The point is that a payment for agri-environmental measures not only has to compensate direct costs and the opportunity land value but also the capitalized value of another payment scheme with a different policy aim. Many farmers have indeed shown reluctance to take up agri-environmental measures, although their reasons are not yet clear.

Nitrates Directive

The Nitrates Directive is applied as a nation-wide measure. The whole of Denmark has been declared nitrate vulnerable. The direct implication of the Directive is a ceiling, on the application of animal manure on areas prone to nitrate leaching, of an average of 170 kg nitrogen per hectare. Some relaxations are accorded to the dairy sector in the proposed national implementation allowing 230 kg of nitrogen per ha on grassland areas. Some standards for livestock densities in harmonization rules already apply, but some of these are due to be further tightened gradually over a seven-year period towards the year 2003 as a result of specific national measures. In a recent assessment (DIAFE, 1997) of possible impacts of the Nitrates Directive, it was concluded that it only had limited consequences for the livestock sector compared with existing national requirements. The Nitrates Directive, indeed, resembles the existing national rules to a large extent, reflecting the fact that it was prompted in large part by Danish efforts to curb nitrate leaching.

The main direct impacts from the national implementation of the Nitrates Directive occur within the dairy sector due to an increased national requirement for harmonization area and a reduction in the allowable amount of nitrogen per livestock unit from 115 kg to 100 kg. For pork production there is no change in the existing national requirement for harmonization, and the Nitrates Directive does not differ from national policy. It is anticipated, and it has already been experienced, that the price of land in some regions may increase. Increased flexibility, for example by allowing milk quotas to become tradable, could well ease the pressure on structural development. Furthermore it was concluded that some incidence would be shifted towards the cereal feed animal sector (i.e. pigs and poultry), and further investments would be directed to regions without a harmonization problem. No attempts were made in the assessment to investigate the environmental impacts of the Nitrates Directive.

Other EU policies with a potential impact on the environment

The EU Farm Development Programme (Reg. 2328/91; before 1985 named the EU Modernization Act) was recently evaluated in Denmark (Hansen et al., 1997). It is a scheme for investment support in the agricultural sector. The overall conclusion was that the scheme had had insignificant impact on the development of the agricultural sector. Using a general equilibrium model of the Danish agricultural sector it was assessed that abandoning the programme in Denmark while maintaining it in other Member States would only result in minor consequences mainly accounted for by the reduced transfers from the Community. Abandoning the scheme in all of the EU would be beneficial to the Danish economy due to lower costs of the CAP. No assessment of the likely environmental impacts was attempted, but they are likely to be limited.

8.3   Impacts of the CAP

This section reviews some Danish research projects which did attempt to assess the impacts on the agricultural sector from changes in agricultural policy. The focus is on issues which are considered to encompass the most severe impacts on the environment. Emphasis is given to the impacts on land use practice and cultivation intensity, arising from the 1992 CAP reform.

Assessment of productivity growth in Danish agriculture

More than two decades with the CAP has had a dynamic impact on Danish agriculture, e.g. through the accumulated environmental impacts, the investment behaviour in the agricultural sector, capitalization of subsidies and market distortions, and national agricultural policy. The 1992 CAP reform was a shift in the CAP which made a partial assessment of its influence on the agricultural sector possible.

Prior to 1992, the CAP consisted mainly of price and trade regulation (i.e. market interventions) with the unambiguous target of sustaining agricultural production and farmers' income. Figure 1 shows the development in factor productivity until 1992. The chart reveals that more has been produced against lower factor requirements. This is a result of structural and technological development in the agricultural sector relevant for the assessment of the environmental impacts of agriculture, because it indicates that the environmental impacts per unit of production are reduced. Caveats, however, remain and the results are not conclusive.

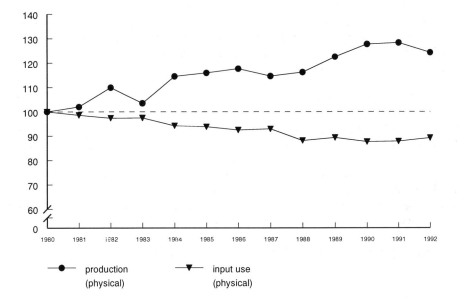

Figure 1   Physical production and factor use by the agricultural sector (index: 1980/81 = 100) (Adapted from Hansen, 1995)

The role of the CAP in this development cannot be identified directly. In theory, a distinction should be made between the impacts of the CAP on productivity and on production including input use, i.e. production intensity, relative distribution of cropping and livestock, and area cultivated. The direct impact of the CAP on productivity (i.e. a shift in the production function) probably has been limited. However, it did include an increase in output (i.e. a movement on the production function and a larger production capacity) through an increased application of input factors and cultivation of arable land in comparison with an unregulated sector. The elevated price level has been capitalized in land values, and it thus has had an impact on the fixed costs at the farm level for new farmers. Productivity and production levels also depend on the structural development of the agricultural sector when there is an increasing scale of production. The impact of the CAP on the structural development of the agricultural sector is not specified.

The consequences for the agricultural sector cannot directly be assessed, but the productivity and terms of trade of Danish agriculture have been analysed (Figure 2).

189

The annual increase in total factor productivity (physical index of production divided by a physical index of factor use) of the Danish agricultural sector from 1980/81 to 1992/93 was 3.2%. This is substantially more than in the 1960s and 1970s where the factor productivity growth was around 1.8%. The factor productivity growth was 3.1% in the crop sector excluding production of roughage, 3.4% in the pig sector and 2.3% in the dairy and cattle sector. The most important source of productivity growth was due to technology; mechanical like buildings, machinery and equipment, chemical like new pesticides and fertilizers, and biological like improved plant material and breeds of livestock. Other sources included increasing farm size and changes towards a more optimal product and factor mix on individual farms. The terms of trade (value index of production divided by value index of factor use) in the same period declined by 3.0% per year for the entire agricultural sector. The average annual decline in factor productivity was 4.0% in the crop sector excluding production of roughage, 2.4% in the pig sector and 1.5% in the dairy and cattle sector. It is usual that the terms of trade decrease when the factor productivity increases. If the decline balances the growth it implies that the entire economic gain from an increased productivity is not captured within the sector but 'down stream' with the processing industry and the final consumers due to lower real prices on agricultural products.

Figure 2   Productivity and terms of trade of Danish agriculture
Based on: Hansen (1995).

Assessing policy implications on the agricultural sector

An assessment of the structural and environmental impacts of the CAP is very much depending on the shifts in productivity and production levels. An integrated assessment of the policy implications on the agricultural sector may, inter alia, be done in a model framework which takes into account the complex relations between agricultural policy, sector production and productivity and the environmental impacts (Figure 3). An analysis of the structure of Danish agriculture and its prospects towards the year 2003 is provided by Rasmussen (1996).

A sector model for Danish agriculture (ESMERALDA) developed at DIAFE has a predominant role in the assessment of the aggregated behavioural relations of Danish agriculture from changes accruing to direct policy measures or price changes. The model is econometric and based on the duality approach of production economics. It is a top-down model with an empirical data source from Danish accountancy, price, quantity data, etc. from 1974/75 onwards.

One analysis estimated the effect of a 10% decrease in cereal prices compensated by a 790 DKK/ha premium for cereal areas. The model showed a rather low impact on cereal production (-1.9%) but a rather large indirect effect on the livestock sector. The reason was a large increase in the pig production sector due to unchanged prices on the production prices of pigs while the cost of input of cereal for feeding declined. In the same study the use of pesticides was estimated to decline about 10% and the use of fertilizers about 5%. The model also revealed that the agricultural sector was over-compensated.

Figure 3   An applied econometric sector model for Danish agriculture
Based on: Jensen (1996).

Integrated economic-environment modelling

The impacts of agricultural policy on the environment were analysed in a joint study by the National Environmental Research Institute (NERI) and the Danish Institute of Agricultural and Fish-

190

eries Economics (DIAFE). The focus of the study was the impacts of policy on the area and crops cultivated and the resulting effects on the use of nitrogen and pesticides.

The economic impacts were modelled using the DIAFE agricultural sector model (ESMERALDA) and the environmental impacts were modelled with the NERI model on leaching of nitrogen into groundwater aquifers, surface waters and inner coastal waters. The study did not take the absolute economic impacts on land rent into account, though relative changes were included in ESMERALDA. The study used 1989 as a base year and included two scenarios. Scenario A was based on the reformed CAP and scenario B was the actual development in prices and outputs observed in 1993/1994. The two scenarios were different since the CAP reform was not entirely implemented until 1995/1996.

The study estimated that the isolated effect of the CAP reform compared with the 1989 baseline amounts to a 19.7% decline in nitrogen emissions to Danish coastal waters, while the estimated decline based on the actual development from 1989 to 1993/94 was 14.4%. The study reveals that the expected environmental impacts of the CAP is to some extent diminished by other developments such as structural changes in the sector. In consequence, the use of fertilizers did not drop as much in scenario B as was expected from scenario A. One of the conclusions of the study was that further measures are required to comply with targets in the national action plan on nitrogen emissions from agriculture, i.e. a reduction of 50% from the level in the mid-1980s by the year 2000 (Miljøstyrelsen, 1996).

8.4 Conditions in environmental policies

Concern over the environmental impacts of agriculture is substantial in Denmark. Agriculture is considered a prime cause of the reduction in environmental quality and specifically of water pollution from excess fertilizers and pesticides. An emerging issue of concern, in which modern agriculture is likewise indicted, is animal welfare. These concerns are driving forces, shaping not only environmental policy, but also national agricultural policy. The CAP, however, is generally not considered as the main cause of the environmental impacts of agriculture which is put down to national policies including environmental legislation.

Although the major focus of public concern is environmental pollution from agriculture, there is also considerable public disquiet over the landscape effects of contemporary agriculture. Current policy action, however, reflects the emphasis regarding the impacts of industrialized agriculture on water quality. In part, this is because the landscape transitions have taken place over a longer period albeit at an accelerating pace. The institutional setting for landscape protection is also far more developed. This is due to the fact that the conservation movement at its outset was more focused on *nature* than on the *environment*, i.e. conservation rather than pollution. Therefore, conservation legislation already to a large extent controls landscape effects. Structural developments in agriculture, however, continue to induce detrimental landscape effects. The spatial scale of issues related to landscape also is considered to be different compared to pollution from excess minerals and pesticides.

National environmental policy: nitrate, phosphate and pesticide regulation

The public focus on the environmental impact of the agricultural sector emerged in the 1980s. By that time, one of the dominant issues was nitrate leaching to groundwater aquifers and to surface waters and subsequently to coastal waters. Legislation to control nitrogen leaching was introduced in the mid-1980s:

(i)   the NPO-paper (on nitrogen, phosphorus and organic compounds) in 1984 revealed that the contribution of the agricultural sector to nitrate leaching was approximately 260,000 tonnes of nitrogen out of a total loss of 290,000 tonnes;

(ii)   the governmental action plan on water resources from 1987, based on the NPO-paper, requested that the agricultural sector should reduce nitrate leaching by 127,000 tonnes by 1992;

Mineral policies

Aim: Nitrate leaching from Danish agriculture should be reduced by 50% and phosphorous emissions by 80% before year 2000 compared to the level in mid-1980s. Point sources should be eliminated.

Measures:
-        maximum application of manure equivalent to 2.3 LU (livestock units) per ha in cattle production; 1.7 LU per ha in pig production and 2.0 LU per ha on other farms (further adjustments are ongoing due to the implementation of the Nitrate Directive);
-        mandatory mineral balances at farm level including a minimum utilization rate in 1997 of nitrogen in manure of 45% in cattle and dairy production; 50% in pig production and 40% on other farms. On farms with deep bedding stables the minimum utilization is 15%. A general increase of 10% is expected by 1998,
-        spreading of manure is in general prohibited from harvest to February 1;
-        minimum storage capacity for 9 months production of animal manure;
-        the proportion of winter crops should exceed 65% on all farms;
-        reporting on cropping (land-use) and fertilizer application at farm level

Pesticide policies

Aim: The use of pesticides in Danish agriculture should be reduced by 50% before the end of 1997 compared to the average level in 1981-85. The use is measured as active ingredients and frequency of application.

Measures:
-        increased research and extension on the use of pesticides;
-        requirements regarding handling and applying pesticides by professionals;
-        accounts of pesticide use on all farms over 10 ha;
-        pesticide use is prohibited in ecologically sensitive areas (e.g. meadows and water catchment areas) and within 2 meters from shores of streams and lakes;
-        all pesticides traded in Denmark should be assessed according to their toxic effects on humans and the environment before approval;
-        a graduated sales tax on pesticides introduced in 1996 (27% for insecticides, 13% for herbicides and fungicides); a further increase is expected in 1998.

Figure 4   An overview of Danish mineral and pesticide policies.
Based on: Olsen et al. (1996).

(iii) the governmental action plan for sustainable agriculture in 1991 included a revised target to achieve the reduction in nitrogen loss from the agricultural sector by the year 2000.

Policies on controlling phosphate leaching and the use of pesticides were also introduced during the 1980s. They are summarized in Figure 4. Figures 5 and 6 give indicators of the progress that has been made.

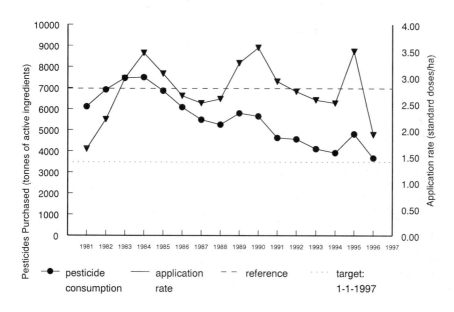

Figure 5   Development of indicators for pesticides (active ingredients and application rate)

An outlook of future policy on nature, environment and the CAP

A government committee in 1996 prepared a report, from the Danish perspective, on the future direction of nature, environment and EU agricultural policy (Ministry of Agriculture and Fisheries, 1996). The report emphasized the need for greater harmony between agricultural policy and considerations for nature and the environment. The report also underlined the role of existing national policies, on reductions in nitrate and phosphate leaching and the application of pesticides. The Committee concluded (p.3, translated) 'there should be a twin strategy, which focuses partly on the purposeful marginalization of particularly delicate areas and partly on a certain general extensification of agricultural production through price policies'. Cross-compliance of the CAP supports to specific environmental requirements through voluntary agreements was stressed. It was pointed out that the tying of agricultural support to environmental protection may be required to justify pure income transfers in coming years. The Committee endorsed an increase in designated environmentally sensitive areas (ESA) from about 50,000 ha to about 350,000 ha.

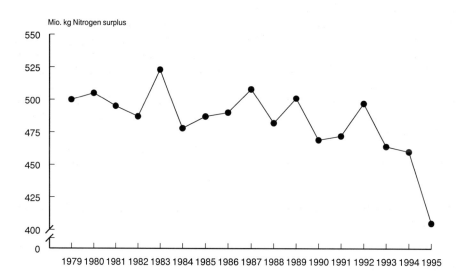

Figure 6   Development in nitrogen surplus

8.5   Interactions between agriculture, CAP and the environment

The impact of the CAP on the environment in Denmark is generally considered through the future prospects for improvements, rather than the damage that may have occurred in the past. It includes future reforms in CAP, e.g. due to EU enlargement and WTO agreements on agricultural trade, which generally are assumed to lead to reduced environmental impacts because agricultural subsidies are removed or diverted towards environmental objectives. In this respect, the CAP as it evolves is seen to complement national environmental policies regarding the regulation of the direct environmental impacts from the agricultural sector. Some would also argue that direct income support makes investments in agriculture for environmental protection more affordable and thus feasible. The emission of surplus input factors, i.e. excess minerals and pesticides, remains the dominant environmental problem of the Danish agricultural sector, while other topics such as ethical concerns over animal production and the socio-economic consequences of changing rural and agricultural structures are emerging.

The apprehension of the Danish agricultural sector in relation to the CAP and the environment can be generalized in a few observations:

(a) the Danish agricultural sector is highly effective and the immediate influence of the CAP on production intensity and cultivated area could be overstated. It is probably for the dairy sector that the CAP has provided the strongest market protection;

(b) the scope of Danish environmental policy with regard to the agricultural sector is considered to be greater than that embedded within EU regulations;

(c) set-aside has provided an acknowledged landscape change, albeit a temporary one;

(d) the accompanying measures to the CAP reform provide a more transparent payment to environmental activities directly related to CAP, although their take-up has been limited. One reason may be the distortion created in land prices from other agricultural and environmental policy instruments;

(e) a recent report considered the benefits from making milk-quotas tradable to ensure a more efficient distribution of dairy production and to ease the sector's adaptation to such measures as the Nitrates Directive;

(f) the pig sector is a dominant and growing sector. Pig production increased by 20% between 1989 and 1996, and continues to expand. The sector is not directly regulated by CAP but its cost side is influenced by the price of cereals as feed stock.

The general observations reveal a composite pattern of linkages between agricultural policy and subsequent environmental impacts.

8.6 Concluding observations

There is a correlation between activities in the agricultural sector and the environment but the quantitative impacts from agricultural policy on the environment are still subject to uncertainty;

(a) agricultural policy to a large extent is based on the CAP, while there is less influence from EU regulation on objectives and implementation of national environmental policy;

(b) the impacts on the agricultural sector from CAP cannot be isolated from other and parallel policy measures. Different tiers of policy are independent and one policy may level out the incentive created by another. Agri-environmental measures, for example, and voluntary cross-compliance instruments have to compete with distorted land values due to the likes of capitalized area premiums and direct payment schemes. The distortion of land rent as a result of environmental constraints or direct income compensation subject to a particular use may put further pressure on marginal areas;

(c) a redirection of agricultural policy towards the inclusion of environmental considerations should enhance the correlation between policy changes and the control of environmental impacts. There are, however, unresolved problems in using such mechanisms as cross-compliance. One problem is the attempt to strive for at least two separate policy objectives with the same instrument;

(d) the integrated analysis of environmental and economic impacts at the sector level still is in progress. Uncertainty remains regarding the extent to which the integration can proceed due to the constraint imposed by a trade-off between reliability and efficiency of the models and their requirements for complex information in order to be accurate.

The environmental impacts of the CAP mainly result from shifts in land-use and the application of input factors. It is nevertheless difficult to separate CAP from other policy and market impacts incurred when interpreting past changes in the agricultural sector and to assess the environmental impacts from changes in the structure of the agricultural sector. The same caveats apply regarding the delineation of future impacts on the agricultural sector and the environment from specific changes in both CAP and national agricultural policy. In this respect, forthcoming policy reforms are anyhow assumed to be assessed in a more qualified manner than was the case, for example, with the 1992 reform. This is mainly due to recent advances in agricultural sector modelling and the prospects for an integration of the environmental consequences. In the 1992 reform the environmental impacts were considered a by-line, while future reforms will include the environment among the headlines.

References [1]

DIAFE (1997) *Om økonomiske konsekvenser af en opstramning af harmonikravene i forbindelse med implementering af EU's nitratdirektiv i Danmark (On the economic consequences of the tightening of the harmonization requirement following the implementation of the EU nitrate directive in Denmark);* Notat, 7 pp. (upublished)

Hansen, J. (1995) *Udviklingen i produktivitet og bytteforhold i dansk landbrug 1980/81-1992/93. (Productivity and terms of trade in Danish agriculture 1980/81 - 1992/93);* Statens Jordbrugs- og Fiskeriøkonomiske Institut, arbejdsrapport, unpublished, 48 pp.

Hansen, J., S.E. Frandsen and P. Steffensen (1997) *Forbedringsstøtten for jordbrugsbedrifter - jordbrugs- og samfundsøkonomiske konsekvenser* (Farm Development Programme - economic impacts); Statens Jordbrugs- og Fiskeriøkonomiske Institut, Rapport, nr. 93, 55 pp., English summary

Jensen, J.D. (1996) *An applied econometric sector model for Danish agriculture (ESMERALDA).* Statens Jordbrugs- og Fiskeriøkonomiske Institut, Rapport nr. 90, 121 pp.

Landbrugs- og Fiskeriministeriet (1996) *Natur, miljø og EUs landbrugspolitik* (Nature, environment and EU common agricultural policy), Betænkning, nr. 1309, 97 pp. + app.

Miljøstyrelsen (1996) *Landbrugspolitik og miljøregulering - 2 delrapport.* (Agricultural policy and environmental regulation), Miljø- og Energiministeriet, Miljøstyrelsen, Miljøprojekt nr. 321, 68 pp.

Ministry of Food, Agriculture and Fisheries (1997) Rapport vedrørende den danske kvægsektors udviklingsmuligheder. (Report on the prospects of the development of the Danish cattle sector), Unpublished report, 201 pp. + appendices

Ministry of Agriculture and Fisheries (1996) *Nature, environment and EU common agricultural policy;* Report from a Committee, Translated summary in English of Report No.1309

---

[1]   An english title in (...) indicates that the reference originally is in Danish.

Olsen, P., J.S. Schou and H. Vetter (1996) *Sustainable strategies in Danish agriculture - A modelling approach;* In: Walter-Jørgensen, A. and S. Pilegaard (eds.); Integrated Environmental and Economic Analyses in Agriculture, Danish Institute of Agricultural and Fisheries Economics. Rapport nr. 89

Rasmussen, S. (1996) *En analyse af strukturudviklingen i dansk landbrug og en fremskrivning til år 2003. (An analysis of the structure in Danish agriculture and prospect towards year 2003);* 'Sustainable strategies in agriculture'; Working paper, nr. 10, 30 pp.

Skop, E. and J.S. Schou (1996) *Calculating the economic and environmental effects of agricultural production;* 'Sustainable strategies in agriculture'; Working paper, nr. 11, 29 pp.

PART 3

MEDITERRANEAN AGRICULTURE

9. SPAIN

Consuelo Varela-Ortega and José M. Sumpsi

9.1 Introduction

In Spain, concern with environmental issues is recent compared to the northern countries in the EU. Although increasing attention is being devoted to these issues, there is still limited information and what there is is generally dispersed. The assessment of the effects of the CAP programmes on the environment in Spain is a difficult task given the great variety and spatial diversification of agricultural systems, cropping patterns, natural areas and ecosystems. So far, very few studies have addressed the links between agricultural policy programmes and their impacts on the environment.

This chapter is based on an exhaustive survey of the available documents. The chapter builds in part on Sumpsi et al. (1996).The first section presents the main environmental problems encountered in the agricultural sector in Spain emphasizing the different perspectives under which these problems are being analyzed. The second section provides an analysis of the effects of CAP measures on the environment. The third section briefly examines relevant environmental conditions in national policies.

Productive sectors versus ecosystems. A different approach to agri-environmental issues

A given ecosystem may extend over an array of crops and livestock and therefore the correspondence between commodity sector analysis and ecosystems analysis is not straightforward. Nevertheless, some penetrating studies have analysed Spanish ecosystems in the framework of the CAP programmes.

Valuable analyses have been made on steppe farming practices and their effects on bird species, including the impacts of intensification processes in cereal lands in the central plateau (Diaz et al., 1993), pseudo-steppes biodiversity and threatened species such as the great bustard (Suárez et al., 1997b) and the relationship between the evolution of extensive grazing systems and the conservation of natural habitats for raptor species (Donázar et al., 1997). Policy options are analyzed in these studies and important considerations are set forth as to whether the roles of the 1992 reformed CAP and the agri-environmental accompanying measures (Regulation 2078/92) have been sufficiently articulated to achieve the desired objective of resource and nature conservation. It is concluded that there is still a lack of integration of these complementary measures.

Traditional steppe farming systems in Spain have also been analyzed from a more general perspective and conservation measures have been proposed to preserve these lands and their natural value (Suárez et al., 1996). The most important conservation problems that affect the Iberian steppes originate from the reduction of their natural extent in response to policy programmes. These programmes are afforestation under EC Regulation 2079/92, development of large-scale infrastructure, and expansion of plantations, mainly olive groves. Policy driven intensification processes, also reported to be responsible for the reduction and loss of natural steppe, include new irrigation projects, reduction in fallow lands and farm size increase in response to structural modernization policies (Suárez et al., 1997a). Proper grazing management, national and regional conservation laws and monitoring of bird species and natural vegetation are proposed as a means to preserve these natural habitats.

Integrated ecosystems analysis has also been applied extensively to the study of the dehesas in the Iberian peninsula, especially in the region of Extremadura in south western Spain and in Andalucia in southern Spain (Campos, 1995a and 1995b; Beaufoy, 1996; Diaz et al., 1997a). The dehesas have been affected by land use changes oriented to short term profitability in competition with irrigation agriculture and forest crops, by increases in livestock density as a result of policy programmes, by the abandonment of tree plantation management and by the abandonment of traditional tillage practices (Diaz et al., 1997b).

Only specific productive sectors, such as cereals, olives and fruit and vegetables have been the objective of specific research that analyzes conjointly the economic and environmental effects of some of the CAP programmes. In other cases more attention has been given to the structural evolution of the CMOs and the related impacts of the extensification or intensification of farming practices. The environmental implications of the application of the accompanying measures of the CAP reform and the structural programmes have been analysed from an integrated approach considering the effects on related ecosystems (e.g. wetland recovery from aquifer depletion) or on specific flora and fauna (e.g. steppe bird preservation in extensive cereal areas). The following section therefore analyses the main environmental problems of Spain in relation to the evolution of the agricultural sector. However, it is generally maintained that the environmental effects of a given policy programme are mainly long term outcomes, and it is still too soon after the CAP reform for a definite assessment of the cause-effect relationships involved.

9.1.1   Features of agriculture

Agricultural modernization in Spain, which started at the beginning of the sixties, gave way to a process of intensification in the best farming areas together with a steady abandonment of increasingly large zones of marginal agriculture. Both processes had a series of negative impacts on the environment (erosion, destruction of habitats, loss of biodiversity, increase in fire incidents in the forest areas, diffuse agricultural pollution, etc.). Later on, Spain's entry into the EEC in 1986 and the application of the European price and market policies reinforced those processes, aggravating the environmental damages.

In-depth analyses on linkages between agricultural policy and the environment remain limited so far. Nevertheless, there are some partial studies that cover specific processes going on in certain zones. Most of these studies come to the conclusion that the process of agricultural modernization in Spain reinforced by the application of the EC policy on prices and markets has produced a dual evolution of Spanish agriculture: on the one hand, the intensification of farming in highly productive zones and on the other hand the abandonment of marginal lands with low productive capacity (Ruiz and Groone, 1986; Baldock and Long, 1988; Sancho Comins et al., 1994; Fernández-Guillen and Jongman, 1994; Vera and Romero, 1994; Egdell, 1993; Baldock et al., 1996).

This dual development is reflected in land use and the distibution of agricultural production across crop types. In Table 1 we can see that of a total surface of 50 million ha in Spain, 42 million is classified as agricultural land of which almost one half is arable land, 38% forest and 15% pasture. Irrigation water is a basic resource for Spanish agriculture and its use marks clearly the differences between farming systems as can be seen in Table 2. Changes during the last few years reflect the impact of the long lasting drought (1992-95) and the effects of the compensatory direct payments under the reformed CAP which have resulted in a distinct farm extensification. A clear decrease in irrigated surface has been observed during this period along with a shift towards less water demanding crops. Farm income reductions as a result of the drought were compensated by the application of the CAP programmes to extensive arable crops and to beef and sheep production as well as by specific national aid programmes.

Table 1   Land use in 1993 (1,000 ha) a)

	Agricultural land			Other land	Total land
arable land	grassland and pasture	forest	total		
19,657	6,494	16,137	42,288	8,192	50,479

a) Last year before the change in the methodology to calculate the surface base.
Source: M.A.P.A. Anuario de Estadistica Agraria 1994 (1996a).

Table 2   Land use by crop type (1,000 ha)

Type of crop	Dry land			Irrigated land			Total		
	1992	1993	1994a)	1992	1993	1994	1992	1993	1994
Herbaceous	8,962	8,542	8,080	2,203	2,198	2,183	11,165	10,740	10,263
Tree plantations	4,000	3,927	3,935	746	749	754	4,746	4,676	4,689
Fallow	3,777	3,949	3,314	258	292	188	4,035	4,241	3,502
Total	16,739	16,418	15,329	3,207	3,239	3,125	19,946	19,657	18,454

a) Variations due to the change in the methodology for the base area calculations. (Anuario de Estadistica Agraria, MAPA, 1994, p.26).
Source: MAPA anuario de Estadistica Agraria 1994 (1996).

Crop distribution in Spain shows the importance of horticulture and fruit products which extend over less than 2% and 5% of the surface respectively but account for circa 30% of total agricultural production (including flowers). The importance of water resources in Spanish agriculture and the high value added of these crops are behind these figures. In fact, it is estimated that irri-

Table 3   Distribution of land surface and agricultural production in 1994

	Surface (ha)	Share of total cultivated land (%)	Production (mill. pta)	Production (mill. $)	Share of total agricultural production (%)
Cereals	6,489,569	24.9	255,303.9	1,906.1	6.85
Leguminous	354,273	1.4	16,164.6	120.7	0.43
Horticulture	429,753	1.7	556,479.7	4,154.7	14.93
Tuber crops	202,902	0.8	193,714.7	1,446.3	5.20
Industrial crops	1,690,302	6.5	80,692.8	602.5	2.16
Citrus	268,175	1.0	212,518.5	1,586.7	5.70
Other fruit trees	943,872	3.6	267,515.6	1,997.3	7.18
Vine	1,235,397	4.7	137,228.0	1,024.5	3.68
Olive	2,177,333	8.4	236,581.8	1,766.3	6.35
Other tree plantations	91,590	0.4			
Forage	1,202,677	4.6	157,408.2	1,175.6	4.22
Flowers	184,475	0.7			
Other crops	3,183,600	12.2			
Grassland and pastures	7,571,600	29.1			
Animal production			1,583,021.7	11,822.4	42.47
New plantations			30,953.0	231.2	0.83
Total	26,025,700	100.0	3,727,582.5	27,834.2	100.00

Source: MAPA Anuario de Estadística Agraria 1994 (1996) and own calculations.

Table 4   land distribution and environmental impact according to development trend

Development process	Surface (Million ha)	Environmental impact
Intensification process (11 mln ha)		
- Highly productive land (5 mln ha)		high
- Irrigated agriculture	3.5	
- Fertile dry agriculture	1.5	
- Extensive farming in non-marginal land (6 mln ha)		medium
- Cereal	3	
- Vineyard, olive	1	
- Dehesas	1	
- Pastures	1	
Land abandonment process (12 mln ha)		
- Extensive farming in marginal land		high
- Cereal	6	
- Marginal pastures	4	
- Pluriannual dry crops (vineyard, olive, almonds)	2	

gated agriculture represents a mere 15% of cultivated land but it generates 60% of total agricultural production and 80% of all farm exports 1).

The dual evolution of agriculture is also central to the environmental effects of Spanish agricultural development. Table 4 presents the distribution of land according to the two ongoing opposing forces, of intensification and abandonment.

9.1.2   Major environmental issues

Judgments about the considerable impact on the environment of agricultural policy measures are based on two considerations. The first is that extensive agricultural systems in Spain have lasted longer and up to more recent times than in the rest of Europe. In general, Spain's agriculture has been preserved from many of the intensification trends of the northern EU countries, resulting in lower yields and more traditional forms of farming (Baldock and Long, 1988). The second is that the rich variety of ecosystems and natural habitats in Spain is far greater than in the rest of Europe, as a consequence of its distinct climatic and topographic characteristics (Baldock and Long, 1988; Sancho Comins et al., 1994; Fernández-Guillén and Jongman, 1994; Vera and Romero, 1994; Egdell, 1993, MAPA, 1995). This richness relies on the relative importance of extensive agriculture, such as the dehesas in South western Spain, the mountain pastures in the North and dryland farming which extends across most of the Castillian central plateau. Moreover, several authors have stressed the vulnerability of Spain's natural heritage and resources as a consequence of the modernization process, which started in the sixties and was reinforced by the application of the CAP (Valladares, 1993; Baldock and Long, 1988).

According to the two opposite evolutionary trends that characterize Spanish agriculture, we can summarize the main environmental impacts (Table 5).

Table 5   Environmental impacts in agriculture according to the development process

Development process	Environmental impacts
Intensification process	Erosion Loss of biodiversity Pollution from intensive farming (dry and irrigated) of surface water and groundwater (nitrates; pesticides; salinization; and over-exploitation of aquifers) Landscape deterioration Pollution from intensive animal production
Land abandonment process	Erosion Fires Loss of biodiversity Landscape deterioration

---

1)   Calculation based on MAPA (1996a): Annuario de Estadistica Agraria 1994 and MAPA (1996b): Plan Nacional de regadios.

Erosion and loss of biodiversity are outcomes of both intensification and abandonment. More than half of the cultivated land extends along arid or semi-arid regions where extensive as well as intensive farming systems are developed depending on altitude, temperature range, rainfall distribution, frequency of torrential rains and availability of water for irrigation. In this geographical context, erosion constitutes a very severe environmental problem. Firstly, erosion is a very specific environmental problem of the Mediterranean countries and a matter of international concern (United Nations Organization - LUCDEME project, 1993). This particular aspect will have to be taken into consideration for policy analysis and policy evaluation at EU level (to ensure a balanced and integrated compromise between the environmental problems of northern and southern EU countries). Secondly, erosion in Spain affects both intensive and extensive agriculture, although the effects are more severe in the extensive lands with high risk of abandonment. Thirdly, erosion affects largely arable land and thus exemplifies how the CAP can affect the environment and the practical and policy responses that have followed. Erosion vulnerable areas of extensive agriculture extend over 12 million ha, mainly situated in Spain's central and southern inland areas and on hillside cropland and plantations. In contrast, erosion vulnerable intensive farming covers only 1 million ha and, strongly affected by unevenly distributed torrential rainfall, is mainly situated along the Mediterranean and south Atlantic coastline where horticulture crops and fruit tree plantations are grown on heavily irrigated farms.

9.1.2.1 Intensification of production

Most of the general studies conclude that agricultural intensification has induced considerable environmental problems. It has been concentrated in certain zones of high productive potential, but has also affected other zones with more limited productive capacity where extensive traditional farming systems have played a very important conservation role.

i)   *Intensification of traditional farming systems*

Three illustrative cases of the intensification of traditional farming systems are as follows:

a)   The intensification of pasture zones and prairies has been directed towards the improvement of pasture land (input use increase), or to reforestation with species of rapid growth. These changes have produced and continue to produce negative environmental impacts (Ruiz and Groone, 1986; Baldock and Long, 1988; Fernández-Guillén and Jongman, 1994).

b)   More than six million ha. in the Western part of Spain are covered with dehesas, which constitute a balanced ecosystem of great ecological, social and cultural importance (Sancho Comins et al., 1994; Fernández-Guillén and Jongman, 1994). A triple exploitation of livestock, agricultural and tree resources is achieved keeping a sustainable integrated system in which farm production coexists with the preservation of natural resources. In response to policy programmes, the dehesa ecosystem has been progressively destroyed and in the past thirty years, about two million hectares have been transformed into arable land for cereal and sunflower cultivation or irrigated crops (Campos, 1993a; 1993b; 1994; Pulido and Escribano, 1994; San Miguel, 1995; Beaufoy, 1996, 1997b). Many endangered species such as lynx, wolf, eagle, vulture and black stork have suffered severe population falls as well as millions of migratory birds coming from Northern Europe that overwinter in the Western Iberian dehesas. The European Commission (through its DG VI-Agriculture), has been sponsoring the initiative of the European Natural Heritage Fund (Fondo Patrimonio Natural Europeo, 1996) to demonstrate the interest and feasibility of recuperating Iberian dehesas.

c) The intensification of extensive cereal crops in the Spanish central plateau (Castilla-León, Castilla-La Mancha, Aragón and Extremadura) has occured in response to high EU prices (Baldock and Long, 1988; Fernández-Guillén and Jongman, 1994; Suárez et al., 1997a). The environmental impacts of a significant conversion of fallow land to agricultural use is difficult to evaluate but the vulnerability of these soils to erosion impacts is very high (Baldock and Long, 1988; Groves-Raines, 1992). The consequences for nature conservation, particularly the impact on bird populations, have been covered in various studies by SEO-Bird Life (Martin-Novella et al., 1993; Naveso, 1992a; Naveso, 1992b; Naveso and Groves-Raines 1992; Sears, 1991; Naveso, 1993). These studies have been used specifically to prepare zonal programmes of support under the agri-environmental Regulation 2078/92.

ii)   Irrigation transformations

During recent years, the increase in irrigated agriculture has been considerable due to the application of structural fund programmes (Sumpsi, 1994b). The transformation from dry land agriculture to irrigated agriculture has one of the most negative impacts on the environment (Baldock and Long, 1988; Sancho Comins et al., 1994; Lopez Bermudez, 1994; Perez Ibarra, 1994; Cruz Villalon, 1994). The Spanish Ministry of the Environment has evaluated for each river basin, the main environmental problems of the irrigated areas in Spain, namely pollution, salinization and over-exploitation of aquifers and wetland destruction (MOPTMA, 1996).

There are several studies that evaluate the environmental impact of certain irrigation zones of intensive cultivation. Among these studies, the most important ones are those related to the irrigated agriculture of the 'La Mancha' plains in the Spanish south central plateau where more than 100,000 ha were put into irrigation by drilling wells during the period 1975-1988. Based on private funding this process was encouraged by the CAP favorable price conditions for crops such as corn, a basic crop in this area, and resulted in over-exploitation of aquifers and increased pollution. This situation has provoked severe water supply problems for the nearby population, as well as the drying of natural wet areas such as the National Park of Tablas de Daimiel (MOPU, 1983; MOPU, 1988; Ocaña et al., 1981; López Camacho, 1987; Garcia Rodriguez and Llamas, 1993; Sancho Comins et al., 1994; Llamas, 1995).

The same situation of aquifer over-exploitation can be found in the strawberry fields of Huelva, the greenhouse and plastic covered plantations of Almeria (region of Andalucía) (González, 1992; Naredo et al., 1993) and in the region of Valencia. In this area of intensively irrigated fruit and vegetables, water comes predominantly or solely from groundwater sources. Pollution, salinization and over-extraction in the aquifers create a severe environmental problem (Sanchis Moll, 1991). Inefficient water management has also been reported to cause important environmental effects, such as salinity problems (Astorquiza, 1994) and nitrate pollution (Fernández-Santos et al., 1992).

iii)   Environmental effects of intensification

The most important environmental problems caused by intensification of agricultural practices are erosion, fires, loss of biodiversity, landscape damage, pollution of surface and ground waters by nitrates, pesticides or slurries, and over-exploitation and salinization of aquifers.

Erosion

Erosion constitutes one of the most severe environmental problems of Spanish agriculture. Spain has been classified by the United Nations as the European country with the highest risk of desertification due to soil erosion. Climatic and geological conditions are only partially responsible, and inappropriate agricultural practices in fact account for the acceleration of erosion in Spain (Diaz Alvarez and Almorox, 1994). Several policy programmes have contributed to the intensification of agriculture, which subsequently have increased erosion due to deforestation, farming of uncultivated land, burning of pastures and straw mulch, overgrazing, intensive tillage operations, use of heavy and overpowered machinery, excessive use of agrochemicals and destruction of natural vegetation cover.

According to official estimates, 18% of the Spanish territory suffers from high erosion and 26% from medium erosion. Thus, medium to extreme erosion affects almost half of the Spanish territory, of which 9 million ha is considered to be under high to extreme erosion impact (annual soil loss of more than 50 tonnes per ha) (Table 6) (ICONA, 1991; LUCDEME, 1993; Rubio and San Roque, 1990). The World Bank has judged erosion in Spain to be very severe. Erosion is concentrated basically in the south-eastern part of the country (regions of Andalucía, Murcia, Castilla-La Mancha and Valencia) where 70% of the Spanish territory suffering from severe erosion problems is located (MOPU, 1989).

Some specific CAP measures have contributed to an increase of erosion. One example is the subsidy given to vineyard uprooting which leaves the uncovered soils more prone to erosion pressures, especially in hilly areas with steep slopes. During the period 1986-1994, about 90,000 ha of vineyards were uprooted. Table 7 shows the distribution of erosion by crop type. Highest

Table 6   Degrees of erosion (annual soil loss, in tonnes/ha/annum)

Erosion degree	Ha	%
Extreme (>200)	1,111,551	2.2
Very high (100-200)	2,561,426	5.1
High (50-100)	5,488,460	10.9
Medium (12-50)	12,922,872	25.6
Low (5-12)	17,308,701	34.2
Very Low (<6)	11,151,334	22.0
Total	50,544,344	100.0

Source: ICONA, 1991.

erosion rates can be found in dryland tree plantations (more than 20 times the erosion impact of forest areas) such as vineyards, olive groves, dry fruit plantations in hill areas (inland zones of the south eastern part of Spain), followed by dryland arable crops.

Table 7   Erosion versus land uses

Land uses	Soil loss (tonnes/ha/annum)	Surface (%)	Erosion (%)
Forest trees cc >0.7 a)	0.216	9.12	1.43
Permanent pastures	0.446	4.3	1.39
Forest trees cc 0.2-0.7	0.452	11.95	3.92
Irrigated crops	0.615	6.04	2.70
Bushes and scrub	0.866	9.9	6.23
Unfarmed lands and scrub	1.006	16.36	11.96
Dry land (arable crops)	1.724	29.88	37.43
Dry land (trees)	4.772	10.07	34.94

a) cc: forest density index (cc: continuous cover; cc >0.7 means that more than 70 % of the forest has a vegetal continuous cover formed by the tree tops).
Source: Diaz Alvarez and Almorox (1994), Soto (1990).

Forest fires

Intensive reforestation with rapid growth species (forestry crops) has contributed to a large increase of fire damage in the forestry zones, since those trees species are particularly combustible (ADENA-WWF, 1994a; Vélez Muñoz, 1991). Poor forestry management, the invasion of scrub and the decrease in extensive animal production are other aspects associated with intensification that have extended fire damage (Valladares, 1993; Vélez Muñoz, 1991).

Loss of biodiversity

Intensive agricultural systems, based on a limited number of species and non-indigenous varieties that are economically profitable, lead to a great loss in biodiversity (Gómez Sal, 1994; Fernández-Guillén and Jongman, 1994). The increasing intensification of agriculture has negatively affected areas which used to be preserved as semi-natural zones including pastures and grasslands, such as the dehesas and the cereal production areas in the steppe regions of central Spain. It has been estimated that in the past 30 years, 60% of the traditional wetlands in Spain have disappeared with the subsequent loss of the related aquatic habitats (Ortega and Rodríguez, 1990). Also, rapid growth timber has caused the elimination of 4 million ha of native forest. The application of the CAP programmes has been considered one of the most serious causes of the degradation of numerous natural areas (Valladares, 1993).

Landscape deterioration

In some cases the CAP has contributed to the deterioration of the landscape also. The transformation of land uses in Spain has resulted in great losses to traditional rural landscapes. There has been a general homogenisation of the landscape, whether through the disappearance of the traditional dehesas (Fernández-Guillén and Jongman, 1994) or the uprooting of vineyards (Sancho Comins et al., 1993). Loss of vegetational cover, the farming of uncultivated land, the increase in erosion and of forests damaged by fire, scrub invasion, monocrop fields of corn in the new irrigated lands of La Mancha plains (Castilla La Mancha), or the 'plastic seas' of Almería (Andalucía) are some examples of the changes which have taken place to the traditional landscapes.

Pollution

The great increase in the use of fertilizers and agrochemicals has led to severe water pollution problems. In general, the quality of surface waters in Spain is satisfactory (De León, 1990) and they constitute the predominant water source that irrigates 70% of all irrigated land. Concerning groundwaters, the situation is quite different. The main problems are pollution and over-exploitation of some aquifers linked to poor management of irrigation water and pollution caused by excessive use and misuse of nitrogen fertilizers. Nitrate pollution in groundwater is concentrated in intensively irrigated zones. The cost of groundwater supplies for irrigated agriculture is very high due to the high energy cost of water pumping from deep wells. For that reason most of those irrigated areas are highly intensive, like the coastline of Huelva (Andalucía), Barcelona (Catalonia) and Almería (Andalucía), the plains of Castilla-La Mancha, and the coastal areas of Murcia and Valencia. It has been reported that in the six river basins in which pollution problems are severe, agricultural activity has a clear responsibility, and is generally related to intensive irrigation practices (De León, 1990).

Nitrate pollution is a severe problem but is localized in very specific zones, especially in heavily irrigated areas from groundwater sources, covering no more than 1 million ha (MINER and MOPTMA, 1994). In recent years, the nitrate content in groundwaters has increased at an alarming rate in several zones of intensive agriculture in which the concentration of nitrate repeatedly exceeds the permitted 100 mg/l. The Mediterranean coastline, specially the zone of Maresme-Llobregat (Cataluña), the coastal plains of the region of Valencia and Mallorca (Balear Islands), the plains of La Mancha (Castilla-La Mancha) and the fluvial valleys of the Guadalquivir river (Andalucía) are the regions in which the pollution process has reached a greater intensity, affecting sources of urban water supply and producing problems of water quality.

Pesticide pollution. Contamination by pesticides in Spain is mainly concentrated in the heavily irrigated areas of parts of the regions of La Mancha in the southern central plateau, Andalucía in the south (provinces of Huelva and Almería in the southern Atlantic coast), and Valencia and Murcia along the Mediterranean coastline where high value added crops of fruit and vegetables are grown in open and greenhouse plantations for the domestic and export markets. Irrigation water comes in these areas mainly from groundwater sources (MINER and MOPTMA, 1994). Vulnerable aquifers are concentrated in those areas where use of pesticides is largely above national averages. Residues of pesticides in groundwater have been reported in the aquifers of the river basins of Tajo (region of Castille-La Mancha), Guadalquivir (region of Andalucia) and Ebro (region of Aragón) as well as in the sub-regions of La Mancha (region of Castille-La Mancha) and La Plana in the province of Castellón (region of Valencia). Concentration of these pollutants may exceed the maximum levels permitted by the Spanish Health Regulations but these water sources are not used for urban consumption (MINER, 1992).

Presently, pesticide contamination in intensively irrigated areas is not considered a primary environmental problem in Spain. Limited as it is in its geographical scope, reliable information can be found in the relevant regional institutions 1). Climatic and geophysical conditions largely determine the low impact of pesticides in these areas in spite of significantly higher levels of pesticide consumption as compared to other regions. These zones are characterized by dry and hot climatic conditions, non saturated soils and very deep water pumping levels of 100 to 400 m. This contributes to a rapid degradation of pesticides and a deep water table not easily reached by pollutants (Gomez de Barreda et al., 1997). For these reasons, pesticide contamination in heavily irrigated intensive agriculture in Spain, especially in citrus groves, has not been catalogued as non-point source pollution and it has not been reported to cause environmental damages to surface water or ecosystems. Conversely, pesticides may be a source of point pollution when residues are sometimes observed at low levels in irrigation wells, following episodes of torrential rain that generate soil preferential pathways that impel pesticide accumulation. (Gomez de Barreda et al., 1997). Pesticide contamination in protected or greenhouse horticulture crops presents particular features that have to be further evaluated (Varela-Ortega, 1997).

Over-exploitation and salinization of aquifers

Salinization of coastal aquifers comes about through the ingress of sea water, as a result of the drop in the groundwater level due to over-exploitation of the aquifers (Llamas, 1995). Out of the 82 aquifers in the coastal areas of the Peninsula and Balearic Islands, 58% have some degree of ocean water inflow caused by over-exploitation. To a great extent, intensive agriculture is responsible for the over-exploitation of the aquifers, since it uses the majority of the available water resources in these areas. Excessive water use for irrigation along with the uncontrolled expansion of irrigation have aggravated the problem. The CAP has encouraged these developments in two ways: on the one hand, through the regulation of prices and markets that favour, through high prices, the expansion of certain crops (corn, sugar beet, oil seeds etc.); and on the other hand, by subsidizing the private development of irrigation through structural programmes.

Pollution from intensive animal production

The problem of excess manure in intensive livestock farms reaches severe levels in the case of the pig sector where the slurry causes important pollution impacts in the nearby waters and bad odours. However this problem is limited to very specific sites in some areas of the Catalonia region (Carballo, 1996).

---

1)   IVIA (Instituto Valenciano de Investigaciones Agrarias). Servicio de protección de los vegetales. Consejería de Agricultura de la Communidad Valenciana.

9.1.2.2   Marginalization and abandonment of production

In the European context, the Mediterranean region accounts for the greatest variety of extensive farming systems including permanent crops (olive and almond trees, holm-oaks, cork trees etc.), annual crops in semi-arid zones, traditional mixed agriculture, and different animal production systems based on the use of grazing land, pastures and traditional farming practices (Beaufoy et al., 1994). In Spain the importance of extensive agricultural systems is very high and, according to reliable estimates, it extends over 25 million ha of a total of 42 million of agricultural land (see Table 1) (ADENA-WWF, 1994b).

The CAP, at least before its reform in 1992, created incentives to intensify agricultural production. Parts of the traditional extensive systems have been gradually transformed into more intensive farming while the rest of the land is threatenend by abandonment. Approximately 60% of the total Spanish surface is classified as mountainous or less-favored areas and is therefore unsuitable for agricultural intensification. Many of these lands suffer from progressive desertification (Ruiz and Groone, 1986). The likely prospect is a dramatic increase in land marginalization in Spain as a response to low productivity (cereal crops and extensive livestock farming), market surplus, production quotas, low demand (olive oil), or a combination of these factors (vineyards) together with the difficulty that many farmers have to adjust to the cost reducing trend of the CAP programmes (Ruiz, 1985).

i) Environmental effects of land abandonment

The main environmental effects of land abandonment are erosion, forest fires, loss of biodiversity and landscape deterioration. However, there has been no systematic evaluation of the overall damage that land abandonment may cause to the environment. There are some scientific studies of specific areas (Llorens and Gallart, 1992), including an informative analysis of the impact of land abandonment on the Spanish central plateau (González Bernáldez, 1991).

Erosion 1)

In Spain, abandoned fields constitute a serious threat to the environment, and have been shaping a characteristic landscape in many areas, like the middle mountains of the Central Pyrenees. The abandonment of agricultural land reduces the vegetation cover, provoking an increase in erosion (González Bernáldez, 1991; Llorens et al., 1992). The abandonment of terrace cultivation in hill areas (abandonment of vineyards, olive groves and almond trees plantations in marginal zones) is another cause of the increase in erosion (Díaz Alvarez and Almorox, 1994). In some of these areas a dense shrub cover has been developing but other areas have been affected by sharp sheet wash erosion that has resulted in completely eroded soils, paved with stone flows (García Ruiz

---

1)   As erision control is heavily herbicide dependent, this part has been extracted partially from: Varela Ortega, C. (1997). Situation and possibilities in Spain for the assessment of additional policy instruments for plant protection products. In: Oskam, A.J. et al. (eds.). Final report for the EU project: Additional EU policy instruments for plant protection products, 2nd phase PES-A programme. DG XI. EU Commission, Brussels.

et al., 1991). Natural processes of plant growth have to be preserved from fires or overgrazing to allow the development of a vegetation cover. In this respect, overgrazing causes severe damaging effects on soils and increases erosion impacts. Human control and management of abandoned pasture lands determines largely the amount of the negative effects caused on the environment in mountain areas (García Ruiz et al., 1991). In Mediterranean arid zones, desertification arises from mismanagement of soils through over-exploitation of fragile ecosystems, over-mechanization and poor farming practices (Rubio and Herrero, 1995; Arnal, 1996).

Conservation tillage practices (minimum tillage or no tillage farming) have increased substantially in extensive agriculture during the past 15 years, which has contributed to reduced erosion impacts and soil loss (Cantero, 1995; Arnal, 1996; Herruzo, 1996). Consequently, herbicide use for weed control has spread. The herbicide dependence of this type of farming is high as production may decrease by almost one third otherwise (Oerke et al., 1994; García Torres, 1996). Conservation tillage practices have been promoted by public, private and research institutions and have been adopted increasingly by farmers based on cost reduction and crop yield sustenance (Hernánz et al., 1995). Cost reduction rates depend on farm productive structure, and fixed versus variable input use - the highest levels of cost reduction are found in farms relying heavily on off-farm contracting of labour and machinery. In general, the increase in herbicide costs are completely overshadowed by reductions in labour, machinery and energy use (Herruzo, 1996). In extensive cereal and legume cultivation areas (a large proportion of the 8 million ha of dry farming herbaceous crops), overall production costs are reported to decrease by 20-25% (basically labour and machinery cost reductions) when conservation tillage is adopted (Hernánz et al., 1995).

Olive tree plantations in dry farming systems extend over 2 million ha and, being largely sited on hillsides, are severely affected by erosion impacts. Among erosion control measures, barley and vetch are grown to develop a soil protection canopy in the sloping plantations, and herbicide applications have been reported to produce clear advantages when conservation tillage is used in olive groves.

Fires

The abandonment of agriculture and forestry induces shrub invasion in forest areas, which in turn increases the vulnerability to forest fires (ADENA-WWF, 1994a). Also the decrease in extensive animal production results in lower shrub control and increasing fire risks (ADENA-WWF, 1994a; González Bernáldez, 1991). Moreover, an expansion of fires causes in turn an increase in erosion (Valladares, 1993; Díaz Alvarez and Almorox, 1994).

Loss of biodiversity

The abandonment of traditional crops and pastures for livestock production has resulted in the reduction of certain plant and animal species with the subsequent loss in biodiversity (González Bernáldez, 1991). Part of the traditional flora and fauna is destroyed by the abandonment of pasture land in high mountain areas due to the decline in extensive livestock production.

Landscape deterioration

Land abandonment can severely damage the landscape through the invasion of scrub, reduction in the vegetative canopy, loss of terrace cultivation, fires, etc. (Sancho Comins et al., 1994). Reducing land abandonment and maintaining extensive agricultural systems are seen as the key factors in preserving the natural landscape (Fernández-Guillén and Jongman, 1994; Morey, 1995). However, preserving extensive agricultural systems as encouraged in the CAP reform, is not enough. It is necessary to identify the farming practices that will produce the highest conservation levels. Research on the interaction between extensive farming practices and nature conservation, though, remains limited (ADENA-WWF, 1994b).

9.2 Impacts of the CAP

9.2.1 Common Market Organizations (CMOs)

Very few sector studies are available with a joint focus on the CAP and the environment. Most studies concentrate on the effects of the CMOs on the agricultural sector without considering the consequences for the environment. In the absence of the appropriate scientific investigations, in some instances, the interconnections between the two processes, the CAP and the environment, can be pieced together by drawing on the opinions and observations of experienced officials and experts. Nevertheless, the environmental consequences of the application of the CMOs are not at the centre of policy concern in Spain, neither in the Central and Regional Administrations nor among most agricultural economists. Lack of understanding of environmental issues in relation to policy programmes, the need for a more regionalized perspective, the necessity of a better integration of policy measures and a more active participation of all actors involved are probably responsible for this situation.

9.2.1.1 The arable crop regime

One of the objectives of the CAP reform has been the extensification of agricultural systems with the aim of reducing surplus production and protecting the environment and the natural resource base. Extensification is however to be attained by means of the two complementary activities, land set aside and cultivated land. In this context, the main questions that arise concerning the arable crops regime can be summarized as follows:

i) Set aside

In the cereal sector one major environmental impact of the CAP programmes is related to the application of the set-aside requirements and the traditional fallow practices. Their impact on soil erosion is one of the most severe environmental problems that affect Spanish agriculture. Nevertheless it is difficult to conclude a clear effect of the reform of the cereal regime on the environment although it is possible to argue that there has been a certain extensification trend since the

CAP reform was applied. Table 8 shows the annual distribution of subsidized surface in dry lands from 1993 to 1997. The set-aside surface has increased from 700,000 ha in 1993 to 1,226,000 ha in 1996, decreasing slightly in 1997. Looking at the distribution of types of set aside one can observe that mandatory set aside showed a steady decrease since 1993 while voluntary set aside increased proportionally so that total set aside remained stable during this period at a level of slightly more than 1 million ha. The area of crops under price support measures increased after 1993 and has remained stable during the past four years at a level of close to 8 million ha of dry land crops.

Table 8   Set aside and crop distribution in dry land (ha)

Year	Set aside			Cereals	Oilseeds	Protein crops	Other crops	Total
	mandatory	voluntary	total					
1993			736,104	4,811,965	1,539,285	19,856	60,200	7,167,410
1994	977,345	248,313	1,225,658	5,204,276	1,023,452	70,030	55,738	7,579,154
1995	999,291	261,954	1,261,245	5,596,124	801,923	77,027	55,093	7,791,412
1996	763,031	463,004	1,226,035	5,647,358	883,025	82,138	30,547	7,869,103
1997	440,499	594,518	1,035,017	5,856,420	821,201	59,469	22,939	7,795,046

Source: MAPA (1998) (unpublished report).

Table 9   Set aside and crop distribution in irrigated land (ha)

Year	Set aside			Maize	Cereals	Oilseeds	Protein crops	Other crops	Total
	mandatory	voluntary	total						
1993			122,560	124,228	222,407	570,132	8,928	3,920	1,052,175
1994	148,122	32,547	180,669	223,812	459,717	360,524	42,265	2,290	1,269,277
1995	142,804	61,170	203,974	238,488	522,087	376,797	27,504	3,415	1,372,265
1996	113,457	43,635	157,092	352,420	506,770	351,122	33,636	3,594	1,404,634
1997	67,745	39,522	107,267	440,527	513,010	265,156	26,780	2,078	1,354,818

Source: MAPA (1998) (unpublished report).

With respect to irrigated crops the extent of land diversion is dependent essentially on the availability of water which in turn affects cropping techniques and intensification levels. Table 9 shows that set-aside land for irrigated crops increased substantially during a dry period (1994-96). The area almost doubled in 1995 compared to 1993, which was a year with average rainfall patterns. Therefore it is difficult to evaluate to what extent the extensification practices induced by the CAP have produced environmental benefits without considering the role that the availability of water for irrigation plays in the technical packages chosen by the farmers. In fact, the CAP programmes have been associated with a reduction in the use of pesticides and fertilizers, but also with observed technological changes which have been the result of the long, severe and wide-spread drought between 1993 and 1996. The application of the CAP programmes compensated to a considerable extent the loss in farm income during the drought period. Input use in 1996-97, however, increased substantially in response to the good climatic conditions and the end of water shortages. Therefore a clear link between the application of the CAP programmes in the cereal sector and an increase in farming extensification in these lands cannot be unambiguously established. The role that natural resources, such as water, play in the level of extensification response to policy programmes has to be taken into consideration to evaluate properly the changing technological structure and therefore the environmental effects of such policy programmes (Varela-Ortega, 1998).

ii) Economic and environmental effects

During recent years Spanish agriculture has been affected by two different factors, a long lasting drought and the application of the reformed CAP. Distinguishing their separate effects is not always easy. Thus it is not possible to conclude that the observed reduction in input use was the response to a policy driven extensification process or to production cost reductions in response to lower income expectations from yield decline during the drought period. It has been observed that input use increased substantially when precipitation recovered. Therefore, natural weather conditions are considered to have a substantial effect on farmers' decisions (in terms of technological patterns and crop choice) (Flichman et al., 1995; Varela-Ortega et al., 1995).

Environmental as well as economic impacts in response to different policy scenarios (pre CAP reform and CAP reform) had been studied for the region of Andalucia in Southern Spain (Varela-Ortega et al., 1995; Flichman et al., 1995). Following these studies it has been argued that CAP reform may induce a clear extensification trend and a subsequent reduction in environmental damage (nitrate contamination) as a response to lower input use (fertilizers and plant protection products). However the environmental consequences of the application of policy programmes in water dependent, highly productive, irrigated agriculture deserves a more careful analysis. Indeed the reduction of environmental damage due to the policy driven extensification process does not depend solely on policy programmes but relies to a large extent on water availability. Farmers tend to chose more intensive techniques in humid years and grow more water demanding crops (with the related increase in nitrate pollution); and they tend to choose less intensive techniques and substitute for less water demanding crops in drought periods (and subsequently reduce pollution) (Varela-Ortega, 1998; Sumpsi, 1996b).

Other regional studies have reported some beneficial effects on the environment of the application of the CAP programmes. In steppe cultivation areas of the region of Extremadura, a traditional dehesa area for winter grazing of transhumant sheep herds, the 1992 direct payments stopped the abandonment of arable lands and the 'arable land-fallow land' reference surface requirement prevented the reduction of fallow (Beaufoy, 1996).

9.2.1.2 The dairy regime

Although there are no specific studies in Spain that evaluate the direct impact of the evolution of the milk quota system on the environment, we can assess the indirect impact of the structural evolution of the dairy sector on the level of intensification and extensification that has been observed across regions and producing areas. The application of milk quotas in Spain has limited the expansion of total production that would have been expected otherwise as a response to a steady increase in productivity. The number of livestock units and number of farms have fallen steadily from 1992 to 1997, total production has levelled off and farm income has been basically maintained with respect to other producing sectors (Castillo, 1992; Sumpsi, 1996b). The application of the milk quota system has accelerated the structural adjustment of this sector in Spain, that started from a less developed and less competitive productive structure than other member states. The result has been a controlled restructuring of the milk sector avoiding the uncontrolled intensification trend that could have been expected in a market pricing system. The quota system provided the possibility to establish regional considerations and therefore allowed for a more balanced spatial allocation of production units, mitigating the excessive concentration in large intensive farms and the abandonment of small scale farms and less intensive production units.

9.2.1.3 The beef regime

There are no specific and direct studies that analyse the relationship between the reform of the beef regime and the environment. The reason for this lack of information is that it is not thought to be a major problem. However, some limited references are made to the ecosystems that are the natural productive environments for the beef sector, such as extensive pastoral systems. The CAP programmes are reported to produce some damaging effects to the natural pastures due to the abandonment of traditional management practices (Zorita and Osorio, 1995). Some beneficial effects of the application of the beef regime have also been reported in relation to the stocking density limits on eligibility for the beef premiums (Maris and Albiac, 1997).

The application of the beef regime in Spain has produced some harmful effects that contradict the objectives of the CAP reform in this sector, namely the support of extensive meat producing systems for beneficial environmental considerations. One of the reasons lays in the unbalanced regional distribution of the beef sector in Spain, together with an extreme regional specialization of intensive and extensive production units. Extensive production systems are located in the north and north western part of Spain (Asturias, Cantabria and Galicia regions) and the intensive fattening units are heavily concentrated in the eastern part of the Catalonia region along the river Ebro valley, to which young livestock are transported from the extensive farms of the northern regions. Eastern Catalonia benefits from a locally developed compound feed industry in the vicinity of important Mediterranean harbours. The negative environmental effects of intensive livestock production in this region are severe (see Section 1.2.1) and considerable animal health problems arise from the transport of the livestock.

Some 70% of meat consumption originates from intensive production units and the reform of CAP was intended to induce the development of more extensive systems by establishing a maximum of 2 LU/ha and induce a more balanced regional production with clear benefits to the environment. However, eligibility for beef premia require a minimum time for marketing beef (10 months) that surpasses the traditional calf production cycle of the extensive northern systems (7 months) that were therefore not affected by the aid payments. In turn, the intensive livestock farms of the Ebro valley, that clearly surpass the maximum stocking rate of 2 LU/ha, do benefit in part from the beef premiums by renting pasture lands in the nearby mountain areas of the Pyrenees. This situation has caused damaging effects on the environment by limiting the desired reduction of intensive production units, and by inducing the abandonment of pasture zones in mountain areas (see Sections 1.2.1 and 1.2.2).

9.2.1.4   The sheepmeat regime

Compensatory payments under the sheep regime have induced a shift from milk producing sheep to meat sheep due to the better returns in the latter type of production. This in turn has provoked a shift in the distribution of herds and a change in herd size. In fact, sheep raised for their meat require less management effort and a lower labour qualification thereby reducing the problem of an increasing scarcity in skilled personnel for hire. The impact on the environment of these changes has still to be evaluated with precision. However, some partial regional studies have been assessing the impact of the sheepmeat sector on the environment.

One of the few studies was conducted in the province of Badajoz region of Extremadura, a traditional winter grazing area of transhumant sheep that evolved to year-round sheep production during the last three decades (Beaufoy, 1996). The CAP is not solely responsible for the environmental effects observed over these years in this cultivated steppe area. On the contrary, the process of modernization of Spanish agriculture has been driven as much by pressures for labour cost reductions and more intensive use of subsidized inputs. This resulted in a clear reduction of arable crops, legumes and fallow as well as the abandonment of traditional transhumance. That led to overstocking and in turn over-grazing, loss of natural feeding habitats for bird species, loss of vegetation and erosion. Although the CAP programmes are reported to cause certain beneficial effects in this area (curbing the trends towards land abandonment and fallow reduction), the CAP headage payments (Sheep Annual Premia) have produced clear incentives for overstocking, not sufficiently compensated with other policy measures (e.g. the CAP accompanying measures 2078/92). Stocking rates continue to increase, as well as the total number of sheep, in response to dairy production under the EU regional labeling system, without any environmental considerations (Beaufoy, 1996; CEAS, 1997b).

9.2.1.5   Fruit and vegetables

Expert opinions regarding the environmental impact of the fruit and vegetables regime indicate a dual effect. There is a negative effect due to the requirement of diverting surplus production by burning or burying the harvested fruit and vegetables. Actually, the recent 1996 modification of this regime eliminates this practice for environmental considerations. On the other hand, the CAP programme applied to certain traditionally extensive fruit productions, like almonds and hazelnuts, has recorded clear beneficial effects for the environment according to the judgements made by governmental experts. This programme includes special payments for cultivation practices and fruit quality enhancement that have halted the trend towards abandonment and weed invasion of these traditional plantations. Appropriate farming practices and lower tree densities in the plantations have resulted in better soil conditions and fire barriers, helping to control erosion and fire hazards which are two of the most severe environmental problems in the arid and semi-arid areas in Spain where these plantations are located (see Section 1.2.2). More than 60% of all almond and hazelnut producers in Spain have joined these programmes, which include payments of 400 ECU/ha for changing to high quality varieties and 275 ECU/ha for improving cultivation practices.

9.2.1.6   Olive

Environmental considerations are being taken into account in the reform of the olive oil sector. One of the most realistic alternatives for reform is partially decoupling the aid payments from production by adopting a direct payment mechanism per olive tree or per hectare. The potential environmental effects of this policy change are still to be investigated. The olive sector is a major traditional sector of great economic, social and environmental value. Given the various types of olive plantations that exist in Spain, it is necessary to distinguish the different environmental effects of such a reform. Broadly, the reform may produce a positive environmental effect by reducing incentives to intensify production or to extend olive plantation in unsuitable areas. Also it may prevent the abandonment of olive groves on marginal lands, but it may also cause a negative effect by favouring more intensive producers in dense plantations (Beaufoy, 1997a). Erosion control is a key environmental issue in olive groves that are planted extensively on steep slopes (see Section 1.2) and therefore require appropriate farming practices and environmentally sensitive production techniques (Beaufoy, 1997a; SEO-RSPB, 1997).

9.2.2   The accompanying measures

9.2.2.1   Agri-environmental policies

In Spain, Regulation 2078/92 includes 4 horizontal measures, which are applicable throughout the whole country (i.e. promotion of organic farming, maintenance of endangered local breeds, agri-environmental training and extensification in cereal crops). There is also a set of zonal measures applicable to specific areas designated by the Central Administration: in areas covered by National Parks, wetlands included in the Ramsar Agreement and Special Protection Areas under the Birds Directive. There are also specific areas designated by regional governments. The total budget of the overall programme (horizontal and zonal measures) for the period 1994-2000 is circa 215,000 million pesetas (approximately 1,380 million ECU) with an average cofinancing by the FEOGA budget of 65% (up to 75% for regions included in Objective 1 and 50% for the rest). The land potentially affected by these programmes is about 5 million ha (MAPA, 1997). The programmes proposed by the central administration amount to 60% of the total cost of the Spanish agro-environmental program. Just over half of this is on horizontal measures.

Among horizontal measures, the one on extensification of cereal crops should be emphasized, due to its expected impact. This programme introduces improved management practices to avoid erosion and establishes a grazing schedule adapted to the biological cycle of each habitat. It covers almost 2.5 million ha (which is equivalent to 98% of the total area of the programme) and an amount of 60,458 million pta. (86% of the total budget for horizontal measures) (Table 10). The objective of this measure is to preserve traditional fallow, regardless of the set-aside requirements, as esthablished by the Spanish Regulation 1765/92. Until now, this traditional fallow had not received any support from the CAP. Soil and climatic limitations in Spanish arid and semi-arid regions have favoured the development of a special extensive agricultural system where fallow has long been a common practice in crop rotations. Potentially, this traditional fallow accounts for about 4 million ha (20% of the total arable land). Although it is observed throughout most of the country, 80% is concentrated in five regions: Andalucía, Aragón, Castilla-La Mancha, Castilla-León and Extremadura. The area has decreased during the last 20 years due to incentives for intensive agricultural practices (Sumpsi, 1994b).

Table 10 Horizontal measures under Regulation 2078/92 (period 1994-2000)

Measure (units)	Beneficiaries	Cost (Million pta.)	% Cofinance FEOGA
Extensification (ha)	2,443,187	60,458	69.3
Training (farmers)	4,900	3,200	69.0
Endangered breeds (LU)	66,187	2,645	69.0
Organic farming (ha)	28,130	4,609	65.0
Total		70,914	

Source: MAPA (1997).

The first two zonal programmes were approved in 1993: first, the programme for irrigation water saving in the National Park 'Tablas de Daimiel' of the Castilla-La Mancha region; and, second, the programme for the protection of steppe birds in two areas of the Castilla-León region, respectively in the southern and northern central plateau. However, the complete application of the agri-environmental Regulation in Spain was postponed until the end of 1994.

The total budget and land area of the zonal programmes are not distributed in an even manner, and 4 of the 13 programmes account for a high proportion of the financial and land resources (Table 11). These programmes are, firstly, flora and fauna protection in extensive arable lands (Measure D1), which accounts for 37% of the total budget of the zonal programme and potentially covers more than 1 million ha; secondly, erosion control (Measure D4), with 17.4% of the total budget, and a coverage of about 400,000 ha; thirdly, the maintenance of abandoned lands (Measure D6), with 12% of the budget and coverage of nearly 200,000 ha; and fourthly, irrigation water saving with 4% of the total budget, and applicable to 90,000 ha.

The implementation of the agri-environmental Regulation was relatively slow during the first years due to administrative difficulties in the management of the programmes (i.e. coordination of the central and regional administrations), the complexity of the legislation involved, the lack of funds from the central and regional governments, the lack of experience in the application

Table 11a   *Implementation of Regulation 2078/92 (period 1994-2000) by region*

Region	Horizontal measures a)				Zonal measures b)			Total horizontal and zonal measures	Regional zone programmes	Total measures
	H1	H	H3	H4	Z1	Z2	Z3			
Andalucía	7,325	510	462	1,014	5,250	405	2,746	17,711	12,550	29,961
Aragón	9,979	383	185	313	3,160	998	1,068	16,085	1,950	18,035
Asturias	0	27	277	50	2,537	0	0	2,891	3,675	6,566
Baleares	100	32	14	152	0	0	0	298	0	298
Canarias	0	255	309	183	3,497	0	344	4,588	1,744	6,332
Cantabria	0	18	139	0	655	0	0	811	0	811
Castilla La Mancha	14,048	319	166	438	17,463	0	285	32,715	8,660	41,375
Castilla y León	12,098	510	462	424	2,437	0	67	15,998	35,138	51,136
Cataluña	582	223	92	721	0	5,031	274	6,923	8,963	15,886
Extremadura	7,095	191	231	356	2,802	275	299	11,250	6,017	17,267
Galicia	0	383	120	120	0	50	0	672	3,674	4,361
Madrid	1,732	19	9	79	0	0	38	1,878	2,975	4,853
Murcia	3,637	96	32	110	0	0	0	3,875	1,095	4,970
Navarra	956	13	19	70	0	0	0	1,057	995	2,052
La Rioja	300	96	69	144	0	0	513	1,122	0	1,122
Valencia	2,607	128	60	439	0	3,775	1,910	8,919	630	9,549
Pais Vasco	0	0	0	0	0	0	0	0	0	0
Total	60,459	3,200	2,646	4,609	37,801	10,534	7,544	126,793	87,766	214,559

a) Horizontal measures: H1=Extensive systems, H2=Training, H3=Endangered breeds, H4=Organic farming; b) ZonAL measures: Z1=National parks, Z2=RAMSAR wetlands, Z3=SPAs (Special Protection Areas EEC Birds Directive 79/409).
Source: MAPA (1997).

of such programmes and the low sensitivity shown by farmers and the administration to agri-environmental programmes. In spite of all these reasons, the agri-environmental programmes in Spain have been gaining considerable momentum and scope. The total annual budget allocated initially in 1993 was 9,895 million pesetas, which more than doubled by 1996 (21,756 million pesetas), involving more than half a million ha and close to 30,000 beneficiaries. This trend is expected to continue in the coming years. However, in spite of this increasing trend, the total budget allocated by mid 1997 had reached only 45,744 million pesetas (actual payments were close to 19,000 million pesetas only), which is less than one fourth of the total available budget. The regional distribution of the agri-environmental programme shows a clear bias towards the inland central regions in Spain, where two regions of the central plateau (Castilla-Leon and Castilla-La Mancha) concentrate almost 70% of the total budget. This is due to the fact that the two most important programmes in Spain - steppe birds protection and wetlands recovery - are mainly concentrated in these two regions.

The main concern of the Spanish authorities, both central and regional governments, with respect to the application of the agri-environmental programmes, is the preservation of natural

Table 11b *Implementation of Regulation 2078/92 by type of measure (million pta)*

Region	H1	H2	H3	H4	A	D1	D2	D3	D5	D6	E	F	G	H3+ D3+ E	H3+ H4	To- tal
Andalucía	21	78	11	28	0	0	0	0	0	0	0			0	0	138
Aragón	704	3	38	20	0	8	0	0	0	0	0		27	0	0	800
Asturias	0	0	1	0	0	0	0	595	0	0	83			620	0	1,299
Baleares	0	0	0	10	0	0	0	0	0	0	0			0	0	10
Canarias	0	0	17	16	0	0	0	0	170	0	0			0	103	307
Castilla La Mancha	0	0	0	0	0	47	0	0	0	12,355	0			0	0	12,402
Castilla y León	131	67	30	66	0	1,391	0	1,395	0	0	0			0	0	3,080
Murcia	98	0	0	0	25	0	0		0	0	0			0	0	122
Navarra	0	74	0	0	0	0	0	0	0	0	0			0	0	74
La Rioja	0	3	22	0	0	0	0	0	0	0	0			0	0	25
Valencia	0	0	0	0	0	0	640	0	0	0	0	22		0	0	662
Total	954	151	120	141	25	1,146	640	1,990	170	12,355	83	22	27	620	103	18,845

H1,H2,H3,H4: Horizontal measures
A: Integrated pest management
C: Stocking rate reduction
D2: Flora and fauna protection in wetlands
D4: Fight against erosion
D6: Irrigation water savings
F: Long-term set-aside
Source: MAPA (1997).

B: Conversion of annual crops into pastures
D1: Flora and fauna protection in extensive crop patterns
D3: Landscape conservation and fire prevention
D5: Environmental practices in the Canary Islands
E: Maintenance of abandoned lands
G: Land management for leisure uses

habitats and agricultural landscapes of high natural value. The programme with the widest take-up has been steppe bird conservation in extensive cereal cultivation areas (Regional Steppeland Cereal Programme) (22.6%), followed by conservation of terrace cultivation, preservation of endangered breeds and strains and organic farming (11.9%) (Suarez et al., 1997a). The environmental effects of these programmes have been substantially hampered by low responsiveness, little time lapse for measuring environmental effects, inappropriate monitoring and control and the need to establish adequate indicators (Suarez et al., 1997a; Oñate and Alvarez, 1997).

One of the most successful zonal programmes has been the programme for the recovery of wetland areas in Castilla-La Mancha and Campo de Montiel in the southern central plateau of Spain (aquifers 23 and 24). Income compensation payments are reported to be beneficial for attaining social, economic and environmental objectives (Viladomui and Rosell, 1997). The first programme approved by the Commission, the aquifer recovery programme of Castilla-La Mancha is the most expensive of all environmental programmes in Spain and one of the most expensive in all the EU with a total budget of 16,200 million pesetas (104 million ECU) for a five year period (1993-97). The environmental objective of this programme was to recover the natural wetlands of the National Park 'Tablas de Daimiel' by reducing water extraction from the aquifers.

The programme's socio-economic objective was to maintain the agricultural activity in the nearby area by compensating the irrigators from the income losses incurred by the reduction in irrigation water in a heavily irrigated zone of about 120,000 ha and 8,400 farms. By mid 1997, almost 3,000 farms had joined the programme covering close to 90% of the total surface area affected. Annual water extraction had been reduced by 299 million m^3; surpassing programme's target (255-270 million m^3 per annum). Meeting largely its environmental and socio-economic objectives, the programme has produced a much larger impact than previously expected. It has contributed to reduced social distress in the area caused by the years of severe drought and has brought about agricultural extensification. It has induced farmers to grow less water demanding crops, change to dry farming and reduce input use, such as machinery and agrochemical products, in an overall switch to less damaging production techniques. However, the programme's potential has still not been fully utilized and the Spanish administration is pressing for an extension in the programme's application in the next period. In part this reflects the success of the initial programme but also recognition that the effects of the reduction in water consumption in the aquifer's recovery will be a long term outcome and that a better management of groundwater resources remains to be accomplished (Viladomiu and Rosell, 1997).

Large differences across regions are observed in the application of the agri-environmental programmes. This is mainly due to the variation in income levels, rather than the relative importance of the agricultural sector in the region. High income regions tend to devote a higher percentage of their resources to environmental programmes than low income regions in spite of the smaller share of the agricultural sector in the region's overall GDP.

9.2.2.2   Forestry measures in agriculture

The afforestation policy developed in Spain in the last decades has been severely criticized by many organizations and experts. Negative effects range from the destruction of soils and natural flora and fauna to the acceleration of erosion, a result of inadequate terrace cultivation. The number of forest fires has also increased drastically over the last two decades, which was mainly due to bad management and uncontrolled shrub invasion. At times there have been larger extensions of burnt forest than of the reforested area, contradicting the policy objective (Ruiz and Groone, 1986).

Since 1986, the Spanish afforestation regulation transposing Directive EEC 937/85 has required the presentation of an environmental impact assessment which has resulted in a better control of the environmental effects. EC 797/85 Regulation introduced for the first time aids to subsidize afforestation of agricultural land. This regulation was implemented in Spain through the Royal Decree 808/1987. However, the low level of grant meant that comparatively few farmers have joined this programme and therefore the effects on the environment have been limited. EC Regulation 2328/91 was clearly oriented towards conservation, and can be considered the base of the new forest regulation introduced by the CAP reform in 1992.

Regulation (EC) 2080/92 of the CAP reform represented a complete change in the farmers incentives, by subsidizing tree plantation costs over the first five years and an income flow during a 20 year time period. This programme affects the whole territory and has an environmental objective (fight against erosion, conservation of flora and fauna, and landscape preservation) and a socio-economic objective (timber production). The estimated cost of the programme for the 1994-1998 period is 276,000 million pesetas with an average cofinancing of 65% by FEOGA (up to 75% for Objective 1 regions and 50% otherwise). However, this programme has been criticized by some specialists who have argued that it does not represent the starting point of a real forestry policy (Fernández Espinar, 1994), as it does not open the possibility for afforestation of mountain areas not previously covered with trees, as is the case with more than 7 million ha in Spain.

The implementation of this Regulation, through the Royal Decree RD 378/1993, is clearly based on environmental considerations rather than on production objectives, and it is aimed essentially to restructure and control soil use changes in marginal land. This objective is reflected in the following points: the programme is directed to reforest cultivated marginal lands that were previously covered by forests; local and slow growing species are preferred to other rapid growing species, favouring environmental aspects rather than productive aspects; regional Administrations have selected species and forest practices in each area according to environmental criteria, and afforestation projects require evaluation of their potential environmental impact.

Spain submitted a very ambitious programme aiming to reforest 800,000 ha and to improve 200,000 ha of existing forests in agricultural lands during the first 5 years. The programme started in 1993 and attained a significant level during 1994. Applications covering 330,771 ha were presented and about half of them were approved. However, unfavourable conditions caused by the drought and low payments reduced participation in the programme to a minimum. As a result, only 27,944 ha were reforested by mid 1995. In 1996, a more productive orientation was introduced (with shorter time limits for growing timber production species), payments were substantially increased and better management conditions in the new plantations were offered. This change brought about expanded participation in the programme. As a result, 94,000 ha were reforested with a total budget of 25,000 million pesetas during 1996. For 1997 an investment increase of 20% is expected.

The distribution by species (Table 12) shows a predominance of local and slow growing species. This trend is particularly important in the regions of Extremadura and Andalucia, where 90% of the dehesa ecosystem is located. In principle, these data may lead to think that the expected environmental impact of this measure must be quite positive. However, since Regional Administrations are responsible for the implementation of this programme, there are marked dif-

Table 12   Distribution of afforested surface by tree species (ha)

Region	Chestnut	Pine	Quercus	Other conifers	Other leafy trees	Walnut
Andalucía		5,292	24,681		6,960	665
Aragón		2,450	2,393	297	207	165
Asturias	850	440	310	130	724	130
Castilla-La Mancha		13,030	19,451		314	236
Castilla-León		10,300	24,520	260	1,250	200
Extremadura		528	18,181		539	15

Source: MAPA (1994).

224

ferences between them. In some regions, monitoring of forestry practices and environmental impact is very strict while in others no environmental assessment is required.

Some specific environmental conditions apply in Spain in relation to the application of the forestry programmes. These programmes are under the competence of the regional governments and in some regions the application of the reforestation programmes is controlled by the regional government's Environment Department to ensure that a given project complies with the region's environmental requirements.

9.2.3 Structural policies

9.2.3.1 Less Favoured Areas

The Less Favoured Areas (LFA) policy was introduced after joining the EC in 1986. Since then, the area qualified for LFA has doubled. At first, this measure was applied to mountain areas as a compensation for low farm incomes. In 1989 the measure was extended to areas at risk of depopulation with a 'need for nature conservation' as established in the current regulation and certain environmental criteria were introduced in the legislation. Lastly, in 1993 the environmental criteria were established with priority status and zones affected by specific restrictions from natural conditions were also included. However, this third category is still of little importance and represents only 0.3% of the total LFA surface but compensatory payments are double. The total number of beneficiaries increased to 200,000 farmers in 1992, but was reduced by 6% in the last years

Table 13 Distribution of LFA and total area by region (1,000 ha)

Region	LFA	Total	Percentage of territory in LFA
Andalucía	3,752	8,787	43.0
Aragón	2,021	4,768	42.4
Asturias	958	1,056	90.7
Baleares	105	494	20.5
Canarias	491	750	65.6
Cantabria	431	528	81.6
Castilla-La Mancha	2,822	7,922	35.6
Castilla-León	3,904	9,400	41.5
Cataluña	1,399	3,193	43.8
Extremadura	605	4,160	14.5
Galicia	1,300	2,947	44.1
Madrid	239	800	29.9
Murcia	182	1,132	16.1
Navarra	530	1,042	50.7
La Rioja	262	503	52.0
Valencia	740	2,326	31.8
Total	19,741	49,748	44.0

Source: MAPA (1993).

due in part to the drought which caused the abandonment of farming as the main income generating activity, particularly in marginal areas with low yields (MAPA, 1993).

In total, close to 20 million ha in Spain are eligible to be included under LFA, which represent 44% of the total national territory and 77% of agricultural land (SAU). The area under LFA classification coincides in many cases with natural areas highly dependent on traditional farming activities. The abandonment of farming activities in these areas may lead to serious environmental problems (see Section 1.2.2). Even though important positive effects of the LFA policy on nature conservation could be expected, there are several factors which weaken its potential benefits. These are the following:

i) Lack of environmental criteria

Although the selection of LFA is made in part by reference to certain environmental criteria, the implementation and modulation of LFA subsidies is based on socioeconomic criteria. In fact, the implementation of the LFA Directive in Spain does not consider the possibility of providing incentives specifically to farmers who use agricultural practices that are compatible with environmental protection. In general, the LFA policy seeks to support extensive farming systems but does not consider if the farming practices involved are beneficial or harmful to the environment. In fact, some extensive farming practices can be very damaging, such as burning of stubble or excessive use of tillage.

On the other hand, strict social criteria exclude part-time farmers from the LFA programme. However, extensive farmers are often underemployed and they either rely on off-farm employment or engage in farm intensification as a way to full-time work and higher incomes. The exclusion of part-time farmers can be considered as a penalty for off-farm employment and may indirectly be damaging to the environment. The inclusion of part-time farmers, though, would involve a further dispersal of the budget. Nevertheless, the implementation of certain environmental priorities, for example through a more restrictive selection of the areas according to their ecological value, would lead to a higher concentration of the budget and would produce greater benefits for nature conservation.

ii) Scarcity of subsidies

At present, the limited budget and the large area to be covered by the programme results in a low level of payment per farmer and therefore a limited policy impact. From a survey conducted of 200 farms in the Segovia province (region of Castilla-Leon in the northern central plateau) it could be concluded that to most beneficiaries, LFA payments did not help to change the productive system on their farms. Compensation under the LFA scheme represents only 2.7% and 5.1% of the farmer's income in depopulating and mountainous areas respectively, and can be considered marginal for most farmers (Sumpsi, 1991). Although the Spanish regulation establishes higher LFA payments for mountainous areas than for depopulating areas, the study concludes that these minor differences do not compensate for the income differences between these two zones, with the result that farmers' income in mountainous zones continue to be significantly lower.

In any case, the LFA scheme cannot be considered, on its own, an efficient instrument to avoid depopulation, because it does not address the inadequacies in infrastructure, services and amenities that afflict these areas.

iii) Budget limitations

The application of this policy is limited largely by budgetary restrictions of the Central and Regional governments as co-financing is required. Marginal regions with greater risk of depopulation face more difficulties in generating the co-financing element. Spain, after Germany, has the largest number of farms in the EU which are eligible for the LFA scheme. However, the budget allocated is one of the lowest in the EU, being about a tenth of that for the UK and a quarter of that for Germany.

iv) Payment scheme

The LFA aid scheme follows a similar pattern to that of other CAP policies, which are also based on headage and area payments. It has resulted in overgrazing in zones where the limit of 1.4 LU/ha exceeds the environmentally sustainable limit. However, positive effects of this policy have been reported in certain areas. In certain low income marginal areas, the LFA payments do play a significant role in farmers' incomes and thus in maintaining certain extensive production practices to provide benefits for nature conservation (Peco and Suárez, 1993). It has also been reported that this policy measure encourages transhumance practices in some specific zones since farmers from non-LFA areas keep their herds grazing 90 days in mountain areas to receive the LFA subsidy (Baldock and Beaufoy, 1993).

Overall, the LFA programme has seen a steadily reducing number of beneficiaries, from 224,000 in 1990 to 173,000 in 1996. Aid payments had been fluctuating, although average payment per farmer increased slightly during the 1990-96 period, especially after 1992, and was around 66,000 pta. However, budget limitations did not permit an increase in the aid payments in 1996 (MAPA, 1997). The regional distribution of the LFA payments shows that the central inland regions of the Castillian plateau receive the highest proportion of the total budget - with around 31% of the total budget and total number of beneficiaries in the north central Castilla-León region, and around 13% of the total budget and total number of beneficiaries in the south central Castilla-La Mancha region.

9.2.3.2   Farm modernization programme

Subsidies directed to the modernization of farms constitute the most relevant structural policy in Spain (Regulation 797/85 and following modifications). This measure had been applied since 1989 (RD 808/87), by granting of investment subsidies. Some 90,220 million pesetas was allocated during the period 1989-1993 to 64,336 farmers, which contributed to a total investment of 275,346 million pesetas (MAPA, 1996a). In general, the investment plans subsidized by this policy implied the intensification of agricultural systems with potentially damaging effects on the environment. Only a small proportion of the modernization projects had a favourable impact on the environment, such as water saving, soil conservation and prevention of erosion, organic farming, afforestation, etc. However, there are no official data or reports concerning the distribution of investments per type of plan and there is no information about the environmental impact of such an important structural policy.

Agricultural policies have played a key role in the development and expansion of publicly funded irrigation networks and private irrigation schemes (Sumpsi, 1994b). During 1989-1993, the modernization policy allocated a total of 8,500 million pesetas to private irrigation plans and promoted the transformation of 48,200 ha into irrigated land. Irrigation water in a large proportion of these lands comes from groundwater sources thus contributing to the increased over-exploitation and pollution of aquifers.

9.3 Environmental conditions in policies

9.3.1 Common Market Organizations

9.3.1.1 The arable regime

Specific set-aside conditions were introduced with the 1992 reform of the arable crops regime. Market set aside was introduced in order to control cereal production. The time span of the different set-aside systems largely determines the objectives of the land diversion and therefore the related environmental considerations. Short term set-aside - such as the mandatory annual, rotational and five-year set-aside regimes - has a clear market orientation and scarce regard for environmental objectives. The environmental effects of this type of land diversion is difficult to assess in any specific country and across the EU member states due to its limited time horizon (OECD, 1997). Conversely, long term set-aside has a defined environmental goal and the effects on the natural resource base and landscape will probably be more specific, as well as more beneficial, across regions and ecosystems.

The treatment of land which is put aside must comply with general environmental rules, applicable to all Member States, including specific indications concerning use of fertilizers and plant protection products, treatment of the vegetation cover and spreading of animal manure. In addition, Member States can stipulate more detailed environmental rules that apply to the specific features of their agricultural systems. These include, for example, a list of permitted cover crops, measures for weed control and certain exceptions to the general regulations based on the specific features of that country.

In Spain, traditional fallow extends over almost 4 million ha (see Table 2) and has been subject to special national regulations apart from the set-aside policy mechanism. Not subject to aid payments, traditional fallow is regionally defined in Spain and considered a mandatory farming requirement to attain support under the arable crop regime. Certain additional environmental conditions on traditional fallow are required of farmers wishing to take up agri-environmental payments (Regulation 2078/92) for long-term set-aside.

The environmental conditions that apply to the set aside scheme, as well as to the traditional fallow practices, have not been changed since Spain established its complementary environmental rules in 1993 through specific national legislation. These include special farming practices and seasonal constraints on tillage operations. However, one of the main problems with the environmental conditions on payments under Regulation 2078/92 for traditional fallow is the restriction on herbicide use that restrains the application of conservation tillage, which is a fundamental instrument to control soil erosion (see Section 1.2). In this respect the CAP reform may limit the expansion of conservation tillage farming, whereas in other arid regions of the world such as the US, Australia and Argentina it is developing at an increasing pace as a means of erosion control.

In general, it is found that production increases when conservation tillage is applied to the whole farm rather than farming the land under a traditional rotational scheme of fallow-cultivated land using conventional tillage practices. Conservation tillage, therefore, is likely to lead to the disappearance of traditional fallow in most zones.

9.3.1.2   The beef regime

There are no additional environmental requirements to beef production, besides the maximum of 2 LU/ha for receiving direct payments. The reason, according to the Spanish Administration (central and regional), is that this sector does not cause significant damage to the environment. It may also be feared that more restrictive national regulations would reduce the competitive position of the Spanish beef sector, which could have severe effects in less favoured areas, particularly in mountainous regions. In these areas it is believed that maintaining traditional extensive systems will be beneficial for the environment and that introducing additional restrictions will conversely provoke the abandonment of pasture lands and farms causing serious damage to the environment and the landscape.

9.3.1.3   Fruit and vegetables

Regulation EC 2200/96 and the subsequent application of Regulation EC 411/97 establish the market organization for fruit and vegetables. These regulations incorporate some environmental conditions to be developed by Member States. The requirements include the implementation of farming practices, production techniques and the handling of chemical residues, to ensure the quality of water, soil and landscape and to preserve biodiversity. In Spain, to attain the desired EU regulation objectives, complementary legislation was advanced to establish the general guidelines for the operation of funds and programmes as well as the financial aid mechanisms addressed to the fruit and vegetable producing sector (basically to the producers' organizations). This national legislation was set out in a Ministry Order of May 14, 1997, under the competence of the Ministry of Agriculture, Fisheries and Food with direct collaboration of the regional governments and the professional organizations. Specific items included in this legislation are:
i)   Rational use of natural resources and inputs, such as soil conservation (i.e erosion prevention with the application of conservation tillage practices, water run-off and percolation protection, wind screens etc.). Conservation of soil fertility conditions (i.e. application of adequate crop rotations, use of organic fertilizers to increase the organic content in soils, reduce the quantity and number of plant protection products). Conservation of soil moisture (i.e. use of plant cover crops,

adequate tillage practices). Reduction of soil contamination, such as control of soil salinization and toxicity, and proper use of fertilizers and pesticides.

ii) Efficient use of water resources, including a reduction in water pollution.

iii) Optimization of energy use, such as reducing energy consumption (e.g. reduction of tractor use and other machinery, utilization of low energy consumption tillage, control of energy pollution). Production and use of renewable energies (e.g. solar and wind energy).

iv) Reduction in fertilizer use by making use of traditional organic fertilization and composted organic residues.

v) Rational use of plant protection products, such as reducing pesticide use (e.g. application of integrated pest management practices), utilization of plant protection products and application procedures that will minimize the environmental impact (e.g. authorized products, low danger and low toxicity products), correct dose treatments, alternative systems of pest and plant disease control (e.g. non chemical traditional pest control and biological pest control).

vi) Use of integrated environmental production systems, such as locally based integrated production systems, ecological agriculture to maintain product quality and reduce environmental impacts with the efficient management of the agri-ecosystem.

vii) Proper handling of residues in the productive process, reducing farm pollution (e.g. air emissions, reduce burning of residues and reuse of organic residues), liquid effluents, use of biodegradable and recycable packaging.

viii) Conservation of landscape by maintaining traditional cultivation, conservation and recovery of ecosystems (i.e. maintenance and restoration of natural vegetation by using traditional farming practices to protect plant and animal species) and ecological processes (i.e. regulation of hydrological flows and erosion control, restoration of aquifers in arid zones, avoid damaging of natural drains and food chains for flora and fauna).

In Spain, these measures have a marked regional and zonal orientation and are to be applied by the Producer's Organizations based on the farm characteristics and the environmental conditions that affect each of the production zones.

9.3.2   Nitrates Directive

The implementation of the Nitrates Directive has been slow and difficult in Spain. The distribution of political and legislative competencies between the central government and the regional governments is one of the major reasons for this delay. The regional governments are still elaborating the Code of Good Agricultural Practice and the identification of vulnerable zones is still under discussion. The Ministry of the Environment in collaboration with the River Basins Authorities and the Mining and Geological Institute (ITGME Instituto Técnico Geológico y Minero) has defined the potential vulnerable zones of nitrate contamination selecting those zones where more than 50% of the samples analyzed exceed a level of 50 mg of nitrates per litre. The selection of vulnerable areas is being discussed by the regional governments. Lack of sufficient experience of how to measure, treat and resolve nitrate contamination problems in Spain is also producing a clear delay in solving this issue. Concentrated in very specific zones (see Section 1.2.1), nitrate pollution in Spain is found in the intensively irrigated fruits and vegetable production zones along the Mediterranean coastline and southern Atlantic areas (provinces of Valencia, Málaga and Huelva, in open and protected crops) and in the maize fields of the region of La Mancha and in the Andalucia rice fields.

Intensive pig production areas are also vulnerable to nitrate pollution, but according to the Administration these areas are less harmful to the environment than the intensive crop production regions. Restricted to very limited areas, slurry emissions in swine intensive farms are controlled differently in landless production units and in land-based farms. The former is controlled by the specific regulations on slurry emissions and the latter is controlled by the application of the Nitrate Directive. There is a strict obligation to apply the Code of Good Agricultural Practice in vulnerable zones where action plans have to be implemented. No special compensation payments of regulation 2078/92 are considered in the intensively irrigated areas of high nitrate contamination because nitrogen fertilizer application doses are excessively high and therefore mandatory reductions in these areas will not result in yield decrease but will cause clear environmental benefits.

9.4 Concluding observations

Integrated perspective in the assessment of agri-environmental policies

The CAP has evolved in the recent years and, especially after the Mac Sharry reform of 1992, environmental considerations have taken a decisive role in the national and regional programmes. But in general it is too soon to evaluate the effects. As we have seen the CAP reform has provoked, in many instances, a dual effect on the environment. Some positive impacts can be identified, like the extensification processes that produce a reduction in the use of fertilizers, machinery and pesticides, the maintenance of traditional farming systems therefore reducing abandonment of marginal lands, and the environmentally managed traditional fallow. But on the other hand, the CAP programmes have provided incentives to increase irrigated crops through the regionally based compensation payments, and have favoured intensive animal production, overgrazing and increased livestock production in some areas through headage payments and low feed prices and the provision of support to silage production. To come to an overall judgment would require studies on a regional basis of the great variety of farming systems, natural habitats and ecosystems in Spain, following an integrated sector- ecosystem multidisciplinary approach.

The need for policy integration (intra-national balance)

The implementation of the CAP reform measures provides evidence that farm support policies and environmental policies can have contradictory objectives. Therefore the evaluation of the overall environmental effects can pose serious difficulties. One example relates to the importance of irrigation for Spanish agriculture, from a productive perspective as well as from an environmental perspective 1). Subsidized investments through structural polices aimed at farm modern-ization (Regulation (EEC) 797/85) and the Spanish parallel programme (Royal Decree 808/87) have caused intensification of production in some areas, with harmful effects on the environment, especially in case of irrigation. Evaluation of the environmental impacts caused by these policies such as over-extraction and pollution in aquifers along the intensively irrigated Mediterranean coastal areas and the La Mancha plains in the central plateau is being conducted lately.

Recently, the Ministry of Agriculture started to develop an ambitious dual programme (1995-2005) aimed to expand irrigation areas and to improve existing irrigation districts covering 1.4 million ha (MAPA, 1996b) which might potentially increase water consumption. In this re-spect, farmers engaged in intensively irrigated agriculture (especially in the Mediterranean areas where export-oriented high value added crops are grown), will probably oppose any policy aimed to reduce intensive farming as it will be viewed as a potential reduction in crop yields. If policies tending to reduce input use and induce sustainable production techniques are to be applied on a voluntary basis, they will have to resort to compensation payments to the farmers for the income loss incurred.

Conflicts between these policies and the objectives of the agri-environmental Regulation (2078/92) are increasingly apparent in many irrigated areas. In fact, environmental policies, deriv-ing from EU, national and regional legislation, have to be adequately coordinated in Spain where there is still a lack of experience, compared to northern European countries. Moreover, a varied and complex agricultural sector and a division of legal competence among many administrative units increase the difficulties for the design, implementation, enforcement and monitoring of such policies (Sumpsi, 1996a). In this context, some of the EU and national policies that affect specific zones and are therefore applied in certain regions, can be under the competence of the regional governments. Examples of these zonal measures are the implementation of the programmes under Regulation (EC) 2078/92 in the region of La Mancha (in the southern central plateau) for recover-ing wetland zones, and in the region of Castille-León (in the northern central plateau) for protect-ing steppe bird species. Both measures focus on extensification of agricultural practices, thus po-tentially reducing input use and causing beneficial effects on the environment. The application of such programmes can be impaired, in some areas, by policies that promote irrigation as well as by the application of the CAP arable crop programmes that may induce the extension of irri-gated crops. In this context, the growing scarcity of water resources and the increasing demand among competing users (agricultural, industrial, urban and 'environmental' users) calls for a more rational and efficient use of water resources and this issue is at the core of the nation's debate. Proper management of water resources is therefore a key issue for policy design and implementa-tion. The links between farm policies and water policies have to be clearly established to avoid damaging effects on the environment. Therefore policy integration and potential contradictions in objectives are of great importance when analyzing farm policy and agri-environmental pro-grammes in Spain and are crucial issues in achieving environmental objectives and in the effec-tiveness of policy instruments.

---

1) The agricultural sector in Spain required 80% of the available hydric resources, it extends only over 15% of total agricultural land (3.5 million ha) but accounts for 60% of total agricultural production and 80% of the exports of agricultural produce.

The need for policy integration (inter-EU balance)

In some cases, the agri-environmental policies may have contradictory effects across regions and countries in the EU. In fact, some of these policies represent two ongoing trends that follow opposite directions with respect to environmentally sound farming practices. As an example we may refer to the opposite effects that two environmental policies may cause on arable land in Spain. On the one land, one of the most widely used erosion control practices has been to apply conservation tillage that requires the use of specific herbicides. On the other hand, reducing the consumption of plant protection products is likely to be beneficial to the environment. The balance of these two apparently contradictory measures has to be carefully studied to avoid unwanted effects on the environment. But it has to be taken into account that in spite of its potential economic and environmental benefits, conservation tillage developments do respond less to policy driven farm management changes than to market forces which are encouraging their increasing adoption in Spain and elsewhere in arid zones. In this respect, curbing the use of herbicides in the EU could impair the application of erosion control farming practices in arid zones and damage the environment. In this context, a crucial policy issue in Spain will be to consider whether these farming practices will increase substantially the use of herbicides and whether this non-reduction in herbicide consumption will be considered as a benefit (along with the prevention of crop losses) or as a cost in the risks-benefit balance of a given policy choice. This example can serve to point out that some attention should be given to potential policy contradictions in the design and implementation of future farm and agri-environmental policies in the EU (i.e. the need to balance the environmental requirements of northern and southern EU countries as well as regional specificity, both in environmental policies and farm policies). In general, a better definition of policy instruments (i.e. the balance between the effectiveness of the instrument in meeting the policy goals and the related enforcement costs) and efficient control and monitoring will be required to increase the overall cost-effectiveness of the policy.

References

ADENA - WWF (1994a) *Incendios forestales*

ADENA - WWF (1994b) *Agricultura extensiva en Europa;* Una paronámica del Proyecto y del Inforne Nacional de España

Arnal, P. (1996) *Diez años de laboreo de conservación en Navarra: balance y expectativas;* In: Garcia Torres et al. (eds.): Proceedings of the National Congress: Agricultura de Conservación; Rentabilidad y Medio Ambiente. (Conservation Agriculture. Profitability and Environment); AELC/SV; Córdoba, Spain, pp. 85-92

Astorquiza, I. (1994) *Transformaciones en regadío de zonas con condiciones limitantes;* Evaluación de la sostenibilidad de Monegros Il. Revista de Estudios Agrosociales 167, pp. 209-220

Astorquiza, I., I. Bardaji, E. Ramos, F. Ramos and J.R. Múrua (1996) *Economic responses facing the CAP reform in the Spanish Cereal Sector;* Poster paper presented in the 8th Congress of the EAAE. Edinburgh, UK, (3 - 7 September 1996)

Baldock, D. and A. Long (1988) *The Mediterranean environment under pressure: the influence of the CAP on Spain and Portugal and the IMPs in France, Greece and Italy;* Report to WWF

Baldock, D., G. Beaufoy, F. Brouwer and F. Godeschalk (1996) *Farming at the margins. Abandonment or redeployment of Agricultural Land in Europe;* Institute for European Environmental Policy (IEEP) and Agricultural Economics Research Institute (LEI-DLO); London-The Hague

Baldock, D. and G. Beaufoy (1993) *Nature conservation and new directions in the EC Common Agricultural Policy;* Institute for European Environmental Policy, London

Beaufoy, G. (1996) *Extensive Sheep farming in the steppes of LA Serena, Spain;* In: Mitchell, K. (ed.). The Common agricultural Policy and Environmental Practices. Proceedings of the seminar: Nature Conservation and Pastoralism. European Forum; WWF. COPA, Brussels; pp. 34-48

Beaufoy, G. (1997a) *Environmental considerations for the reform of the CAP olive-oil regime;* Unpublished report

Beaufoy, G. (1997b) *Assessing the effects of the CAP accompanying measures and rural development measures on biodiversity and landscape in southern Europe;* Case study: Extremadura. Spain. Report of the project of Lierdeman et al. (1997). DG XI. Commission of the European Communities. Brussels

Beaufoy, G., D. Baldock and J. Clark (1994) *The Nature of Farming-Low Intensity Farming in Nine European Countries;* Institute for European Environmental Policy (IEEP); London

Birdlife International (European Agriculture Task Force) (1996) *Nature conservation benefits of plans under the Agri-environment regulation (EEC 2078/92);* RSPB-EU Project. DG XI. Brussels -Bedfordshire

Briz, J., I. De Felipe and M. Merino (1994) *Set-aside programme in Spain;* Aspects of Applied Biology, 40, pp. 575-578

Brouwer, F. and S. van Berkum (1996) *CAP and environment in the European Union: analysis of the effects of the CAP on the environment and assessment of existing environmentol conditions in policy;* Wageningen, Wageningen Pers.

Campos, P. (1993a) *Valores comerciales y ambientales de las dehesas españolas;* Agricultuira y Sociedad, 66, pp. 9-41

Campos, P. (1993b) *Sistemas agrarios. Análisis aplicado al monte mediterráneo;* In: Naredo, J.M. and F. Parra (eds.); Hacia una ciencia de los recursos naturales. Siglo XXI, Madrid pp. 281-304

Campos, P. (1994) *Economía y conservación del bosque mediterráneo en la Península Ibérica;* In: Cadenas A. (ed) Agricultura y desarrollo sostenible. Serie Estudios, 97; MAPA. Madrid; pp. 441-457

Campos, P. (1995a) *Análsis técnico y económico de sitemas de dehesas y montados. Final report;* EU project (CAMAR CT-90-28) DG VI, Brussels, EU Commission

Campos, P. (1995b) *Dehesa forest economy and conservation in the Iberian Peninsula;* In: McCracken, D.I., E. Bignal and S.E. Wenlock (eds.); Farming on the Edge: The Nature of Traditional Farmland in Europe. Joint Nature Conservation Committee Peterborough, UK; pp 112-117

234

Cantero, C. (1995) *Laboreo de conservación en cultivos herbáceos extensivos en Cataluña;* Vida Rural, 19-20; pp 36-43

Carballo, M. (1996) *Problemática ambiental generada por explotaciones porcinas;* El impacto medioambiental de las explotaciones de porcino, Porcis 31, pp. 11-18

Castillo, M. (1992) *La políticas limitantes de la oferta lechera. Implicaciones para el sector lechero español;* Serie estudios, MAPA. Madrid

CEAS-EFNCP (1997a) (Centre for European Agricultural Studies, Wye and European Forum on Nature Conservation and Pastoralism) *Possible options for the better integration of environmental concerns into the various systems of support for animal products. Volume I;* Final Report for the EU Commission, DG XI. CEAS, London

CEAS-EFNCP (1997b) (Centre for European Agricultural Studies, Wye and European Forum on Nature Conservation and Pastoralism) *Possible options for the better integration of environmental concerns into the various systems of support for animal products. Volume II: Case Studies;* Final Report for the EU Commission, DG XI. CEAS, London

Cruz Villalon, J. (1994) *La agricultura en las zonas h·medas mediterráneas;* Agricultura y Sociedad, 71, pp. 183-209

De León, A. (1990) *Nitric pollution of waters in Spain: present situation and prospects. Nitrates, Agriculture, Water;* International Symposium. Paris 7-8 Nov. 1990. INRA pp. 527-554

Diaz Alvarez, M.C. and J. Almorox (1994) *La erosión del suelo;* El Campo, 131 Agricultura y Medio Ambiente; pp. 81-93

Díaz, M., R. Carbonell, T. Santos and Tellería (1997a) *Breeding birds communities in pine plantations in Spanish plateaux: biogeography, landscape and vegetation effects;* American Journal of Applied Ecology

Díaz, M., M.A. Naveso and E. Rebollo (1993) Respuestas de las comunidades nidificantes de aves a la intensificación agrícola en cultivos cerealistas de la Meseta Norte (Valladolid-Palencia, España). Aegypius 11, pp. 1-6

Díaz, M, P. Campos and J. Pulido (1997) *The Spanish dehesas: a diversity in land-use and wildlife;* In Pain, D.J. and M. Pienkowski (eds.) Farming and birds in Europe. Academic Press, London

Donázar, J.A., M. Naveso, J. Tella and D. Campión (1997b) *Extensive grazing and raptors in Spain;* In: Pain, D.J. and M. Pienkowski, (eds.). Farming and birds in Europe. Academic Press, London. pp. 117-149

Egdell, J.M. (1993) *Impact of Agricultural Policy on Spain and its steppe Regions;* Studies in European Agriculture and Environment Policy. RSPB, Sandy

Fasola, M. and X. Ruiz (1997) *Rice farming and waterbirds: integrated management in an artificial landscape;* In: Pain, D.J. and M. Pienkowski (eds.) Farming and birds in Europe. Academic Press, London. pp. 210-235

Fernandez Espinar, L.C. (1994) *El sector forestal y la reforestación de terrenos agrícolas;* El Campo, 131. Agricultura y Medio Ambiente. pp. 49-59

Fernandez-Guillén, M. and R. Jongman (1994) *Diversidad y Agricultura;* El Campo, 131; Agricultura y Medio Ambiente pp. 65-81

MOPU (1988) (Ministerio de Obras Públicas) *Evolución de las extracciones y niveles piezom étricos en el acuífero de la llanura manchega;* Report 10/88. Madrid

MOPU (1989) (Ministerio de Obras Públicas) *La degradación del suelo;* Medio Ambiente en España,. pp. 23-54

Morey, M. (1995) *Paisajes mediterráneos rurales amenazados;* Ecosistemas, 14, pp. 26-31

Naredo, J.M., J. Lopez Salvez and J. Molina (1993) *La gestión del agua para regadío;* El caso del Almería; El Boletín

Naveso, M.A. (1996) *The nature conservation benefits of the programmes under Regulation 2078/92;* SEO-RSPB, Cambridge

Naveso, M.A. (1992a) *Proposal to declare the steppeland of Madrigal-Peñaranda as an environmentally sensitive area;* SEO-RSPB, Madrid

Naveso, M.A. (1992b) *Proposal to declare the steppeland of Villafafila as an environmentally sensitive area;* SEO-RSPB, Madrid

Naveso, M.A. and M. Groves-Raines (1992) *Proposal to declare the steppeland of Tierra de Campos as an environmentally sensitive area;* SEO-RSPB, Madrid

Ocaña, L., S. Niñerola and C. Ruiz Cela (1981) *La problemática de los contenidos de nitratos en las aguas subterráneas de la llanura Manchega;* Proceedings of the seminar: Análisis y evolución de la contaminación de las aguas subterráneas en España. 1, pp. 113-118, Barcelona

OECD (1997) *Environmental Performance Review of Spain* (Draft document)

Oerke et al. (1994) *Crop production and crop protection, estimated losses in major food and cash crops;* ECPA, Elsevier. pp. 12-20

Oñate, P. and P. Alvarez (1997) *Transcendencia socioeconómica de las medidas agroambientales: el caso de Castilla y León;* Revista Española de Economía Agraria, 179

Ortega, F. and F. Rodríguez (1990) *El medioambiente;* In: Bosque, J. et al. (eds.) Geografía de España. Planeta, Barcelona

Pastor, M. and A. Guerrero (1990) *Influence of non-tillage on olive grove production,* Acta Horticulturae, 286 (pp. 283-286); In: Pastor, M. et al. (eds.) Uso de herbicidas en la formación de cubiertas vegetales con crecimiento reducido en olivar. ITEA, 65, pp. 35-44

Peco, B. and F. Suárez (1993) *Recomendaciones para la gestión y conservación del medio natural frente a los cambios de uso relacionados con la PAC;* Report ICONA (Instituto para la Conservación de la Naturaleza), Madrid

Peco, B., F. Suárez, J.J. Oñate, J.E. Malo and J. Aguirre (1997) *Report on the Implementation of Regulation 2078/92 in Spain (Part B);* In: Biehl, D. (coord.). Implementation and Effectiveness of Environmental Schemes Established under Regulation 2078/92. First Progress Report. Parts A and B. EU Project FAIR 1 CT95-274. EU Commission; Brussels

Perez Ybarra, C. (1994) *Alteraciones ambientales en las transformaciones en regadío;* El Campo, 131 Agricultura y Medio Ambiente, pp. 117-133

Pulido, F. and M. Escribano (1994) *Análisis técnico y económico de sistemas de dehesas;* Junta de Extremadura. Servicio de Investigación y Desarrollo Tecnológico. Badajoz, Spain

Rubio, J.L. and P. San Roque (1990) *Water erosion and desertification in Mediterranean Europe;* In: Rubio and Rickson (coord.) Commission of the European Commities, Directorate-General for Agriculture, Brussels

238

Rubio, J.L. and J. Herrero (1995) *La desertificación del litoral mediterráneo español;* Quercus, 18, pp. 8-11

Ruiz, M. (1985) *Impacto ecológico de la entrada en la CEE;* Quercus, 21, pp. 4-5

Ruiz, M. and H. Groone (1986) *Spanish agriculture in the EEC: A process of marginalization and ecological disaster?* FFSPN Rencontres Internationales de Toulouse; Agriculture-Environment, pp. 456-461

San Miguel, A. (1995) *Aprovechamiento sostenible del monte mediterráneo;* Ecosistemas, 14, pp. 40 - 47

Sanchis Moll, J. (1981) *Los nitratos en los acuíferos costeros de Valencia. Su distribución en el espacio y en el tiempo;* Proceedings of the seminar: Análisis y evolución de las aguas subterráneas en España; Barcelona

Sancho Comins, J., J. Martinez Vega, J. Garcia-Abad, P. Navalpotro and A. Santaolalla (1994) *La tradición e innovación en el paisaje agrario: Los efectos de la PAC en la región central española;* El Campo, 131; Agricultura y Medio Ambiente, pp. 215-235

Sancho Comins, J., J. Bosque, F. Moreno Sanz, (1993) *Crisis and permanence of the traditional lands agromediterranean in the central region of Spain;* Landscape and Urban Planning, 23, pp. 155-166

Sears, J. (1991) *Case study of European Farmland Birds: Mediterranean Steppelands;* The Dehesa System Research. Report RSPB Sandy

SEO-RSPB (1997) *Consideraciones ambientales para la reforma de la O.C.M. del aceite de oliva. Comments to the report of the European Commission to the Minister's Council and to the European Parliament;* 1st Draft. Cambridge

Soto, D. (1990) *Aproximación a la medida de la erosión y medios para reducir Ésta en la España peninsular;* Ecología, 1, pp. 169-196

Suárez, F., P. Oñate, J. Malo and B. Peco (1997a) *La aplicación del Reglamento Agroambiental 2078/92/CE y la Conservación de la Naturaleza en España;* Revista Española de Economía Agraria 179

Suárez, F., M. Naveso and E. De Juana (1997b) *Farming in the drylands of Spain: birds in the pseudosteppes;* In: Pain, D.J. and M. Pienkowski (eds.). Farming and birds in Europe. Academic Press, London

Suárez, F., J. Herranz and M. Yanes (1996) *Conservación y gestión de las estepas de la España Peninsular;* Unpublished Report, Madrid

Sumpsi, J.M. (1991) *Metodología para la evaluación del impacto socio-económico de las ayudas socio-estructurales de la CEE;* Secretaría General de Estructuras Agrarias, MAPA, Madrid

Sumpsi, J.M. (1994a) *El regadío como instrumento de la Política Agraria;* Symposium Nacional: Presente y Futuro de los regadíos españoles, CEDEX, MOPTMA, Madrid

Sumpsi, J.M. (1994b) *Las ayudas a los sistemas extensivos de cereales de secano;* Secretaría General de Estructuras Agrarias. MAPA, Madrid

Sumpsi, J.M. (1996a) *El nacimiento de la política agroambiental en España;* Revista Economistas, 64, pp. 398-406

Sumpsi, J.M. (1996b) *La agricultura española ante los nuevos escenarios de la PAC;* Revista Española de Economía Agraria, 176-177, pp. 265-301

Sumpsi, J.M. and C. Varela-Ortega (1995) *The Common Agri-environmental Policy and its Applications to Spain;* In: Albisu, L. M. and Romero, C. (eds): Environmental and Land Use Issues. An Economic Perspective. Wissenschaftsverlag Vauk, Kiel, pp. 119-133

Sumpsi, J.M., C. Varela-Ortega and E. Iglesias (1996) *CAP and the environment in Spain* (internal document)

United Nations Organization (1993) *Lucha contra la Desertificación en el Mediterráneo;* LUCDEME project. UNO

Valladares, M.A. (1993) *Effects of the EC policy implementation on natural Spanish habitats;* The Science of the total environment, 129, pp. 71-82

Varela-Ortega, C. (1997) *Situation and possibilities in Spain for the assessment of additional policy instruments for plant protection products;* In: Oskam, A. et al. (eds). Additional EU policy instruments for plant protection products, 2nd phase PES-A programme. DG XI. EU Commission, Brussels

Varela-Ortega, C. (1998) *The Common Agricultural Policy and the environment: Conceptual framework and empirical evidence in the Spanish agriculture;* In: Antle, J., Lekakis, J. and Zanias, G. (eds.): European Agriculture at the Crossroads: Competition and Sustainability. Edward Elgar; UK

Varela-Ortega, C., A. Garrido and M. Blanco (1995) *Analysis of the Socioeconomic and Environmental Impacts of Different Policies in the Spanish Region of Andalucia;* Regional Report. (EU contract number 8001-CT91-0306-4706A); Commission of the European Union, DG VI. Brussels

Velez Muñoz, R. (1991) *Los incendios forestales y la política forestal;* Revista de Estudios Agrosociales, 158, pp. 83-107

Vera, F. and J. Romero (1994) *Impacto ambiental de la actividad agraria;* Agricultura y Sociedad, 71, pp. 153-183

Viladomiu, L. and J. Rosell (1997) *Gestión del agua y política agroambiental: el Programa de Compensación de Rentas por reducción de regadíos en Mancha Occidental y Campo de Montiel;* Revista Española de Economía Agraria, 179

Zorita, E. and K. Osorio (1995) *La utilización del territorio mediante sistemas pastorales;* Producción de carne de vacuno de calidad (II). Bovis 67

10. ITALY

Alessandro Bordin, Luca Cesaro, Paola Gatto and Maurizio Merlo

10.1 Introduction

Farming and forestry in Italy are often linked to the state of the environment and the quality of rural landscapes. Conservation groups, the mass media and the public in general are all involved in the debate on the relationship between agriculture and the environment. Farmers also participate in the debate. Both positive and negative aspects of the relationship are raised, according to the issue and the arguments posed by interested groups.

The positive externalities of farming and forestry are mainly evident in upland areas and include, for example:
- landscape quality due to attractive land uses, such as vineyards, orchards, citrus groves, pastures, meadows, trees and hedgerows;
- environmental conservation due to low intensity pastures and meadows and uneven-aged, mixed forestry.

The negative externalities are mainly evident in the lowlands, particularly in the Po valley and the coastal plains where intensification is most acute. These include, for example:
- pollution of groundwater and surface water from nutrients and pesticides;
- disappearance of traditional farming patterns and their replacement with new ones involving loss of landscape features and negative environmental impacts.

Another harmful effect of lowland intensification is the increased income gap with marginal hilly and mountainous areas. The consequence is often land abandonment, degradation of landscapes and rural life, and unpredictable environmental outcomes. The unknown question concerning large areas across the country is whether they return to wilderness, providing at least some sort of positive externalities, or are just facing scenic and ecological decline, losing their traditional positive externalities while, perhaps, gaining new negative externalities from agricultural and forest land abandonment. It must be stressed that the boundaries between positive and negative externalities are far from being well understood, let alone defined (Hodge, 1991). The lack of a positive externality to which people are accustomed, is often perceived as a negative externality. The established cultivation and management of mountain meadows and pastures have traditionally provided such wider benefits as watershed management, beautiful scenery and biodiversity. Abandonment reverses the situation and may lead to negative externalities, including floods, avalanches, monotonous landscape, and loss of biodiversity.

Both positive and negative externalities are thus related to agricultural technologies and farming practices, but also to the circumstances of the farming community and the social and economic viability of rural areas. However, compared to some EU countries, the state of rural areas and the environment in Italy is not so much seen as a consequence of the CAP. In fact the technological tendencies and the polarization between intensive production systems in the lowlands, with their harmful effects on the environment, and the abandonment of mountain and hill farming, are considered background trends which started with 19th century industrialization and its associated economic developments (Sereni, 1961; De Benedictis 1996, and Merlo, 1991).

The evolution of agriculture since the beginning of this century is summarized in Tables 1 and 2. The intensification trend became pronounced in the '50s and '60s, as did land abandonment, the reduction of pasture and meadows and the expansion of woodland. The overall rural land use pattern has been impinged upon by urban growth, which during the past few decades has been around 50,000 ha per annum. Such a large increase in urban land has not been observed in other countries. This change has been particularly felt because of its suddenness and because it has included more than two million ha of the more fertile peri-urban areas with considerable landscape effects. Agricultural intensification and urban growth have mainly taken place in the lowlands.

The other paramount features of Italian land use change - abandonment and afforestation - were mainly observed in mountainous and hilly areas (Table 2).

Table 1   Land use evolution in Italy, 1910-1990 (1,000 ha)

Year	Agricultural employment (% of total)	Agricultural area	of which: crops including orchards and vineyards	meadows and pastures	Idle and abandoned land	Forests	Urban areas	Non productive land	Total land area
1910	58.4	20,773	15,193	5,580	1,035	4,564	611	1,003	27,986
1929	51.7	20,586	14,873	5,713	1,832	5,295	639	1,054	29,406
1955	38.8	20,908	15,760	5,148	1,110	5,761	727	895	29,401
1965	25.8	20,438	15,320	5,136	1,011	6,089	1,131	766	29,453
1975	15.3	17,527	12,313	5,214	3,220	6,309	1,655	754	29,465
1985	11.1	17,090	12,114	4,976	3,056	6,429	2,130	757	29,462
1990	8.9	16,850	11,972	4,878	2,881 a)	6,751 a)	2,219	753	29.454

a) Taking into account the 2 million ha of abandoned farmland, now under natural afforestation, this area increases to 8,750,000 has as shown by the 1985 Forestry Inventory (IFNI-ISAFA, 1988).
Source: Elaborated from ISTAT (various years) and IFNI (1988).

Table 2   Farmland evolution in mountainous, hilly and lowland areas of Italy, 1961-1990 (1,000 ha)

Year	Mountainous areas	Hilly areas	Lowlands	Total
1961	4,687.7	8,497.5	5,337.9	18,523.1
1970	4,402.2	7,998.0	5,090.0	17,492.0
1982	3,914.0	7,206.0	4,723.0	15,843.0
1995	3,639.2	6,849.0	4,558.0	15,046.0

*) Estimated area.
Source: elaborated from ISTAT (various years).

Concentration of production has taken place in the lowlands: the richest areas in terms of water availability and soil fertility. The rising intensity of production has required, however, an increasing amount of chemical fertilizers and pesticides, as shown in Tables 3 and 4.

Table 3   Agricultural use of chemical fertilizers: total sales (1,000 tonnes) and application (kg per hectare of agricultural land)

					Sales					
	1950	1955	1960	1965	1970	1975	1980	1985	1990	1994
Nitrogen	316.2	319.2	347.8	461.8	550.4	710.1	953.2	1,011.2	757.1	842.8
Phosphates	454.3	490.2	397.0	452.8	486.3	441.0	625.8	609.8	603.4	585.4
Potash	106.7	109.8	127.3	167.6	194.5	218.9	366.3	340.0	355.2	335.3

					Application					
	1950	1955	1960	1965	1970	1975	1980	1985	1990	1994
Nitrogen	15.1	15.3	17.0	22.6	31.4	40.5	55.8	59.2	44.9 a)	50.0
Phosphate	21.2	23.4	19.4	22.1	21.7	25.2	38.2	35.7	35.8	34.7
Potash	5.1	5.2	6.2	8.2	11.1	12.5	21.4	19.9	21.1	19.9

a) The apparently sharp decrease in nitrogen fertilizer application was the result largely of the adoption of new survey methods. However the declining consumption trend is widely accepted and is confirmed by subsequent data.
Source: ISTAT (various years) and FAO/ECE (1990).

National figures on the use of agrochemicals per hectare of land do not reveal the actual concentration of consumption in the lowlands, as shown by INEA-TECNAGRO (1991) (Figure 1). It is clear that the environmental situation of several Italian lowland areas, particularly the Po valley in Northern Italy and the coastal plains, is as bad as other intensive farming areas of Europe. These are also the most urbanized parts of the country, which are affected most by industrial pollution.

The concentration of intensive crops in lowland areas has been paralleled by livestock developments, particularly pig farming. Whilst the number of cattle has decreased since the 1960s, the number of pigs rapidly increased from 3.3 million in 1961 to almost 9 million in 1982. The factor which has had the most significant impact on the environment is the increased size of the production units. In 1961 all pig farms had fewer than a thousand pigs, whilst in 1982, 57% of pig numbers were concentrated on farms with more than thousand pigs. Although the trend towards larger production units has recently slowed down, nevertheless, in 1990, two-thirds of pig production was concentrated in units with more than a thousand animals (ISTAT Agricultural Census, 1991).

The most significant indicator of agricultural intensification, and the underlying technological development, is given by the Gross Product (GP). From the 60s to date the GP has increased from 44 to 55 billion lire (ISTAT, various years, constant prices 1989) notwithstanding a remarkable loss of 4 million ha (about 20%) of productive farmland. The increase of production has taken place above all in lowland areas (despite urban growth) while abandonment and extensification have taken place above all in mountainous and hilly areas.

The polarization trend can be summarized by the Concentration Index of agricultural GP which distinguishes the contribution of lowland, hilly and mountainous areas to GP. The Index increased from 0.26 (in 1960) to 0.38 - 0.40 (in 1989) as shown by Figure 2. The trend is even more impressive given that some 2 million ha of highly productive lowland have been lost to urban development.

The role of the CAP

It is almost impossible to isolate the environmental effects of the CAP from the overall influence of agricultural technologies and the (world) market prices of production factors and commodities. A number of aspects need to be considered. First, the intensification and polarization of Italian agriculture started well before the full operation of the CAP. Secondly, the trend towards polarization peaked around 1980, which was well before the CAP was substantially reformed. Thirdly, the production trends for pork, which is outside the main domain of the CAP, have been rather similar to those for products which are regulated by CAP measures. Finally, the deterioration of the farmed environment is registered in many other parts of the world including in countries that have never adopted price support policies.

Therefore, the role of the CAP must not be exaggerated. However one has to acknowledge that the CAP has been effective in pursuing, and accentuating, existing trends: of increasing productivity during the '60s and the '70s; and then the turn towards less intensive farming from the second half of the '80s onwards. In other words, as in all 'successful' policies, the CAP has followed the existing 'wave' in terms of technological developments, and the demands of farmers and rural societies for support. More recently, the agricultural sector is being reoriented to the requirements from society to maintain the environment, recreation and conservation together with the provision of safe, high quality food.

The rise in productivity of course is a well established fact, but the turn towards less intensive production methods is a trend which started only recently and is still in its infancy. Therefore the major issue seems to be to what extent did the CAP contribute to enhanced linkages between agriculture and the environment. What can be said is that, in Italy, the sectors where the CAP has had the more negative environmental impact, such as creals and beef, are those where price policy has been more effective.

As is well known, however, the bulk of Mediterranean products have been less protected, and therefore the CAP must not be the only reason for the declining relationship between agriculture and the environment. Therefore, accepting that the CAP has only largely followed the trend wave, at least as far as Italy is concerned, then a commodity-based analysis is called for, considering the specific effects of different CAP regimes. The analysis will also consider the effects of more general policies, such as those determined by the accompanying measures to the CAP re-

246

form, the structural measures, including the Less Favoured Areas (LFA) scheme, and the Nitrates Directive.

10.2 Impacts of the CAP

10.2.1 The arable regime

The area of arable crops has shown a remarkable reduction since the 1950s (Table 5). This has been largely due to the polarization process in Italian agriculture, including the almost total abandonment of mountainous and hilly arable areas, offset by the ploughing of lowland permanent meadows. That the total surface of meadows and pastures has not decreased is purely due to the fact that the abandonned arable land in the mountains and the hills is treated statistically as meadows and pastures (Table 1). To characterize the situation, the term 'maize-desertification' is often used to refer to the intensive, monocultural farming pattern which involves the abandonment of old farmhouses and the enlargement of holdings and fields. This phenomenon is often based on the so-called farm-destructuralization and the use of contractors who undertake all operations from ploughing and sowing to spraying and harvesting, over a number of farms.

Table 5   Total surface area, yield per hectare and sales price of the main arable crops during the period 1950-1993

Crop	1950	1955	1960	1965	1970	1975	1980	1985	1990	1993
					Surface area (1,000 ha)					
Wheat	4,720	4,852	4,554	4,288	3,694	3,545	3,258	3,032	2,612	2,205
Maize	1,241	1,237	1,188	1,028	1,026	897	998	923	877	936
					Yield (tonne per hectare)					
Wheat	1.65	1.96	1.49	2.28	2.34	2.71	2.73	2.82	2.97	3.59
Maize	1.55	2.59	3.21	3.23	4.63	5.94	6.84	6.95	7.71	8.71
			Price (ECU/tonne with reference to 1993 lira/ECU exchange rate)							
Wheat	440	455	435	392	341	278	265	214	167	179
Maize	359	322	269	280	296	285	259	233	199	169

Source: ISTAT (various years) and FAO/ECE (1990).

The economic rationale behind these developments is shown by yield trends (Table 5) which have increased considerably from the 50s to the 90s. The CAP price policy is often blamed for being the main force behind intensification of arable farming. However the prices of maize and wheat in real terms (Table 5) have been substantially reduced. Intensification must therefore be related to technological development, mainly induced by the factor-price ratio. The high cost of labour, compared to all other inputs, is certainly the key driving force behind intensification. Of course the role played by CAP's guaranteed prices must also be acknowledged, being very important for explaining farmers' choices of crops.

The overall environmental impact of these trends are rather negative. Mountain and hilly areas have lost, amongst other things, several traditional crops, landscape beauty, and the so-called 'mosaic' pattern made up by *ager* (arable), *saltus* (meadows and pastures) and *silva* (wood-land) landscapes. Lowlands have seen the disappearance of mixed farming based on meadows, animal husbandry, arable crops and crop rotation, in addition to the various environmental losses stretching from biodiversity, landscape beauty to traditional rural cultures. In other words, the overall diversity of rural areas has been diminished.

Some parts of the CAP may have in themselves remarkable effects on the state of the rural environment and landscape, for example set aside. Its introduction, as a voluntary measure in the agricultural year 1988-89 affected almost 100,000 ha of land, mainly located in the Central-Southern part of the country (Sicily and Tuscany). The area increased in the following year to

Table 6   Set aside scheme on arable land in 1995 (ha)

	5 years scheme (Reg. 1094/88) (voluntary)	Rotational scheme (Reg. 1765/92) (compulsory)	Non rotational scheme (Reg. 1765/92) (compulsory)	Non food (compulsory)	20 years scheme (Reg. 2078/92) (volunary)	Total
Piemonte	2,664	11,764	6,676	6,412	124	27,640
Valle d'Aosta	0	0	0	0	-	0
Lombardia	1,618	25,788	10,939	11,050	80	49,475
Trentino A.A	0	0	1	0	-	1
Veneto	836	22,284	3,926	10,652	54	37,752
Friuli V.G.	1,400	8,355	3,973	4,089	8	17,825
Liguria	1	4	4	2	-	11
Emilia Romagna	6,588	20,755	4,387	9,409	1,410	42,549
Toscana	36,694	20,856	12,675	3,634	457	74,316
Umbria	5,104	9,892	5,897	4,006	414	25,313
Marche	4,843	12,663	2,612	8,420	13	28,551
Lazio	8,161	11,605	4,075	1,810	176	25,827
Abruzzo	1,441	2,303	1,125	557	0	5,426
Molise	1,226	5,265	164	2,807	6	9,468
Campania	272	1,770	275	136	-	2,453
Puglia	26,894	17,506	3,645	2,210	-	50,255
Basilicata	22,812	4,166	909	325	934	29,146
Calabria	4,733	1,176	1,232	137	-	7,278
Sicilia	31,787	4,437	1,954	653	807	39,638
Sardegna	11,560	1,822	447	77	-	13,906
Total	168,904	182,411	64,916	66,386	4,483	486,830

Source: INEA (1996).

168,904 ha. Rotational compulsory set aside introduced under the 1992 CAP reform has affected 182,411 ha mainly located in Veneto, Tuscany and Emilia Romagna regions, whilst the long term, 20 years, set aside, according to the accompanying measure EC Regulation 2078/92, has been applied to 4,483 ha, mainly located in the regions of Emilia Romagna, Basilicata and Sicily. Table 6 gives the overall situation of set aside application in 1995 according to the various measures adopted since the 1988-89 agricultural year.

Empirical results indicate that voluntary set aside has affected primarily the less productive and less intensive agricultural systems of Central-Southern Italy. Participation rates for the voluntary set aside schemes remained very low in the intensive arable production regions of the Po valley. This scheme has therefore failed to achieve its main objective: production containment and promotion of less intensive farming. To a certain extent the 168,904 ha affected by the first generation of voluntary set aside have created more environmental-landscape problems than they have solved. In fact the existing trends towards land abandonment has been favoured, the landscape has lost in terms of beauty, and therefore the rural situation has worsened from various points of view favouring the existing trends of rural out-migration and deprivation, particularly in the southern part of the country.

It has been shown how set aside and the subsequent lack of tillage and/or vegetation cover, increases run-off with general loss of water and soil. According to various cases and situations, this brings about leaching of nitrates where lack of vegetation cover is accompanied by a heavy rainfall regime - as it is the case of Northern Italy - with consequent pollution of underground water (Giardini and Borin, 1996; Covarelli and Montemurro, 1993). In Southern Italy, another negative effect is given by evaporation of soil moisture - and therefore loss of water - where set-aside implies lack of hoeing in the summer season 1).

The shift from voluntary to compulsory and rotational set aside has certainly improved environmental impacts of the set aside scheme. However, of the 182,411 ha of rotational set aside, only 90,000 ha are located in the most intensive farming areas of the Po valley. Long term set aside (20 years period - EC Regulation 2078/92), which potentially has the most important favourable environmental impacts, has been applied to only 4,483 ha (in Emilia Romagna, Basilicata and Sicily).

All in all, the application of set aside to date in Italy seems therefore to be rather positive as far as rotational and long term set aside are concerned, whilst it is rather negative with voluntary set aside measures. The various positive/negative impacts of set aside are listed in Table 7 with reference to the Italian situation. Income and production effects are not, however, considered by the Table e.g. the slippage effect and the elasticity of set aside acreage with respect to the rate of compulsory set-aside (Boatto and Pilati, 1997).

---

1)   It has always been stated by agronomy textbooks that 'one hoeing' is equivalent to 'one watering'.

Table 7   A typology of environmental impacts of set aside

Set aside typology	Positive impacts	Negative impacts
Fallow set aside - no vegetation cover	Correct solution in the following cases: - soil structure recovery when damaged by harvesters that have operated in wet soil; - as part of the management of low productivity marginal areas; - accomplishment of land improvement works.	The lack of soil vegetation cover exacerbates: - leaching of nitrates because of the missing crop interception (especially during the winter); - soil erosion especially in hilly and mountainous areas (wind and surface); - diffusion of weeds.
Set aside with "spontaneous" vegetation cover	Reduces the negative impacts of "naked" fallow set aside. Amongst various positive effects must be mentioned: - recovery of biodiversity (plant and animal species which no longer can survive in 'man-made farming ecosystems'); - habitat for nesting and migratory birds.	Negative impacts less relevant: - intrusion in the landscape (untidiness); - some nitrogen leaching can occur during the winter; - diffusion of weeds.
Short term set aside	-----------------------	It increases negative impacts: - erosion - leaching of nitrogen and other nutrients
Regulation 2078/92: long-term (20 years) set aside	It allows the re-establishment of a more natural environment, favouring afforestation of farm land, creation of wetland and landscape patchwork. Positive effects of set aside are therefore felt to the greatest extent, and negative effects are reduced	It is the best solution for reducing the negative effects of fallow set aside.
Fixed set aside	-----------------------	It favours abandonment of the less productive parts of the farms, where stewardship is more needed. Production effects are not felt.
Rotational set aside	It allows flexibility in a farm's organization. Positive and negative effects are spread around the farm, expected production reduction is more evident, slippage effects are limited	It reduces some negative effects like abandonment and lack of stewardship.

Source: elaborated from Giardini and Borin, 1996.

250

10.2.2   The dairy regime

There is no doubt that milk price support since the '60s has been influential in shaping the sector, now mainly based on medium sized family farms. Many small dairy farms (less than 10 cows), as well as the larger farms employing hired labour have disappeared. Large dairy farms had disappeared well before milk quotas were introduced, due to increased costs of hired labour and the effects of technological progress in crop mechanization. These holdings have been converted to arable farms. Another remarkable consequence is the maintenance of farming in Less Favoured Areas (LFAs) including mountainous areas, where no alternatives exist other than dairy farming (Table 8).

Table 8   Location of dairy farms, 1994-95

	Number of dairy farms	Total milk production (1,000 tonnes)	Average milk production per holding (tonnes)
Mountains regions	51,299	2,071.5	40.4
Other LFAs	10,256	620.0	60.5
Lowlands	43,792	7,475.8	170.7

Source: Pieri and Rama (1996).

To some extent, dairy farming has helped safeguard more diversified farming patterns, including cereals and, particularly in mountainous areas and LFAs, the management of meadows and pastures. The consequences for landscape quality and the maintenance of rural life are therefore remarkable. In general, dairy farming has not induced intensive farming systems with major pollution effects, as has been the case for the cereal and the beef sector.

The milk quota regime since 1984 was able at least to maintain a 20% share of national milk production in the mountainous areas and other LFAs (Figure 3). This is particularly remarkable given that previous CAP measures, together with the background trends, have systematically penalized mountainous areas and LFAs when compared to the lowlands.

The milk quotas regime, including milking cows abatement premia, could certainly have been an opportunity to support mountainous areas and LFA milk production through the re-assignment and transfer of quotas. Environmental objectives could also have been achieved by applying quotas to farmers with sufficient land applying environmentally friendly techniques. Unfortunately, the application of milk quotas in Italy has never been conceived in terms of LFA support or environmental benefit; the huge economic and social pressures from lowland farmers would not have permitted this. Markets and production have been the main concern of the various measures. It is nevertheless remarkable that abatement policy and dissuasion measures have been relatively more effective in the lowlands (-45% milk production) than in mountain areas (-40%) and LFAs (-35%).

It was only recently, though, that some explicit emphasis was given to LFAs and environmentally related issues. In particular Act 768/92 stated that quota transfers must be made locally

251

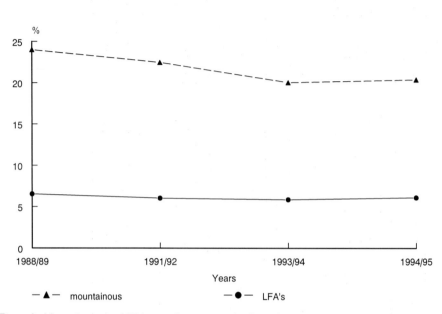

Figure 3 Mountain area and LFAs contribution to total milk production

within LFAs and mountainous areas, preventing the transfer of quotas towards the lowlands. It is also stated that transfer and re-assignment must be oriented towards extensification and environmentally friendly production techniques. Acts 463/96 and 464/96 are indicative of the new attitude giving priority in bidding for new quotas to mountain and LFA farmers. Finally it must be underlined that this new orientation has found its strength in the appellation d'origine cheeses (and other dairy products) policy and in organic milk production (see Section 2.5.

10.2.3 The beef regime

The concentration process during the last decades, paramount in the pig sector, has also been evident in beef production (Table 9). The growth of cattle production units with more than 500 stock (un-heard of in 1961 but 11.5% of total stock in 1990), and also of those with more than 100 stock (from 6.1 to 31.0% of total stock) is mainly due to beef production. In fact the bulk of milk production remains confined to farms with fewer than 100 cows.

As with other agricultural sectors, it is difficult to say to what extent the development is due to general technological-economic trends or to the CAP regime. It must be pointed out, however, that the trend is much more pronounced in the pig sector where the price support policy has been

Table 9 Development of farms with livestock and number of animals by farm size for cattle and pigs, 1961-1990

Cattle

Size (heads)	1961 farms (x 1,000)	%	animals (x 1,000)	%	1970 farms (x 1,000)	%	animals (x 1,000)	%	1982 farms (x 1,000)	%	animals (x 1,000)	%	1990 farms (x 1,000)	%	animals (x 1,000)	%
≤ 5	1,048.3	68.2	2,658.4	27.9	567.4	59.0	1,525.4	17.5	230.5	46.0	650.5	7.5	122.3	38.3	348.3	4.5
6-9	283.0	18.4	2,130.9	22.4	199.9	20.8	1,526.3	17.5	86.8	17.4	627.9	7.2	50.4	15.8	366.6	4.7
10-19	143.2	9.3	2,060.0	21.6	117.7	12.2	1,715.0	19.7	88.4	17.7	1,180.1	13.6	59.8	18.7	804.9	10.4
20-49	51.3	3.4	1,499.7	15.8	57.1	5.9	1,776.8	20.4	61.3	12.3	1,813.5	20.9	52.1	16.3	1,568.9	20.2
50-99	8.1	0.5	556.2	6.3	13.8	1.4	956.2	11.0	20.3	4.0	1,356.4	15.6	20.5	6.4	1,376.1	17.8
100-499	3.5	0.3	579.8	6.1	6.0	0.6	1,019.6	11.7	12.1	2.4	2,170.9	25.0	13.5	4.2	2,405.4	31.0
≥ 500	0.0	0.0	0.0	0.0	0.2	0.1	177.0	2.1	0.9	0.2	886.8	10.2	0.9	0.3	888.8	11.5
Total	1,537.6	100	9,485.1	100	962.2	100	8,696.4	100	500.4	100	8,686.1	100	319.6	100	7,759.1	100

Pigs

Size (heads)	1961 farms (x 1,000)	%	animals (x 1,000)	%	1970 farms (x 1,000)	%	animals (x 1,000)	%	1982 farms (x 1,000)	%	animals (x 1,000)	%	1990 farms (x 1,000)	%	animals (x 1,000)	%
≤ 19	996.9	98.7	2,039.8	60.8	891.2	96.4	2,066.8	34.9	534.1	96.4	1,203.3	13.4	342.9	96.0	799.2	9.5
20-49	8.2	0.8	241.7	7.2	21.5	2.3	649.2	11.0	9.2	1.7	260.6	2.9	6.1	1.7	173.7	2.1
50-99	2.2	0.2	162.7	4.8	5.1	0.6	366.4	6.2	2.9	0.5	196.0	2.2	1.9	0.5	125.5	1.5
100-499	1.8	0.2	264.5	7.9	5.7	0.6	1,287.7	21.6	4.7	0.8	1,100.1	12.3	3.2	0.9	771.8	9.2
500-999	1.4	0.1	644.0	19.2	0.8	0.1	570.9	9.7	1.6	0.3	1,129.3	12.6	1.4	0.4	959.9	11.4
≥ 1,000	0.0	0.0	0.0	0.0	0.4	0.1	996.3	16.8	1.8	0.3	5,061.4	56.6	1.9	0.5	5,576.4	66.3
Total	1,010.4	100	3,353.0	100	924.8	100	5,928.3	100	554.3	100	8,950.8	100	357.4	100	8,406.5	100

Source: ISTAT, General Agricultural Censuses.

253

far more limited, compared to beef and milk. It is the case, yet again, that the role of CAP must be considered within its limits and not be exaggerated.

Beef production, together with milk, pigs and poultry, is mainly concentrated in the Po valley regions, the most densely urbanized and industrialized part of the country. It is not surprising that national health laws, and more recently environmental laws, have been passed, independently and well before changes to the CAP, to reduce the problems of intensive animal breeding and the related impacts. Also Local Authorities, responsible for regional and physical planning, play a key role. The so-called Merli Act n. 319/76 on water pollution passed in 1976 is fundamental in this field. In regard to livestock it lays down a minimum acreage for animal waste disposal. The limit was set at 4 tonnes of cattle per hectare, i.e. 6-7 dairy cattle or 7-8 beef cattle (young stock). All production units exceeding this threshold - the majority of beef and pig production - were to be considered as potentially polluting industrial plant whose waste has to be treated, and effluent discharge has to comply with appropriate standards.

However, from the 80s it has been the EC, also including the CAP, that has assumed the lead in preventing environmental problems created by beef and other animal production. The issue will be examined in Section 2.7 concerning the Nitrates Directive. It must be stressed, however, that farming, in order to be eligible for compensatory payments under the beef regime is limited to a certain stocking density (livestock unit per hectare) that is lower than that specified by the 1976 Italian legislation: 2 LU per hectare as of 1996. The limit was even lower for the so-called sensitive areas (Regulation 797/85, Article 19). Of course it must be underlined that a substantial part of Italian beef and pig production units remain outside 'farming', in other words farms and production units are considered industrial plants, and producers are therefore obliged to apply the general environmental rules (from the previously quoted Merli Act of 1976) implying higher costs due to waste treatment.

However, it should be noticed that the CAP also has adopted a 'carrot' approach. This approach is aimed to prevent pollution from beef production, and indeed all other main agricultural

Table 10   Maximum number of Livestock Units (LU) per hectare of farmland

	Sensitive areas (Article 19, Reg 797/85) (LU/ha according to various regions average) a)	CAP Reform 1992 (Animals/ha)	Merli Act (319/77) (Animals/ha)
Dairy cattle	1.1	2	6-7
Young stock or beef cattle	1.1	2	7-8
Fattening pigs	1.1	16	26-40
Sows with piglets	1.1	5	16
Turkeys or ducks	1.1	100	260-400
Laying hens	1.1	135	1,333
Young hens (0-16 weeks)	1.1	285	2,000

a) The various Italian Regions applying Art 19 have adopted different limits. In particular in terms of LU: Toscana 0.75, Trento 0.6, Bolzano 0.4, Piemonte 0.5 (Alpine pastures), Lombardia 2.0, Veneto 1.0 (pastures) and Emilia 1.0. Of course the different kind of animals have to be transformed into LU according to appropriate coefficients.

254

production. First of all it must be stressed that farms with less than 15 LU are exempted from the acreage limits. An additional premium of 30 ECU is provided to those holdings with stocking density which does not exceed 1.4 LU/ha. It is also foreseen that sheep compensation rights cannot be transferred outside LFAs. This environmentally oriented policy is even more evident when considering Regulation 2078/92 where extensification and conservation of traditional breeds is supported by specific aids. In other words, the EU has introduced policies with payments which are conditional on environmental requirements. However, the trend towards intensification in beef production achieved its peak in the 80s with the 90s showing a slow down.

A new trend may have started during the 1980s with policies to interact with new emerging clean technologies that do not necessarily mean less production but often a more careful and sensible adoption of modern technical means. As shown by Giardini et al. (1989) in relation to soil types and farming techniques, there is scope and room for technical adjustments, i.e. adoption of new updated environmental-friendly practices that do not substantially affect productivity. The future of this technological escape (of course, common practically to all agricultural productions) has to be seen and applied in the context of more environmentally oriented payments. Such payments may gradually replace the temporarily adopted compensation scheme. It must not be forgotten, however, that the 'payments policy' means, almost always, 'bribing' the producers with the aim to prevent pollution or undesirable environmental effect. This means that the 'polluter pays principle' is openly contradicted notwithstanding being often advocated by the EU.

10.2.4   Mediterranean crop regimes, appellations d'origine controllées and organic farming

Mediterranean crops (wine, olive oil, citrus fruit, vegetable and other fruit) represent a limited application of market support compared to cereals, milk and beef. Market regulations came later, tending to assume different forms than simply guaranteed prices. Wine, for instance, has been mainly subject to measures aimed at uprooting vineyards, outside the traditional quality areas. Subsidies for stocking and distillation have been applied only in certain years, without becoming routine. Also support provided to olive groves, orchards and citrus gardens has been aimed above all to maintain a certain land use, rather than the market of a commodity.

Many Mediterranean crops, including cheeses, have been subjected to *appellation d'origine*, a well established tradition in Italy. In order to protect the *appellation d'origine* against frauds, safeguarding both producers and consumers, legislation has been passed since the beginning of this century including codes of practices for each *appellation d'origine*. Techniques and therefore stewardship practices, in line with tradition, are imposed on producers wishing to be part of the *appellation*, including maximum production per hectare, traditional style vineyards, meadow conservation for producing quality hay, pastures in the Alps, reduced fertilization and spraying, etc. The positive effects in term of prices and income have been shown by several studies.

There is now some evidence of a correlation between the quality of certain products, like wine, cheese and fruits, and that of the landscapes where they are produced. This fact has been shown by studies in Italy during the 1970s which analysed the income effects of agricultural products appellation d'origine (Favaretti and Merlo, 1973). The price of these differentiated products was at least 10-20% higher than that of similar products not entitled to appellation d'origine. Of course, the higher the intrinsic quality, the more pronounced the price effects.

What is even more interesting from an environmental point of view is the extent to which higher prices are also effective in maintaining high-cost quality landscapes and other amenities. This applies particularly in mountains and hilly areas where the alternative could be abandonment. In other words, differentiation of agricultural products according to quality and origin seems to have a series of positive consequences, not only for farmers' income, but also for the landscape and other rural amenities. In fact, a virtuous circle is started:

- the product is guaranteed and a specific market segment is created;
- consumers are reassured and stimulated to show their true willingness to pay higher prices for the product and its image;
- higher prices allow producers to meet the higher costs inherent to traditional practices, including lower productivity, as imposed by the code of each *appellation*;
- higher prices also allow areas, otherwise neglected, due to their ultra-marginality, to remain cultivated, as in the case of steeply sloping vineyards and mountain meadows/pastures;
- the application of traditional practices and the survival of farming in marginal areas jointly contribute to the conservation of rural amenities in the form of upkeep of the landscape, maintenance of traditional land uses including soil conservation, footpaths and other rural infrastructures;
- within the geographical coverage of appellation d'origines, the farm areas not directly involved in the scheme are also generally better maintained;
- food processing of local quality products can take place within rural areas at a generally small scale, as sometimes required by the code of production, thus contributing to the overall rural economy;
- favourable circumstances for agri-tourism are also created, giving a further impulse to farm income and rural life and culture.

Overall evidence of the beneficial environmental effects of *appellation d'origine* has been given by early studies during the 70s aimed to analyse the income effects of wine *appellation d'origine controleé* (Favaretti and Merlo, 1973). Close attention to these experience has also been paid by the OECD (1996).

The policy of guaranteed origin and quality production was originally integrated at the EC level during the 1970s. The first policies focused on wine and cheese and were introduced in response to the established legislation of Mediterranean countries. More recently, they have been extended to cover many other agricultural products, from fruits to vegetables. Important EU Regulations apply: Reg. 2081/92 and 2082/92 on products origin and specificity. Thanks to Reg. 2092/91 on organic farming, the issue of healthy food has also been, to some extent, included. It should be noted that the cost of these policies remains generally low: on the one hand, it is the consumer who pays for the quality of products and therefore for the landscape; on the other hand, administrative costs remain low compared to subsidies and other support. Control is often delegated to producers associations (consorzi di tutela) that have a vested interest in protecting the quality of the product together with the label of origin.

Given the increasing interest by consumers in environmentally friendly products, along with the sophistication and willingness to pay by European consumers, policies guaranteeing the origin and quality of production would seem to have a promising future. Until now, the positive environmental impacts have not been directly considered as a policy objective. However, it is interesting to note that the recent Italian legislation on appellation d'origine wines (1994) clearly makes reference to the positive environmental and rural development impacts. Along the same lines, the progress of EU legislation on Ecolabel (EC Reg. 880/92) and the related stewardship certification could be mentioned. At the moment, they are widely discussed in relation to timber and forestry - though applicable to many other products - which should certify the benefits associated with agro-forestry sustainable management.

10.2.5   Structural measures and LFA policies

Structural measures and LFA policies include the following instruments:
-    European Agricultural Guidance and Guarantee Fund (EAGGF)
-    European Social Fund
-    European Regional Development Fund

It is not easy to give clear evidence of their effects on rural areas and the environment. The compensation from Directive 75/268, which is aimed to support farmers in mountainous areas, certainly has had positive landscape effects in terms of conservation and maintenance of pastures and meadows. However, it has not modified the trend towards land abandonment. More radical interventions were required in terms of land mobility: even with 100 or 200 ECU per hectare one cannot think to maintain any viable farming in a context of highly fragmented and dispersed holdings - the average area of an Italian farm is 6-7 ha, fragmented in 4-5 pieces. Measures were needed, first, to consolidate pieces of land. For various reasons, Italy has not been able to produce appropriate legislation in this field. The regions meanwhile have not always been able to channel the EC support, however limited, foreseen by Directive 75/268 and other structural interventions. In fact some regions did not implement the LFA compensation for budgetary or administrative problems, or in the best cases the implementation was not continuous.

Another EC structural measure was given by Reg. 269/79, the first in the field of afforestation. Some 43,000 ha have been planted mainly in Central and Southern Italy, with funds made available by the EC regulation, practically absorbing the already existing Italian programmes. The basic remark, common to all Italian afforestation policies, is the questionable choice in terms of species and environmental orientation. Production forestry too often has been the main objective. Meanwhile, afforestation has taken place using the only available land- obviously, that which is marginal. The policy has therefore failed both its objective of increasing timber production (because of the marginality of the site), and of creating an environmentally-oriented forestry (because of the species used and the techniques applied). The purpose of landowners too often has been the exploitation of the financial opportunities offered by the EC, the state and regions. In the meantime, no guidance has been given on the maintenance and restoration of existing forests, for instance coppices to be converted to high forests and chestnut groves to be restored. Similar misunderstandings have also afflicted the agro-forestry part of Reg. 2088/85 (Mediterranean Integrated Plans) that have sometime failed to achieve their environmental objectives. All in all, though, the results have been more positive than with Reg. 269/79.

Regional/structural policies have substantially evolved since the 70s. The most significant and positive changes have taken place since the second half of the 80s:

- incorporating new objectives into traditional structural policies, including environmental protection;
- revising traditional tools and criteria of eligibility;
- incorporating traditional structural measures into programs for rural development (Objective 5b) and regional development (Objective 1);
- focusing financial support on specific rural/less developed areas;
- more involvement of local administrations, e.g. regional authorities, with substantial improvement in the capacity to assemble projects and increasing consistency between regional and EU programmes.

However, it has been remarked (Mantino and Pesce, 1996) that the main objectives of the reform of the Structural Funds have still to be met because of difficulties in programme definition and co-ordination, in implementation, and in monitoring and evaluation. Now, the major problems concern the following aspects:

- the complexity of normative and institutional references;
- the lack of territorial focus (i.e. the links between specific areas and problems);
- the predominance of top-down approaches;
- the need for greater modulation in financial aid between different sub-regional areas;
- the need to evaluate socio-economic and environmental impacts;
- the compatibility, and competition, with price and market policies.

In this context, the Leader Programmes can be considered really innovative, because they promote:

- intersectorial participation and local partnerships;
- local co-ordination agents (GAL);
- decentralized decision-making;
- building capacity at the local level.

However, it is surprising how the environmental impacts of structural measures are not always considered, even if the role of well-kept landscapes and environments is now considered essential for any land-based rural development. From this point of view, the CAP and the EU environmental policy do not always fully recognise that Mediterranean rural areas have been, and still are, liable to:

- farmland neglect (abandonment) in marginal conditions with depopulation, degradation, erosion, landscape decay etc;
- congestion, pollution, landscape impoverishment etc. in highly developed areas.

In both cases, there is a net loss for individual landowners, the rural inhabitants and society as a whole. The best way to prevent such undesirable evolution is given by agricultural-environmentally oriented policies. However these should be linked and consistent with local physical planing, and regional development policies. The post-reform CAP has started to play a key role in these fields, as discussed in the following sections.

Analysis of the environmental impact of recent structural policies is not available. The only possible remark is that environmental considerations have to be taken into account in plan preparation and also in the ex-post evaluation.

10.2.6   The accompanying measures

The objectives of two of the accompanying measures focus on the rural landscape and the environment. Regulation 2078/92 includes agri-environmental measures and Regulation 2080/92 measures for the afforestation of agricultural land. In both cases implementation has taken quite a long time, being delegated by the Ministry of Agriculture to the Regions, as is usual in Italy. Regulation 2078/92 was implemented in 1994 in 13 regions; and 5 other regions started in 1995. In 1996, Campania was the only region still waiting for official approval by the EU. Thus, the programmes have been running for three years at best, and for only 1-2 years in most Italian regions. This time span is not long enough for allowing an assessment of the real effects of the Regulation on the environment. Data are only available in terms of uptake (i.e. number of farms, hectares and reared animals), and farmers' perceptions and attitudes towards the accompanying measures.

Regulation 2078 has two main objectives: on the one hand, the minimization of negative agricultural impacts through reduction in the use of chemicals and the adoption of eco-compatible practices; and on the other, countryside stewardship and environmental conservation. The first objective has been acknowledged and pursued in almost all Zonal Plans, while the countryside stewardship objective is recognized as an important factor in land use management only in those areas where this has always been an accepted feature of physical planning. Table 11 illustrates the broad range of objectives of the different Regional Zonal Plans applied in Italy.

The anticipated uptake of the Zonal Programmes in terms of area and livestock units is summarized in Table 12 for the whole country and for geographic regions. It was projected that the Regulation would cover 11.5% of the total farmed area - with considerable differences between Northern, Central and Southern Italy - and 2.5% of of the total cattle population.

The data also reported in Table 12 show the rather limited impact that Regulation 2078/92 has actually had: after the first three years, only about one million ha (around 6.5 % of the total farmed area) and 26 thousands units of livestock (LUS) have been involved, with a total of 77,414 applications funded. Although an upward trend can be observed, the forecasts of the Zonal Programmes seem too ambitious, the projected livestock numbers being as much as seven times higher than what has been achieved. Many reasons have been put forward for the shortfall, most noticably the considerable bureaucracy involved, the approval procedure for the zonal programmes and the delay in the availability of the national co-financing. Also of importance are farmers' claims that the premiums are too low to meet the increased costs of the environmental requirements. In addition, there is a certain competition with other support policies. The different codes of practice and environmental standards between Italian regions have also been matters of debate.

More details about the potential environmental impact come from the analysis of the uptake of the different measures (Table 13). Measures A to C relate to the reduction of negative agricul-

Table 11   Objectives stated in the Regional Zonal Programmes of Regulation 2078/92

Region	Objectives
Piemonte	mountain areas: solving the problems of abandonment of land and animal breeding; lowlands: reduction of chemical inputs
Valle d'Aosta	maintenance of existing balance between farming and natural environment
Lombardia	mountain areas: solving the problems of abandonment of land and animal breeding; lowlands: reduction of chemical inputs
Bolzano	maintaining rural population
Trento	maintaining rural population; reduction of chemical inputs
Veneto	improving food quality; environmental protection through reduction of chemical inputs
Friuli V.G.	protection of mountain environment and groundwater
Liguria	maintenance of the balance between marginal farming and natural environment; prevention of soil erosion
Emilia Romagna	prevention of soil erosion; protection of water resources and biodiversity
Toscana	enlargement of protected areas; development of organic farming; conservation of soil fertility
Umbria	upland: multi-purpose forestry, prevention of soil erosion; traditional farming; lowland: reduction of chemical inputs
Marche	internal areas: conservation of nature and landscape; coastal plains: food quality; groundwater areas: reduction of chemical inputs
Lazio	upkeep of abandoned land, decrease of livestock density, organic farming, extensive farming
Abruzzo	mountainous and hilly areas: protection of landscape and woodlands; coastal and Fucino: reduction of chemical inputs
Molise	mountain areas: prevention of soil erosion; upkeep of abandoned land; coastal plains: reduction of agricultural impacts
Campania	coastal plains and lowlands: reduction of chemical inputs; hills: conservation of landscape and maintenance of extensive farming
Puglia	income support for full time farmers in order to avoid depopulation
Basilicata	mountains and hilly areas: upkeep of marginal areas, improvement of animal husbandry; lowlands: reduction of chemical inputs
Calabria	protection of water resources; fire prevention
Sicilia	food quality; control of flood disasters and pollution
Sardegna	increasing the value of local products, upkeep of marginal areas

tural impacts and refer mainly to the more fertile areas, while measures D to G promote active stewardship practices for the conservation of the natural environment and landscape in marginal areas.

Measure A1+A2 for the reduction of fertilizers and pesticides has accounted for the largest share of participants (43,397) and area (402,415 ha). Measure C, on the other hand, has been quite unsuccessful. Measure D1, referring to environmentally friendly farming practices and maintenance of countryside and landscape elements, covers 36,792 ha on 10,921 farms. Then follows organic farming (A3+A4) with 9,377 applications and 157,508 ha. As expected, participation in measures E, F and G has been limited. It is interesting to note that measure D1, is largely confined to the Northern Regions (95% of the total area involved) while the Southern Regions have focused mainly on organic farming.

Table 12 Uptake of the Regulation 2078 Zonal programmes 1994-97 in Italy

	Projected		Actual, 1996	
	ha	% a)	ha	% a)
Italy	1,930,701	11.5	1,011,822	6.0
North	1,140,133	21.9	589,433	11.3
Centre	368,992	13.6	202,396	7.5
South	421,576	5.9	209,993	2.9
	Projected		Actual, 1996	
	LU	% b)	LU	% b)
Italy	196,206	2.5	25,575	0.3
North	129,229	3.2	20,572	0.5
Centre	24,737	2.8	4,463	0.5
South	42,240	1.9	540	0.02

a) Of total farmed area; b) Of total cattle head as reported in Table 9.
Source: INEA, 1997.

Table 13 Reg. 2078 - Uptake by measure, 1996 (funded applications, hectares and livestock units (LU))

	Measures								
	A1+A2	A3+A4	B	C	D1	D2	E	F	G
applications	43,397	9,377	8,401	20	10,921	2,943	1,323	618	150
ha or LU	402,415	157,508	45,357	979	36,792	24,596	25,179	7,341	1,952

A1+A2: Reduction (A1) or maintenance of reduction (A2) of the use of fertilizers and plant protection products.
A3+A4: Change to (A3) or maintenance of (A4) organic agriculture.
B1: Change to more extensive forms of crops or maintenance of existing extensive production methods.
B2: Conversion of arable land into extensive grassland.
C: Reduction of the density of livestock per forage area.
D1: Use of other farming practices compatible with the requirements of protection of environment or natural resources as well as maintenance of the countryside and landscape.
D2: Rearing of local breeds in danger of extinction.
E: Upkeep of abandoned farmland or woodlands.
F: Long-term set-aside.
G: Management of land for public access and leisure activities.
Source: INEA.

In conclusion, the implementation of EC Regulation 2078/92 has had positive results only where the aid scheme has been built on an already existing environmental experience, e.g. the presence of a network of organic farming associations, of extension services for integrated pest management, of pedigree books for local breeds, or where incentives for the maintenance of alpine pasture and meadows had already been promoted at a local level. Positive hints for further developments also come from the measures strictly connected with the production of environmental goods, such as hedgerow conservation and planting when accompanied by information and technical assistance, as well as projects concerning the 20-year set aside and dealing with the restoration of wetland and wildlife shelter areas.

Like Regulation 2078/92, Regulation 2080/92, giving funds for reafforestation and wood improvements could also have direct influence on the environment once fully implemented. So far, however, and even more than with 2078/92, projections of Regional Programmes have not been confirmed by the actual take up (Table 14). This is especially true for wood improvements (e.g. forestry roads, fire prevention and improvement of existing woodlands). The structural organization of forest property and its fragmentation on the one hand and the length of investments and the connected risks on the other, have been put forward as the main reasons for this failure (INEA, 1997). There is also a certain ambiguity between stated objectives and measures. If a more naturally-oriented management of the environment is the goal, it is not clear that this can be attained with species such as Sitka spruce, Eucalyptus and Poplars. Distinctions between species, silvicultural techniques and forest management should be drawn in order to make clear the different impacts. Perhaps incentives should be differentiated with reference, first of all, to forest typologies rather than to tree species (Gatto and Merlo, 1997).

Table 14 Fund assigned and areas under Regulation 2080/92

Funds assigned, 1994-97 (ECU)	731,792
Area to be reafforested (projection)	100,500
Area to be improved (projection)	109,500
Area under reafforestation (actual)	47,841
Area under improvement (actual)	14,175

10.3 The Nitrates Directive

The EC Nitrates Directive 91/676 is specifically aimed at reducing pollution of water by nitrates originating from agricultural sources. It follows a previous Directive 75/440 fixing maximum allowed levels of nitrate in the water at 50 ppm. In the last 20 years, this limit has proved to be very difficult to meet in several European countries, particularly in regions with intensive agriculture and animal husbandry.

In Italy, the concentration of nitrates in internal and sea waters is of considerable concern, since the eutrophication associated with high concentrations of nutrients leads to the proliferation of algae and the consequent demise of aquatic life. The Adriatic Sea, receiving waters from the entire Po Valley and other coastlands, is particularly affected by algal eruptions. Since the second half of the 80s, public opinion is very sensitive to the problem and the tourist industry has been seriously affected.

Directive 91/676 has been incorporated within Italian legislation through Act 146/1994 (Disposizioni per l'adempimento di obblighi derivanti dall'appartenenza all'Italia alle Comunità Europee, G.U. 4 marzo 1994), wherein Art. 37 provides for a Code of Good Agricultural Practice, which is aimed at reducing nitrogen releases. The Code, issued by the Ministry of Agriculture, was published in March 1995 and paid particular attention to the following measures to be undertaken in order to avoid nitrate run-off and leaching (MiRAFF, 1995):

(i)   fertilization should not exceed the physiological requirements of crops, in terms of the quantity and timing of applications;
(ii)   the most appropriate field drainage should be applied;
(iii)   animal wastes must be appropriately stored, managed and distributed in the fields;
(iv)   crop residues should be ploughed in;
(v)   irrigation has to be applied appropriately, according to soil typology;
(vi)   vegetation cover like catch-winter crops, should always be considered;
(vii)   chemicals and manure should be carefully spread;
(viii) specific care has to be paid to cultivation near water courses, such as along rivers bank and lakes. Buffer strips and other appropriate devices should be applied.

Following the above guidelines, fertilization plans have to be developed according to soil typology, rainfall regime, cultivation techniques and of course the individual crop specificity. The relationships and combinations between manure and chemical nitrogen are crucial, and this require careful attention particularly on the sandy soils as in various parts of the Veneto Region in the Po Valley (Giardini et al., 1989). It has been shown that often nitrate releases are due to poor knowledge and poor management of fertilizer and manure particularly in a slurry form. In this context, what is reckoned to be important is, on the one side, the role of extension services able to inform and guide farmers, on the other the employment of slow-release nitrogen fertilizers, able to provide crops with their exact need. The industry has therefore been forced to provide a new generation of nitrate chemicals.

The Nitrates Directive must be seen, however, in the context of existing Italian legislation, particularly Act 319/1971 (the so-called Merli Act, already mentioned in Section 2.3). This legislation dealt with nitrates over the past 20 years. As already shown, intensive animal husbandry was to be considered as any other industrial plant obliged to introduce anti-pollution devices and treat the wastes. This is the case for the majority of pig producers as well as intensive beef producers. Act 319/1971 fixed the maximum limit for waste water at 20 ppm of nitrogen for disposal in rivers. Limits for drinking water were fixed by Presidential Decree 236/88 at 4 ppm of nitrogen.

Although Italy has been able to comply with the deadlines set by the Directive for issuing the Code of Good Agricultural Practice, it has not yet designated the so-called vulnerable zones that should be subjected to a stricter regulation according to Art. 3 of the Nitrates Directive. This is, of course, a serious shortcoming. It can be related to various measures like A1 and A2 of Reg. 2078/92 that are not so easily applicable because of the shortcomings in the implementation of Directive 676/91 (Fabbro, 1997).

Apart from the designation of vulnerable zones, the Code of Good Agricultural Practice is seen rather positively in Italy. In certain circles an extension is advocated to include pesticides, herbicides and soil conservation management, which in the Italian Mediterranean context are even more serious problems than nitrates. Previous failures in the application of the so-called Quaderno di campagna, a sort of register where farmers had to report all operations employing chemicals, pesticides and herbicides, must however be recalled. Mandatory bureaucratic means are not easily applicable particularly in a fragmented structure of several hundred thousand farmers and millions of rural properties, as in Italy. Persuasion and cross-compliance with other measures are perhaps a more efficient means, with a key role for the extension services which it is starting to play. Transaction costs must therefore be taken into account both for the public authority concerned with Codes of Good Agricultural Practice and for the farmers themselves who cannot be obliged to keep account of all those potentially harmful operations, which are part of their daily routine work.

10.4 Concluding observations

It is almost impossible to draw definite conclusions about linkages between CAP and the environment in Italy. What can be said as concluding remarks is that further adaptations, or reforms could make even more effective the role of the CAP with regard to the environment if:
-    compensatory payments were aimed at maintenance and improvement of rural landscapes and environments according to the 'cross-compliance' principle. Of course possible agri-environmental measures should be consistent with GATT and the green box of the WTO;
-    adequate acknowledgement (such as Reg. 2081, 2082 and 2092 on product origin and quality) and support were provided to high value products based on quality and/or technological innovation where EU agriculture can still be competitive in a world context. These products very often are the driving force of local economies - e.g. 'agro-industrial districts' based on *appellation d'origine* products;
-    markets were developed as far as possible for environmental goods and services provided by agricultural and forestry - for example, by means of additional services (parking places, guided visits and other structured recreational activities attached to environmental goods);
-    policies were balanced between 'top down' and 'bottom up' approaches, i.e. subsidiarity - however in a context where overall directions are provided and local participation promoted, including integration with local physical planning;
-    incentives were 'modulated' or 'differentiated' in order to avoid undue rents. Environmentally oriented subsidies should also be 'differentiated' or 'modulated' according to the value-type of goods and services provided by agriculture and forestry: use value (recreation, tourism etc), option and existence values.

The main conclusions could be that rural areas, or non-urban areas, need an integrated policy-mix based on local physical planning, regional-development policies and agricultural-environmental policies. The emphasis of the different policies should be balanced according to the context of individual areas. Policy recipes are not easily available. However, if attention is paid to existing

EU policies and real world developments, one should choose the 'bottom up' LEADER approach - being the most close to the Italian experience of 'agro-industrial districts' (Garofoli, 1991; Saraceno, 1993) or the American 'agropolitan development' (Friedman, 1988) or, again, 'renaissance rural' (Cavailhés et al., 1994) as variously defined. Some of the key considerations are as follows:

- the state of rural areas is mainly in the hand of Local Authorities through physical planning instruments, provision of services and landscape/environmental conservation;
- regional policies, and development policies, have often overlooked rural realities and problems such as landscape/environmental conservation, services and welfare, and agriculture and forestry needs;
- agricultural policies have always had other primary objectives different from rural people's welfare - farmers' income support, a minor section of rural inhabitants, albeit caring for the overall rural landscape;
- environmental/landscape policies can rather easily clash with regional-agricultural policies and indeed with physical planning when aimed at infrastructure improvement - e.g. roads and other public facilities.

References

Boatto, V. and L. Pilati (1997) *The elasticity of set-aside acreage with respect to the rate of compulsory set-aside;* In: What future for the CAP (Ferro, O. ed.); Kiel: Wissenschaftsverlag Vauk

Cavailhès, J., C. Dessendre, F. Goffenette-Nagot and B. Schmitt (1994) *Change in the French countryside: some analytical propositions*; European Review of Agricultural Economics; (21), pp. 429-449

Cesaro L. (1996) *Observed environmental impacts of set-aside schemes and other arable support measures in nine EU member countries, Country: Italy;* In: ' Possible options for better integration of environmental concerns into support for arable crops'; Wye College, University of London

Covarelli, G. and P. Montemurro (1993) *Gestione della vegetazione spontanea nei terreni a set-side*; In: Atti del convegno: La gestione della vegetazione nel contesto della nuova politica agricola comunitaria; Società italiana per lo studio della lotta alle malerbe; BA

De Benedictis, M. (1996) *Agricoltura ed ambiente: interazioni tecniche ed economiche*; Università La Sapienza, Roma

Fabbro, D. (1997) *Direttiva nitrati: l'Italia "buca" le aree vulnerabili*; Terra e vita n° 17/97

Favaretti, G. and M. Merlo (1973) *Effetti economici della legge sulla denominazione d'origine dei vini*; Inea, Roma

Friedman, J. (1988) *Life space and economic space: essays in Third World planning;* Transaction Books, New Brunswick, New Jersey

Garofoli, G. (1991) *Modelli locali di sviluppo;* Angeli, Milano

Gatto, P. and M. Merlo (1997) *Issues and implications for agriculture and forestry: a focus on policy instruments;* In: W.N. Adger, D. Pettenella and M. Whitby (Eds.); Climate Change Mitigation and European Land-Use Policies. Wallingford, CAB International, 295-312

Giardini L., M. Borin (1996) *Effetti del set aside annuale sul contenuto di azoto nitrico nella falda sottosuperficiale;* Riv. Di Agronomia, 30, 3

Giardini, L., C. Giupponi and A. Bonini (1989) *Stima dei rilasci di nutrienti (azoto e fosforo) dei terreni della pianura veneta;* Venezia, ESAV

Hodge, I. (1991) *The provision of public goods in the countryside: how should it be arranged?;* In: N. Hanley *(Ed.),* Farming and the countryside: an economic analysis of external costs and benefits, Wallingford, CAB International, pp. 179-196

IFNI-ISAFA (1988) *Inventario Nazionale Forestale 1985;* Temi (TN)

INEA (1997) *L'applicazione del Regolamento CEE n. 2087 in Italia;* Campagna 1996, Ministero per le Politiche Agricole

INEA-TECNAGRO (1991) *L'impiego dei prodotti chimici in agricoltura;* Studio su incarico MAF, Draft

ISTAT (1961, 1971, 1982, 1991) *1st, 2nd, 3rd, and 4th General Agricultural Census;* Rome

ISTAT (Various years) *Annuario Statistico Italiano;* Roma, Istituto Poligrafico dello Stato

Mantino, F. and A. Pesce (1996) *I programmi per lo sviluppo rurale in Italia: risorse, strumenti ed effetti;* Paper presented at the XXXIII Conference of the Italian Society of Agricultural Economics (SIDEA), Napoli, September 26-28, 1996

Merlo, M. (1991) *The effects of late economic development and land use;* In: Journal of rural studies; Vol. 7, n° 4

MIRAFF (1995) *Codice di Buona Pratical Agricola per la protezione delle acque da nitrati;* Collana del progetto Finalizzato PANDA Edagricole

OECD (1996) *Amenities for rural development;* Paris, Organisation for Economic Co-operation and Development

Pieri, R. and D. Rama (1996) *Quote latte: vincolo o strumento di gestione? - La situazione nei Paesi dell'Unione Europea;* Angeli, Milano

Saraceno, E. (1994) *Alternative readings of spatial differentiation: the rural versus the local eco nomy;* European Review of Agricultural Economics; 21, pp. 451-474

Sereni, E. (1961) *Storia del paesaggio agrario italiano;* Laterza, Bari

11. GREECE

Helen Caraveli

11.1 Introduction

Greece is a combination of mountains and plains. Some 43% of its surface is mountainous, 27% is semi-mountainous and only 30% comprises plains (see Table 1). The mountainous regions run from the north to the south of the country. Only 30% of the country's total area is cultivated, and 40% of this is situated in mountainous and semi-mountainous communes (Ministry of Environment, Spatial Planning and Public Works). The Peloponese, Western Sterea, Crete, Ipiros and East Aegean Islands (see Figure 1) are the most mountainous regions.

Table 1   Land use classification according to terrain in 1.000 km²

Land use type	Plains	Semi-mountainous	Mountainous	Total
Agricultural land	22	10	8	40
Pasture land	11	15	27	53
Forest land	3	8	19	30
Other uses	4	2	3	9
Total	40	35	57	132
	(30%)	(27%)	(43%)	(100%)

Source: National Statistical Service of Greece, 1996.

The country's climate is marked by long, hot and dry periods in the summer. The limited rainfall and its concentration during the winter months, in combination with high temperatures in the summer, result in prolonged periods of drought in many parts of the country. The duration of dry periods is greater in the Aegean Islands, Eastern Sterea and the eastern coast of the Peloponese; dry periods are less common in Northern Greece and very rare in mountainous regions. The irregularity of rainfall is often the cause of severe erosion, which is one of the country's most serious environmental problems.

The agricultural sector is further characterized by the predominance of small and fragmented farms (Greece has the lowest average farm size among EU countries: 4 ha against 16.4 ha for the EU as a whole - Eurostat, 1996), with the smaller farms being concentrated in mountainous and semi-mountainous communes. Table 2 shows trends in farm size and fragmentation. The structural improvements that have taken place have been modest.

Figure 1   Regions of Greece

268

Table 2   Basic structural characteristics of the Greek farm sector

Year	Average size (ha)	Parcels per farm	Economically active population in agriculture as % total	Irrigated land (share of total agricultural land)	Percentage distribution of agricultural holdings by group size (ha)	
					<5	>20
1971	3.5	6.5	40.6	22.2	79.1	1.0
1981	4.6	6.1	29.1	26.2	72.0	1.9
1987	5.3	5.8	27.0	n.a.	69.4 b)	3.0 b)
1993	4.3	5.9	20.4 a)	33.6	75.3	2.8

Source: National Statistical Service of Greece and Commission of the EC, *The Agricutural Situation in the Community,* Brussels, 1996.
a) 1995; b) 1991.

A high proportion of the agricultural land - around 83% - falls within Less Favoured Areas (LFAs), mostly located in the mountainous regions of the mainland or the islands. These regions with small-scale farming are also characterized by an elderly population. The share of farm holders aged 55 or over is 60% in LFAs, which is much higher than in other types of regions (Baldock et al., 1996). The Mediterranean type of production and the small size of holdings are closely linked to each other. The greatest proportion of farms is found in the smallest size classes (Table 3), mainly growing permanent crops, fruit, olives and citruses. The proportion of farms in the largest size classes is highest for cereals and other field crops. Table 4 connects the type of farming to the geographic region. In Macedonia/Thrace and Thessaly - regions with higher proportions of plains - the majority of farms are engaged in the production of cereals. In the other, more mountainous regions, types of farming characterizing Mediterranean agriculture (e.g. olives and permanent crops in general) are more prevalent, with certain areas specializing, say, in vineyards or in olives.

Table 3   Percentage distribution of farms by economic size class, for the main types of production 1993

Type of production	Size class (ESU)		
	1-2	2-4	4-8
Cereals and other field crops	46.2	25.1	28.7
Fruit excluding citrus	50.5	27.4	22.1
Citrus	66.9	21.0	12.1
Olives	78.9	14.3	6.8
Permanent crops various	56.6	24.8	18.6

Source: Ministry of Agriculture, Farm Accountancy Data Network.

Table 4   Distribution % of the number of farms by type of farming and geographic region, 1993

Type of production	Thrace-Macedonia	Ipiros Peloponese-Ionian Islands	Thessaly	Sterea Ellas -Crete -Aegean Islands	Total of Greece
Cereals and other field crops	56.6	3.8	49.3	12.4	26.3
Horticulture	2.0	1.9	1.3	2.6	2.1
Vineyards	1.0	3.9	3.2	6.4	3.9
Fruit excluding citrus	12.3	1.7	12.7	0.9	5.6
Citrus	0.0	17.5	0.0	1.4	2.6
Olives	5.0	36.4	6.6	34.0	23.7
Permanent crops, various	1.2	17.3	2.4	14.2	10.1
Beef	1.7	0.2	0.8	0.1	0.7
Sheep	4.1	6.1	9.4	5.8	5.8
Grazing livestock, various	0.3	0.4	0.5	0.3	0.4
Horticulture and permanent crops	8.3	12.4	6.8	12.5	10.7
Mixed livestock and permanent crops	7.5	8.4	7.2	9.4	8.4
Total	100.0	100.0	100.0	100.0	100.0

Source: Ministry of Agriculture, Farm Accountancy Data Network.

Table 5 shows the changing structure of agricultural production during the past two decades (see also Caraveli, 1993). Clearly, there has been a restructuring in all categories of products, mainly after accession to the EU during the 1980s. In terms of *annual crops*, major changes include reductions in the share of soft wheat and increases in the share of hard wheat, maize and cotton. For *permanent crops*, there has been an increase in the share of fresh fruit, including grapes and citruses, and a decrease in the share of wine. Finally, a decrease in the share of beef has been offset by an increase in the share of pig and poultry production.

These developments reflect an adaptation to EU markets and have been in accordance with policy objectives and in response to specific policy measures (e.g. favourable prices or aids for specific products).

Changes in the structure of production went hand in hand with a process of production intensification in the plains, following the abandonment of traditional, mixed and extensive systems of multicultivation (of annual or perennial crops) and livestock farming, and the movement of the population away from mountainous communes (of the mainland and islands) to the plains. This whole process also reflects the transition from subsistence agriculture to market agriculture - a process very much strengthened by the adoption of the CAP.

A picture of the movement from the mountains to the lowland can be obtained from Table 6, where the proportion of the population in mountainous communes is seen to decrease, while the proportion of the population of lowland communes appears to increase steadily. The process of depopulation in mountain areas has mainly taken place during the period following the Second World War.

Table 5   Percentage shares of each product in the total gross value of agricultural (two-year averages)

	1969-70	1979-80	1988-89	1992-93
ANNUAL CROPS				
Cereals				
Soft wheat	6.4	6.4	2.8	2.2
Hard wheat	1.4	1.9	3.3	4.3
Animal feed				
Maize	2.1	3.2	5.4	5.8
Barley	2.6	2.6	1.7	1.1
Industrial crops				
Cotton	4.1	3.3	6.6	7.6
Tobacco, oriental	4.1	4.3	4.3	3.3
Sugarbeet	1.0	1.2	1.4	1.5
Vegetables				
Vegetables fresh	6.2	6.2	5.8	6.0
Tomatoes for processing	0.7	1.2	1.0	0.8
PERMANENT CROPS				
Wine and must	2.2	1.7	1.4	1.1
Citruses	1.8	1.4	2.3	2.3
Other fresh fruit and grapes	4.0	4.2	5.0	6.5
Olives and dried fruit	4.8	4.1	4.3	3.6
LIVESTOCK PRODUCTS				
Beef	6.0	4.8	3.5	3.3
Sheep	3.9	3.8	3.4	3.4
Goat	1.9	1.8	1.9	1.9
Pig	2.5	4.8	4.7	4.5
Poultry	2.6	4.2	4.0	4.5
Meat, total	17.2	19.5	17.7	17.6
Milk	10.1	8.9	8.8	8.4
Total	27.3	28.4	26.5	26.0

Source: Ministry of Agriculture (Own calculations).

Intensification generally occurred in arable farming. Some types of livestock production have also undergone marked intensification, since EU accession, namely the poultry, pig and dairy sectors. However, less intensive farming practices still remain. The most common form is extensive rearing of sheep and goats for milk and meat, with the livestock often herded in large mixed flocks. This system covers much of the mainland and is especially significant in maintaining the nature conservation value of mountainous areas (Beaufoy et al., 1994, p.24). Furthermore, in the Ionian islands, where farms are very small, mixed farming carried out on a subsistence basis is still widespread. Even though most arable farming is of high intensity, certain tree crops, such as olives, almond and mastic varieties, are still traditionally managed - with pruning and grafting in winter, picking by hand in autumn and early winter, and weed control during the year. About 95% of

served in drip irrigation. Finally, on a regional level, three different irrigation patterns have been adopted, depending on local water availability, crop mix and land type. Thus, in Western Sterea, the Peloponese, Macedonia and Ipiros, the dominant irrigation technologies are sprinklers, and drip irrigation; in Thessaly and Thrace, where arable crops prevail, sprinklers and self-propelled sprinkler systems dominate; and in Crete, where horticulture and tree crops are the main production types, the most widely adopted irrigation system is drip.

Some further evidence of the process of production intensification in the Greek lowlands is given by the rise in the degree of both mechanization and fertilizer consumption. The former, measured by the number of tractors per km^2 of agricultural land, shows an increase from 0.7 in 1970 to 2.3 in 1990. Over the same period fertilizer consumption doubled from 37 to 75 kg/ha.

The degree of intensification and mechanization of Greek agriculture is however quite limited, relative to the rest of the EU 1). This is mainly due to the geomorphological structure as well as to the small scale of farming. The majority of plantations are surrounded by natural areas (hills, ravines, mountains, lakes, gulfs, etc.) that are beneficial to preserving biodiversity and to forming important cultural landscapes (OECD, 1997; Caraveli, 1998).

It is worth referring to the reasons for the overall increase in fertilizer consumption. These have been:

-   the general intensification of Greek agriculture in the post-War period, which is expressed by the increase in irrigated land, the use of improved and higher yielding varieties by farmers (which require larger rates of fertilizers) and the improvement and expansion of industrial crops (sugarbeet, cotton, tobacco) 2), whose production is favoured by intensive fertilization;
-   the heavy subsidization of chemical fertilizers by the Greek state - aiming, on the one hand, at the increase in farmers' incomes (through the reduction in input costs), and, on the other, at the protection of domestic industries.

In the 1950s and 1960s, fertilizer use increased by an average 9-10% per annum. In the following decade, 1971-80, the annual rate of increase fell to 4.5%, and fell again in the decade 1981-1990, to only 2.3% per annum (Beopoulos, 1996).

With the accession of Greece to the EU, subsidies on inputs were deemed incompatible with the CAP regime, and a 5-year transitional period was agreed for their gradual abolition. Subsidies remained in place for a further six years and were abolished only in 1992, as a consequence of which fertilizer prices increased dramatically and consumption fell substantially.

It should finally be mentioned that very high fertilizer consumption per hectare - exceeding the 'optimum' level of fertilization by 15-20% - has been observed in the regions of intensive cultivations (mainly, Thessaly & Macedonia) whereas sub-optimum levels are typical in mountainous areas where extensive (non-irrigated) cultivations prevail (Ministry of Environment, Spatial Planning and Public Works, 1995).

---

1)   The lower degree of production intensification is also evident in the low share of purchased inputs (i.e. intermediate consumption) in farm production, which in Greece is 26.2% against 46.3% in EU-15 (CEC, 1996).

2)   As was mentioned in section 1, this expansion was particularly evident during the 1980s, due to the favourable prices, and other types of supports, for these products, as a result of the application of the CAP.

The use of pesticides, in contrast, has increased tremendously since the country's accession to the EU. Previously, their use had actually decreased during the 1960s and had increased by only 16% during the 1970s. But over the period 1981-92, their use almost doubled, with the intensification and specialization of production encouraged by the high guaranteed prices. Even though pesticide costs still comprise only 3-4% of total production costs, their increased use has caused environmental problems, as well as posing significant risks to farmers.

11.3 Interactions between agriculture and environment

Modernization and intensification of agricultural practices in Greece have not yet induced severe negative implications or the degradation of traditional local ecosystems on a wide scale. This is mainly due to the still limited degree of intensification of production (OECD, 1977; Caraveli, 1998).

However, the intensification process in the lowlands has led to the uncontrolled increase in the level of pollution and soil erosion and the exhaustion of water resources. Irrigation has led to serious problems of water management and supply - with the exhaustion of water resources, salination and eutrofication being the main impacts - and fertilizers have been the major source of groundwater and surface water pollution from nitrates. 'Exhaustive irrigation of groundwater, abuse of pesticides and fertilizers are the most obvious manifestations of the problem' (Skourtos, 1995). Over-abstraction of water for drinking and irrigation purposes often leads to deficits in the balance of ground waters. In coastal areas, this may lead to the ingress of brackish water leading to the salinization of the groundwaters, rendering them unsuitable for either drinking or irrigation purposes. Regions with salinization problems are mainly the coastal regions of the provinces of Pieria and Kavala (in Macedonia), as well as those of the Eastern and Southern Peloponese and the islands of the East Aegean and Crete (Ministry of Environment, Spatial Planning and Public Works, 1995).

A deterioration in the quality of water courses and lakes has also been observed, where the presence of nitrate, phosphate and ammonia emissions over accepted values has been found (Ministry of Environment, Spatial Planning and Public Works, 1995). Where the soil is sloping and can take little infiltration, nitrate run-off accumulates in the water basins where they cause eutrophication of the surface waters (Ministry of Agriculture, 1993). These difficulties have necessitated the construction of dams which, by storing surface waters, limit the exhaustion of groundwater, prevent salinization and preserve the quality of water.

In mountainous regions, depopulation and the collapse of traditional, mixed and extensive, production systems have had serious negative environmental impacts. The existence of traditional farming systems permitted a sustainable exploitation of environmental resources by contributing to the maintenance of soil fertility and the protection of crops. Their collapse, following the mechanization in the plains of the mainland, led to a marginalization of these areas, through a dramatic decline in the arable area, with an increase in the desertification process and the conditions leading to soil erosion, the invasion of scrub, increased incidence of forest fires and major floods, and the reversion of the countryside to wilderness (OECD, 1997; Caraveli, 1998).

Erosion has been the main cause of degradation of the natural terrain in Greece and it is estimated to affect, to a greater or lesser extent, one-third to one half of the total land area of the country. The main factors in the increasing erosion of agricultural and forest lands are: fires (the country as a whole is considered as a high risk belt in respect to fires), grazing after a fire, overgrazing of pastures, cultivating sloping land without safeguards against erosion and the exhaustion of soil fertility (Ministry of Environment, Spatial Planning and Public Works, 1995). Erosion can lead to the loss of valuable soil resources, increase the frequency of destructive floods (that damage good agricultural land and infrastructure), curtail the grazing capacity of mountainous pastures, increase pollution of surface waters, and lead to the loss of scarce water resources. It has been estimated that, due to erosion, 30 million m³ of soil is lost annually (Ministry of Environment, Spatial Planning and Public Works, 1995).

There is also a tendency to abandon or under-graze scrublands. This has been largely the outcome of the decline in transhumance in sheep and goat production (see Table 7) - a traditional land-use practice - which has had significant impacts on the farmed landscape, by leaving large areas of sub-alpine pasture neglected.

Table 7   *Evolution of goat population (%) (1956-1985) according to type of farming*

Type of farming	1956	1959	1961	1970	1975	1980	1985
Home-fed	10.8	11.7	17.4	20.7	19.2	18.6	19.3
Sedentary	60.8	65.1	69.5	71.3	73.4	73.2	75.5
Transhumant	28.4	23.2	13.1	8.0	7.4	8.2	5.2

Source: Hatziminaoglou et al. (1995).

Not all marginal areas are affected by abandonment of agricultural land, and overexploitation of marginal land is a problem in many regions of the country, mainly as a result of the increased incentives for production in such regions (LFAs). This applies particularly to the subsidies under the CAP sheep and goat regime (such as the Sheep Annual Premium), as well as the special LFA payments. Both forms of support have led to overstocking and overgrazing and, therefore, to environmental damage in these areas (Ministry of Environment, Spatial Planning and Public Works, 1995, p. 60). The decline in transhumance, the introduction of new animal breeding structures and the reluctance of breeders to exploit the distant alpine grasslands have further contributed to an excessive increase of the stocking rate of the rangelands 1), a situation which is more evident in the low and middle altitude areas (Nastis, 1985; MAICH, 1997).

Abandonment or overexploitation of marginal grazing land does not allow the rational management of pastures so that their feeding capacity appears to be in decline. In particular, it has been estimated that 25% of the mountainous and semi-mountainous soils have irreversibly lost their productive capacity (Alexandris, 1985). Also, as much as 40% of all extensive grazing lands

---

1)   Characteristic marginal land, located in middle and high altitude zones.

has been degraded to a dangerous level, due to deforestation, under-grazing or over-exploitation and is now unsuitable for grazing (Baldock et al., 1996, p. 64). Yet, even under these conditions, production in rangelands - including scrublands - helps to cover a sizeable part of the nutritional needs of goats, transhuming or not, and constitutes a decisive factor in maintaining the prevailing traditional extensive system of exploiting ruminants (Hatziminaoglou et al., 1995; MAICH, 1997).

The environmental situation of a third category of regions - the coastal regions of a great part of the mainland or islands - should finally be mentioned. In these regions, the urban and tourist expansion has reduced traditional low-input cultivations (such as grains, legumes and other vegetables) - thus, reducing the productivity of valuable agricultural land - as well as important wetlands. These phenomena have led to a significant rise in the level of environmental pollution, over-use of water resources and a notable deterioration of the coastal landscapes (Louloudis, 1996).

11.4 Impacts of the CAP and environmental policies

In Greece, no special environmental provisions have been proposed for the protection of the environment from agricultural activities, with the exception of certain regulations referring to the hygiene standards that agricultural businesses are expected to meet, or some general and non-systematic advisory measures such as appeals for rational use of fertilizers and pesticides, made by the Department of Agricultural Extension of the Ministry of Agriculture. Greece, like some other EU countries did not take up the opportunity offered by Article 19 of the European Union's agricultural structures Regulation 797/85 to provide financial support for environmentally friendly agricultural practices (Beopoulos and Louloudis, 1996).

However, various environmental policy measures have been adopted by the Greek authorities. A distinction is made between national measures and others taken within the framework of the CAP reform. The former type of measures are:
- for the protection of soils from erosion, there have been attempts to adopt improved irrigation methods which would limit the loss of soil and the desertification of land. Moreover, due to the implementation of various programmes (which form part of the Public Works Programme), the use of highly fertile land for agricultural purposes has increased (Ministry of Environment, Spatial Planning and Public Works, 1995);
- for the better management of water (see previous section), the Ministry of Agriculture has, since 1990, been implementing a programme for the construction of big dams - with priority to be given to the islands, to be followed by programmes for the mainland. Furthermore, the same Ministry has adopted special annual programmes, since 1990, to assess the quality of surface waters and measure the degree of salinization of groundwaters due to excess abstraction of water. The impacts of these programmes have not been assessed yet.
- for the protection of waters from nitrogen pollution, caused by the application of fertilizers and by cattle-raising, i.e. for the implementation of the Nitrate Directive (91/676/EEC), the Ministry of Agriculture has, since 1993, adopted the Code of Good Agricultural Practice. These include rules that refer to methods of using nitrogen fertilizers and livestock manure,

especially in some sensitive zones. More specifically, 'the Code aims at the prevention of groundwater and surface waters pollution and nitrogen accumulation due to water percolation or surface run-off' (Ministry of Agriculture, 1993).

The following measures were taken within the context of the CAP reform:
- for the reduction of pollution caused by pesticides and the promotion of environmentally friendly methods in agricultural production, in 1993, the Ministry of Agriculture developed a programme for the implementation of Regulation 2078/92 in Greece. Within the framework of this Regulation, a regime was set, which is co-financed by FEOGA by 75% and which has the following targets (Ministry of Agriculture, 1994):
  - to favour or preserve methods of agricultural production, which reduce pollution caused by agricultural activity;
  - to promote the application of environmentally friendly intensive production and cattle raising methods, and the transformation of arable land to extensive pasture;
  - to favour the exploitation of agricultural land in accordance with regulations which aim at the protection and improvement of: the environment, natural areas, landscape, natural resources, soils and genetic diversity;
  - to encourage the maintenance of agricultural land and forest areas, wherever it is considered necessary for ecological reasons or to prevent fires, and to prevent in this way the danger of rural depopulation;
  - to encourage farmers' acceptance of agri-environmental measures 1).
- the Greek Ministry of Agriculture submitted to the European Commission three Programmes within the framework of Regulation 2078/92. These are: The National (horizontal) Programme for the protection and preservation of biodiversity and genetic diversity; the National Programme for the reduction of the level of environmental pollution caused by agricultural activity; the National Programme for the long-term withdrawal of agricultural activity (Louloudis, 1996). The three Programmes have now been approved and their monitoring has been undertaken by various committees of the Ministry of Agriculture. For their implementation a sum of 52 billion Drachmas have been made available for a three year period covering 1995-1998 (EU contribution 75%).

The first Programme concerns mainly remote semi-mountainous and mountainous communities of the hinterland and the islands, where, due to natural handicaps, intensive agriculture has not expanded and where the cultivation of threatened species should be supported (i.e., through special aids aiming at the maintenance of the extensive nature of agricultural activity or through biological cultivations).

The second Programme consists of two sub-Programmes. The first of these, on biological agriculture, aims at the reduction of pollution caused by agricultural activity, as well as the protection of flora and fauna and of public health. This sub-Programme concerns the whole country and initially covered 2,000 ha of organic cultivations, 50% of which are found in the Peloponese.

---

1) The results of relevant field-work reveal that 'farmers' views are, on the whole, not favourable to environmental problems stemming from agriculture' (Beopoulos and Louloudis, 1996).

Thereafter, for the period 1995-1997, the lands that will be included in this programme cannot exceed 6,000 ha. It will later be expanded to ecologically sensitive areas - belts near lakes and coasts and LFAs of the mountainous hinterland. The support approved for growers of organic products for 1995 and 1996 is presented in Table 8. The second sub-Programme, on the reduction of nitrogen pollution caused by agriculture in the plain of Thessaly, aims, among other things, at the reduction of cotton production in this area.

The third Programme aims at the creation of biotopes and biological parks, the protection of hydrological systems from pollution caused by agricultural activity and the protection of water resources. This target works supplementary to the areas promoted through their inclusion in the NATURA 2000 network through the Habitats Directive (92/43).

Table 8   Support approved for the growers of biological products (period 1995-1996)

Cultivation	1995		1996	
	ecologically sensitive areas	other areas	ecologically sensitive areas	other areas
Intensive olive groves	1,141	1,091	1,220	1,167
Extensive olive groves	490	468	523	501
Korinthian raisins	2,137	2,045	2,285	2,188
Table grapes	2,464	2,355	2,636	2,519
Sultana	1,978	1,894	2,116	2,025
Vineyards	2,443	2,338	2,613	2,500
Tree crops (except citrus)	2,464	2,464	2,636	2,636
Citrus trees	3,520	3,520	3,765	3,765
Garden crops, Vegetables and Edible pulses	880	880	941	941
Cereals - Grains	528	528	565	565

Source: Ministry of Agriculture.

It is expected that the implementation of the Programmes will substantially contribute to the reduction of the level of environmental pollution, as well as to the protection of biodiversity and the better management of water resources.

Even though no assessment of the impact of the 1992 reform on the state of environment and landscape in Greece has been attempted, views have been expressed that the reform will significantly contribute to the alleviation of environmental pressures, through the direct and indirect effects of lowered farm prices and the removal of the gap between EU and world prices.

The direct effect will be the reduction of the incentive to use purchased inputs and thus the discouragement of intensive farming. This effect, in combination with the abolition of subsidies on fertilizers, the liberalizing of the fertilizer market, the reluctance on the part of the EU to finance big irrigation works and the new irrigation techniques, implies less environmentally burdening production (Louloudis, 1996).

12. PORTUGAL

João Castro Caldas

12.1 Introduction

Pressures of agriculture on the environment, although not a major priority in Portugal, have been the subject of scientific research and some public concern. There is no systematic data collection, nor systematic analyses of the environmental effects of agriculture. Portugal has remained agricultural modernization and intensification processes that have taken place in Northern European countries over the past fifty years. This does not mean that the restructuring of agricultural production has not caused problems but that, in general terms, these problems have not had the visibility and ramifications found in other sectors. In addition, the country is characterized by strong contrasts between agricultural production structures which have not improved in recent years. This chapter, which reviews the current state of knowledge, starts with a concise of agricultural production structures. Section 2 makes reference to the contrasts to be found and stresses the principal dynamic forces affecting them over the past few years, namely those arising from European integration. Based on available information, Section 3 outlines the most significant interactions between agriculture and the environment in Portugal. Section 4 provides an evaluation of the environmental impact trends arising from the application of the principal CAP measures, particularly those arising from the 1992 reform. The chapter finishes with a summary and conclusions.

12.2 Structural characteristics of agriculture

12.2.1 Regional differences

The Portuguese territory extends over an area of 88,994 km². It is marked by the contrasts between its northern mountains and extensive southern plains. Notwithstanding a relatively mild degree of relief, 95% of the land above 400 metres is concentrated north of the River Tagus whereas 63% of the territory below 200 metres is located in the southern part of the country. In addition, along much of the coast a narrow coastal strip is separated from the inland regions by a range of hills.

These contrasts in relief are the cause of a regional heterogeneity of restrictions and resources in respect of productive activity. Special reference should be made to the diversity of climatic conditions. This diversity is particularly visible in the abundance of irrigation water along the northern coast and in the major degree of annual and inter-annual irregularity of precipitation in the south. Southern regions, particularly the Alentejo, often suffer acute water shortages during the growing season.

In addition to a marked contrast in settlement patterns (with population densities ranging from 327 inhabitants/km² in the Entre Douro e Minho to 20 inhabitants/km² in the Alentejo), the economic and social history of the occupancy and use of this contrasting territory has had the effect of creating a wide range of vegetation types, agricultural structures and landscapes.

Table 1   Major land uses in the early 1990s (share of total territory, %)

Region	Utilized agricultural area	Forested area	Uncultivated land	Agricultural surface with sub-forested cover	Irrigated land in total agricultural area (%)
Northern Coast					
- Entre Douro e Minho	31.5	37.9	19.9	9.0	58.4
- Beira Litoral	19.8	47.6	12.4	0.1	50.4
Northern Interior					
- Trás-os-Montes	38.8	15.9	14.6	0.7	14.2
- Beira Interior	36.0	31.3	20.5	7.8	17.7
Lisbon and Tagus Valley	38.2	33.6	6.2	12.2	26.0
Alentejo	68.4	38.5	5.6	38.9	5.6
Algarve	25.5	12.7	18.8	1.9	18.2
Total	43.3	33.4	12.0	21.7	20.6

Source: Rolo (1996).

Table 1 shows the high share of forested land and uncultivated land in the different regions of the country. Also it shows the marked contrast between the high share of cultivated land on the plains of the Alentejo (a considerable part of which has a sub-covering of cork-oak groves) compared to the low share in the rest of the country. In many northern zones, in which land cultivation is conditioned by considerations of relief, the agrarian landscape is characterized by narrow terraces. These terraces were painstakingly constructed in the past with the objective of overcoming natural restrictions and extending the cultivated land, which is currently being progressively abandoned.

More than 50% of the agricultural land in the northern coastal area is being irrigated. Almost all of it is in small plots of land under farmer-managed irrigation systems. In contrast, only 6% of the land is irrigated in the south and is mainly reliant on large public irrigation systems.

12.2.2   Farm structures

Utilized agricultural area in total amounts to some 3.8 million hectare. There are some 410,000 farms, with major regional differences in farm structure. Small fragmented farms on small plots of land are mainly located in the north. The largest holdings are observed in Alentejo. The contrasts are also significant, albeit less so, between the coastal and northern regions (see Table 2).

The average agricultural area of holdings in the Alentejo conceals a structure in which land ownership is highly concentrated. In this region about 16% of holdings exceed 50 ha. This group of holdings, however, represents 86% of the land used agriculturally. Table 3 shows that the regional heterogeneity in the physical dimension of the holdings and concentration of land ownership also correspond to a concentration of the capitalist sector in the south and in the Tagus Valley. The northern region is almost exclusively farmed by family labour with specific local exceptions such as vines for the production of Port Wine.

Table 2   Regional variation in farm structures

Region	Share of number of holdings (%)	Share of UUA (%)	Farm size (ha)	Number of parcels per holding
Northern Coast				
- Entre Douro e Minho	19	6	3.0	4.3
- Beira Litoral	21	5	2.3	7.2
Northern Interior				
- Trás-os-Montes	18	13	6.9	10.1
- Beira Interior	11	12	9.6	6.4
Lisbon and Tagus Valley	17	13	6.8	4.1
Alentejo	9	47	50.5	2.3
Algarve	4	4	7.5	5.8
Total	100	100	9.2	6.0

Source: INE, Survey on the Structure of Holdings - 1995.

12.2.3   Agricultural labour force

In the early 1990s the farm population accounted for some 20% of the national population, but as much as 57% (Trás-os-Montes) and 43% (Beira Interior) of regional populations in the Northern Interior, and as low as 10% in the Lisbon and Tagus Valley region (Rolo, 1996).

A current but recent trend suggests an increasing linkage between income generated from agriculture and additional income earned by families from outside the holding (Baptista, 1996). Significantly, the majority of farm family workers spend less than half of their working time on the farm and the major part of family income is usually generated outside the holding (see Table 4). In the coastal region, north of the River Tagus and in the Algarve, income from agricultural activity (self-consumption and sales of farm products), is typically complemented by wages and income from non-agricultural activities of members of the household. In the more sparsely populated northern interiors in which there are fewer alternatives in the labour market and in which the farm population is older, it is more often external transfers (in the form of social security payments) that complement the earnings from the cultivation of small plots of land.

Table 3   Distribution of annual work units (AWU) by type of labour force (%)

Region	Family labour	Hired labour
Northern Coast		
- Entre Douro e Minho	95	5
- Beira Litoral	97	3
Northern Interior		
- Trás-os-Montes	94	6
- Beira Interior	96	4
Lisbon and Tagus Valley	86	14
Alentejo	73	27
Algarve	84	16
Total	92	8

Source: INE, Survey on the Structure of Holdings - 1995.

12.2.4   Types of farming

Indices of production conditions (see Table 5) show small family farm zones in the northern coastal region to have a much higher degree of labour intensification, mechanization and livestock densities, in comparison with an increasing trend towards extensification in inner regions. The Tagus Valley region has the highest concentration of agrarian capitalism zones. Here, the intensification indices, particularly for labour and mechanization, contrast sharply with those of the extensive farming systems in the Alentejo.

These various elements and the regional distribution of types or farming (see Table 6) provide an outline of the principal farming systems in Portugal. Thus, in the Northern Coastal Region, notwithstanding specialization trends in several areas (namely dairy production), irrigated mixed-farming is the norm, spread out over small, fragmented family farms. Given the small dimension of such farms and notwithstanding their relatively high degree of intensification, these farms do not occupy the whole of the available family work force nor, in the overwhelming majority of cases, are they capable of providing the total family income.

In the northern interior, in which the irrigated area is a much lower percentage, systems which combine diverse permanent crops or specialized vine and olive systems as well as sheep and goat breeding are prevalent, together with mixed-farming and more extensive livestock systems. Notwithstanding the fact that these are much larger holdings than those to be found in the northern coastal area, the conditions are such that the income from the farm must usually be supplemented by transfers from outside.

Agrarian capitalism is mainly observed in the Lisbon and Tagus Valley area where there is a greater degree of diversity of specialized farming systems, including general cropping, cereals, viticulture, fruit-growing, mixed permanent crops as well as horticulture.

288

Table 4   Agriculture's share in Farm Family Incomes and in Family Working Time

Region	Principal source of family income (% of number of holdings)			Working time spent on the holding (% of number of family workers)		
	exclusively earned from the holding	principally earned from the holding	principally earned from other sources	<50% of working time on the holding	50-100% of working time on the holding	all of working time on the holding
Northern Coast						
- Entre Douro e Minho	9	34	57	48	36	16
- Beira Litoral	9	20	71	56	32	12
Northern Interior						
- Trás-os-Montes	13	26	61	69	26	5
- Beira Interior	7	16	77	65	26	9
Lisbon and Tagus Valley	13	19	68	70	16	14
Alentejo	13	28	59	70	15	15
Algarve	8	22	70	75	19	6
Total	10	24	66	61	27	12

Source: INE, Survey on the Structure of Holding - 1995.

Table 5   Several technical indicators on production conditions: labour, mechanization and livestock

Regions	Indices of production intensity		
	labour (UAA/AWU, in ha)	mechanization (No. of tractors / 100 ha of UAA)	livestock (Livestock Unit/ha UAA)
Northern Coast			
- Entre Douro e Minho	1.7	12.05	0.59
- Beira Litoral	1.8	14.70	0.47
Northern Interior			
- Trás-os-Montes	7.0	3.41	0.19
- Beira Interior	10.2	2.97	0.19
Lisbon and Tagus Valley	6.3	7.07	0.21
Alentejo	49.2	0.96	0.18
Algarve	8.4	5.48	0.13
Total	7.6	3.89	0.22

Source: INE, Survey on the Structure of Holdings - 1995.

The Alentejo is characterized by a higher concentration of capitalist agriculture but, with the exception of areas irrigated by public irrigation systems (less than 6% of the total agricultural surface), it is dominated by non-irrigated cereals, oak and cork-oak woods and extensive livestock systems.

Finally, in the Algarve, specialized fruit-growing systems and horticulture clearly predominate. Tourist development in the region also provides major opportunities for the employment of family members outside the farm.

12.2.5   Changes in production structure

There has been a progressive fall in the cultivated agricultural surface in Portugal since the drastic reduction in the available labour force, upon which Portuguese agriculture had been based, before the major upsurge of emigration in the 60s. Land, which is no longer needed for agricultural purposes, when not used for afforestation or urban development, adds to the adds to the growing area of the territory left uncultivated. Such evolution has, on the other hand, been accompanied by gains in productivity from labour and the land. These gains in productivity which have taken the form of accrued final production, have been achieved via technological change with consequent accruals to intermediate consumption.

Table 6   Regional distribution of the number of holdings and UAA by farming type (%)

Farming type	Portugal		Northern Coast		Northern Interior		Lisbon and Tagus Valley		Alentejo		Algarve	
	no.	UAA	no.	UAA	no.	UAA	no.	UAA	no.	UAA	no.	UAA
Cereals	1.9	8.9	1.1	2.9	0.6	0.6	2.4	9.3	9.7	15.2	0.9	2.1
General cropping	8.2	5.3	8.2	7.0	8.7	4.6	9.8	13.8	4.4	3.0	5.3	3.7
Horticulture	2.7	0.9	1.7	0.9	0.1	0.1	8.1	4.9	4.7	0.3	3.6	2.3
Viticulture	8.1	3.1	3.6	3.9	13.5	5.6	14.3	7.1	2.2	0.6	1.8	1.0
Fruit-growing	6.4	3.7	1.3	1.5	5.7	4.4	12.0	8.0	3.3	0.8	42.9	29.9
Olive growing	5.6	3.7	0.2	0.2	10.5	7.0	3.2	1.2	21.5	3.7	1.1	0.6
Mixed permanent crops	6.5	4.5	1.7	2.2	11.7	10.4	9.0	5.8	3.8	1.2	11.1	9.7
Dairy	3.3	2.7	6.6	11.4	1.4	2.4	0.8	2.2	1.2	1.0	0	0
Other bovine animals	2.7	6.5	3.8	4.2	2.3	4.3	1.0	6.1	3.7	8.6	0.7	2.2
Sheep and goats	5.1	16.7	4.0	12.0	6.2	19.6	2.7	7.4	13.0	19.8	2.2	4.2
Granivores	4.5	2.9	5.9	2.4	1.9	0.6	5.8	3.2	5.7	4.3	1.9	0.8
Mixed farming/ Others	45.0	41.1	61.9	51.2	37.3	40.4	30.9	30.9	26.8	41.6	28.5	43.5
TOTAL	100.0	100.0	100.0	100.0	100.0	100.0	100.0	100.0	100.0	100.0	100.0	100.0

Source: INE, Survey on the Structure of Holdings - 1995.

In global terms, the direction of change suggests a growing degree of mechanization, the introduction of new crop species and new selected animal breeds and an increase in the use of chemical fertilizers, phytosanitary products and animal feedstuffs.

Table 7 shows the increases in final production during the past few decades. This was accompanied by a progressive decline in the relative importance of the farming sector in the economy as a whole, both as regards employment as well as the level of contribution to Gross Domestic Product (GDP).

The processes of change have had different consequences between regions, types of agricultural activity and farming systems. As such, the small family farms in the northern coastal area of the country have zones in which there has been a progressive degree of specialization in mixed-farming towards dairy farming or vineyard restocking. Given the structure of the small farms, the changes in farming systems and technologies have, to a large extent, used external processes such as the hire of tractors and collective milking and milk collection equipment.

In the northern interior, which has experienced a larger drop in population, there has also been a steeper reduction in the cultivated area. Land has been left uncultivated, and there has been an increase in the forested area and in the number of vineyards and orchards.

In the Lisbon and Tagus Valley region, reference should be made to zones along the coast to the north of Lisbon and to the plains of the Tagus Valley. In the first of these zones, there has been a significant reinforcement of specialization in viticulture, fruit-growing, horticulture and intensive livestock breeding. In the plains of the Tagus Valley (the frontier of the north-south contrast in production structures), in addition to the reinforcement of viticulture and fruit-growing, there has also been an intensification and capitalization of holdings producing irrigated crops (including maize, grain, rice, tomatoes for paste production, tobacco) and cattle breeding.

Table 7   Economic features of the agricultural sector compared to the rest of the economy

Years	Share of working population employed in agriculture (%) a)	Share of agricultural and agro-industrial sectors in the national economy (GDP) (%) b)	Final production of agro-forestry sector (1950=100) c)	Intermediate consumption in agro-forestry sector (1950=100) c)
1960	42.2	20.6	112	199
1970	30.5	14.9	156	635
1980	18.4	9.5	224	1,460
1990	10.1	5.0	254	1,609

Source: a) INE, Population Surveys; b) Rolo (1996); c) prepared on the basis of Rolo (1996)- 1950 market prices.

The Alentejo, after the progressive abandonment of extensive non-irrigated crops on the poorest quality land and an intensification/capitalization of the more fertile land, went through a short period in which landless workers who had seized holdings experimented with agrarian reform. At the time in question, the workers returned to an increase in the cereal crop produced cooperatively. With the return of the land to its former owners, the former evolutionary trends were resumed. Areas irrigated by public irrigation systems were intensified and expanded. The area of forests planted with pine trees and eucalyptus groves has been increased at the expense of the former cork-oak plantations, with an expansion also in the area of hunting reserves.

Finally, in the Algarve, reference should be made to the expansion of the irrigated surface through the use of underground water and a major intensification and specialization in productive systems, particularly in the case of fruit-growing and horticulture.

12.2.6   Impacts of the CAP

Major funding was made available for agricultural investments and income support during the first years following membership of the European Community, to modernize holdings and adapt agricultural structures to market expansion. The enormous growth of investment in agriculture accelerated existing trends (Avillez, 1991). There has been a concentration of the total amount of aid to investment in the Lisbon and Tagus Valley region, the Alentejo and Entre Douro e Minho (Northern Coast), and aid to individual investment has been preferentially targeted at the poles of concentration of agrarian capitalism (Alentejo and Tagus Valley) and on the larger holdings. A recent survey (Baptista, 1997) has shown that, by comparison with Northern Europe, the Portuguese agricultural model (like that for other Southern European countries - Spain, Italy and Greece) provides lower income for labour, a higher percentage of the working population, and a lesser degree of vocational employment (with less full-time farming and a major part-time labour

Table 8   Distribution of support per region in 1996

Regions	Income support (%)	Investment support (%)	Other support (%)	Total support (%)	Share of income support in total support (%)
Northern Coast					
- Entre Douro e Minho	11.7	8.7	1.2	9.8	70.7
- Beira Litoral	8.1	7.1	1.8	7.3	68.3
Northern Interior					
- Trás-os-Montes	11.4	12.3	1.8	10.9	64.0
- Beira Interior	9.1	6.8	0.6	7.7	72.1
Lisbon and Tagus Valley	17.7	19.9	7.3	17.6	62.1
Alentejo	36.2	14.1	3.1	26.9	83.0
Algarve	1.5	3.8	-	2.1	42.6
Not possible to regionalize	4.3	27.3	84.2	17.9	13.3
Total	100.0	100.0	100.0	100.0	61.5

Source: GPPAA, Support for agriculture 1996.

component). The same author has also noted that the CAP, in the case of Portugal, has 'particularly concentrated on distributing money to a social group which is not interested in permanent population settlements nor in the ordered and diversified use of the territory but in maximising the value of rents with the least possible degree of commitment to society' (Baptista, 1997).

Irrespective of specific local conditions, the impacts of the CAP on agriculture can be summarized in the following manner. The CAP has contributed towards modernization of farm holdings and intensification of production. This trend applies to the case of dairy intensification and vineyard upgrading in the Entre Douro e Minho region, and of horticulture and permanent crops in the Algarve and the coastal zone north of Lisbon. In addition large amounts of funds have been made available to the largest holdings of the Alentejo and Tagus Valley (see Table 8), and which have not paved the road to the 'orderly and diversified use of the territory', having permitted the maintenance of marginal farming systems in the Alentejo, while not eliminating bottlenecks or profound contrasts in production structures.

12.3 Interactions between agriculture and environment

Pressures for agricultural intensification in Portugal are much less in their extent and their degree than in Northern European regions (Machado, 1996). Most of Portuguese territory supports an extensive agriculture with a low degree of pollution. In terms of the consumption of nitrogen based fertilizers (37 kg/ha in 1990) and the total consumption of fertilizers (69 kg/ha in 1990), Portugal is in penultimate position among European Union countries. In terms of the number of tractors per 100 ha of UAA Portugal lies in tenth position. Such figures are indicative that there is no general problem of agricultural intensification. They may, however, conceal the existence of specific problems which need to be analysed on a local scale and which not only involve physical parameters of climate, relief and hydrology but also others related to the structures of holdings and to the farming systems.

The areas of most intensive agriculture are limited in territorial terms but they do contribute towards problems of water and soil pollution which in several cases have reached worrying proportions. However, it is not always easy to evaluate the culpability of agriculture as against other pollution sources, particularly when dealing with zones which are subject to high population densities and a major presence of other polluting activities. For example, in one zone of the Algarve the nitrate content in water for human consumption was found to be one of the highest in the country and judged of major concern, even in comparison to other worldwide locations. The causes of this situation have been attributed to the intensification of agriculture and the proliferation of dispersed settlement (Fernandes et al., forthcoming).

In addition to the case of the Algarve, which has been singled out as the Portuguese region with the highest consumption of fertilizers per hectare, reference can also be made to several specific situations in which agricultural can be blamed for water and soil pollution. There are, accordingly, areas of pollution caused by chemical fertilizers used on the irrigated maize crop of the Tagus Valley (a zone whose consumption of fertilizers is also above the national average) and in irrigated areas of the Alentejo. As regards pesticide contamination, the most worrying situations have once again been detected in the Algarve, in fruit-growing and horticultural zones north of

Lisbon, in the rice crop in the Baixo Mondego (Beira Litoral) and in vineyards zones in the Northern Coastal area. Effluent from intensive pig breeding, particularly in parts of the Lisbon region and the Tagus Valley, contaminate surface water (e.g. in the municipal district of Leiria) and underground soil and water (e.g. in the municipal district of Setúbal) (Quintela, 1995). In the Entre Douro e Minho (Northern Coast) there has also been pollution caused by effluent from dairy farming. Lastly, various problems of soil salinization in several of the irrigated areas of the Alentejo, Algarve, Beira Litoral and in zones north of Lisbon (DGA, 1995) have been detected.

As regards problems of soil erosion, according to the 1994 Report on the Status of the Environment (DGA, 1995) the situation in Portugal is worse than in the south of the European Union as a whole (Spain, Southern France, Greece, Italy and Portugal). Soils with a high degree of erosion risk account for 30% of the territorial surface. Some 57% are considered to be of intermediate risk of erosion, with 13% being considered reasonably safe from erosion. Equivalent percentages for the south of the European Union as a whole indicate that 19% of soils involve a high erosion risk, 36% are at moderate risk, 37% under low erosion risk and 8% of the total surface is excluded from the estimates (e.g. urban areas, lakes, missing data). The highest percentages of soils in danger of erosion in Portugal are located in the Lisbon and Tagus Valley and Alentejo regions, where such risks are, to a great extent, the result of the continuation of cereal cropping.

Finally, the abandonment of cultivated areas, although corresponding to a certain return to nature, has an obvious effect on the environment, both owing to the loss of nature friendly ecosystems that depend on human intervention but also through the loss of the farmed landscape (Madureira, 1994). The 'montados' (oak and cork oak groves) of the Alentejo and the 'lameiros' (permanent pasture and meadow) of the Northern Interior are examples of ecosystems which are at risk from the disruption to traditional crop practices (Baptista, 1996). The abandonment of contour farming in many zones north of the River Tagus is an example of the loss of farmed landscape. However, it is the increase in pine and eucalyptus tree plantations and the areas destroyed annually by forest fires, which represent the most dramatic alterations to the landscape.

We now turn to an assessment of the impact of the CAP on agricultural pollution, the exposure of soils to erosion, the conservation of ecosystems and the maintenance of the landscape.

12.4 Impacts of the CAP on the environment

In addition to the productive capacity of agriculture, the 1992 CAP reform emphasized concern over environmental protection. Therefore along with sectoral measures for the reorganization of production and alterations to Common Market Organizations, there were introduced the accompanying measures, such as for agri-environment (Regulation 2078/92) and forestry for holdings (Regulation 2080/92).

12.4.1 The Common Market Organizations

The reform introduced compensatory payments for the price reduction in cereals, oilseeds and protein crops. Payments are based on areas which are effectively cultivated and average regional productivity figures (which, in the case of Portugal, are significantly higher than the actual average productivity). The result, in less competitive regions, has been an incentive to continue production in order to receive the subsidy (Avillez, 1997). On the other hand, the payment of compensation has forced the reduction of the sown area and has enforced the need for a 'set-aside' although producers producing less than 92 tonnes of cereals have been exempted from this requirement (Ministry of Agriculture, 1992). As the average statistical productivity in Portugal is 1.6 tonnes per hectare, the set-aside is only obligatory for holdings with more than 50 ha of cereals, which is practically only the case in the Alentejo and the Tagus Valley. In the case of oleaginous products, such as sunflower seed oil, the past few years have recorded a trend towards an increase in the sown areas together with a marked decline in output per hectare. This provides clear evidence of the referred to changeover from earning returns from production to subsidies (Pereira, 1997). Arguably, through a logic of maximising the amount of rent on the part of the major landowners, area payments have discouraged the reorganization of production and continue to expose high-risk soils to the possibility of further erosion.

As regards the beef regime, the objective of the introduction of the incentive system was to encourage extensive production. Eligibility for livestock premia has thus been limited on the basis of stocking density, though not for small producers on holdings with total Livestock Units of 15 or less. In fact, 90% of cattle holdings in Portugal are exempt from the stocking density norm (Ministry of Agriculture, 1992). The objective of encouraging extensive production has therefore been restricted to zones such as the Alentejo, in which production was already being carried out on these lines. Elsewhere and in the overwhelming majority of cases, there are no limitations on cattle density and the incentives system once again encourages the maintenance of livestock numbers, not for production purposes but as a means of receiving subsidies.

12.4.2 The Structural Funds

According to the criteria arising from the Community's 1988 reform of the Structural Funds, the whole of Portuguese territory is eligible under Objective 1 for EAGGF support targeted at the development and structural adjustment of agriculture in regions with a GDP/inhabitant ratio less than 75% of the EU average. After the 1992 CAP reform and changes to the structural fund regulations had incorporated environmental concerns, the agricultural sub-programme of the second Community Framework Support Programme (for Portugal) was renamed as the Programme for the Support of Agricultural and Forestry Modernization (PAMAF). This programme brought in changes on zoning and the use of instruments for the 1994/1999 period, and did put emphasis on environmental issues. In addition to the structural measures set out in Objective 1, PAMAF also included Objective 5a instruments – to accelerate the adaptation of agricultural structures as part of the reform of the CAP - for which all EU farmers are eligible if at least 25% of their income is earned from agriculture and 50% is earned from agriculture, tourism and forestry of a professional type, when not contributing to excess production and/or prejudicial to the environment.

PAMAF comprises six measures:

- measure 1 is targeted at providing support to the development of agricultural infrastructures, such as irrigation infrastructures, agricultural and rural roads, drainage and soil conservation, the electrification of holdings and land restructuring;

- measure 2, which is targeted at support for individual holdings, encompasses Objective 5a actions for the restructuring of olive groves and vineyards or the modernization of diverse strands (modernization of crops, local breeds and the introduction of innovative cropping activities);
- measure 3 is the forestry development programme for tree planting and the replanting of forests which have been destroyed by fire, improvements to existing forests, the establishment and improvement of forestry nurseries, studies and maintenance, infrastructures and multiple forest use;
- measure 4 involves research and development for reinforcing agricultural co-operation, the competitiveness of agro-food companies and the financing of strategic studies;
- measure 5 finances investments in companies which process and commercialize products, with due preference being given to products having a certificate of origin;
- measure 6 refers to vocational agricultural training.

In line with the trend of the last few years, the three main measures, in 1996, were measure 2 (support for holdings) with 42% of the assistance, measure 1 (infrastructures) with 21% and measure 5 (product processing support) with 15%. Measure 2 is particularly targeted at major specialized holdings which have received significant finance for increasing the irrigated area under measure 1, and measure 5 was targeted at investment finance for the processing of intensive irrigation products (74% of the funding provided to this measure was spent in the Lisbon and Tagus Valley region). It can be concluded, without establishing a direct relation with the worsening of the situation, that this support was targeted at the more intensive agricultural zones in which, as we have seen, specific problems of water and soil pollution have been observed. Measure 3, targeted at the replanting of forests destroyed by fire and which, by means of replanting and the planting of burnt down forests, could have a positive effect on soil erosion, received 6% of the funding.

12.4.3   Agri-environment policies

In Portugal, the full agri-environmental measures programme has been divided up into three major groups of measures and a fourth group referring to vocational training actions.

Group I is targeted at agricultural pollution and comprises 5 measures:

1) Protection of water from agricultural pollution (awaiting the transposition of the Nitrates Directive to Portugal);
2) Recommended chemical treatment;
3) Integrated crop protection;
4) Integrated production;
5) Promotion of biological agriculture.

Group II is targeted at the extension and/or maintenance of traditional extensive farming systems and comprises 14 measures, involving different intervention zones:

6) Traditional mixed farming systems in the north and centre;
7) Non irrigated cereals systems;
8) 'Lameiros' (semi-natural pastures in valleys or hill areas of humid mountain zones in north and central Portugal. They are very traditional and productive systems used for feeding cows and sheep. They are maintained by the farmers who, for example, build water channels to avoid flooding.
9) Extensive forage systems;
10) Traditional olive groves;
11) Torres Novas fig-groves;
12) Contour farming of vines in Dour (Port Wine area);
13) Regional varieties of fruit trees;
14) Traditional non-irrigated orchards;
15) Traditional non-irrigated almond groves;
16) Cork-oak woods;
17) Reconversion of arable land to extensive pasture land;
18) Support for the maintenance of autochthonous breeds in danger of extinction;
19) Extensification of cattle production.

Finally, Group III is targeted at the conservation of resources and the rural landscape and comprises 5 measures:

20) Maintenance of abandoned forested surfaces;
21) Maintenance of forested surfaces to complement holdings;
22) Preservation of areas of arboreal and shrubby species belonging to ecosystems of high biological interest;
23) Maintenance of agricultural land inside forest areas;
24) Maintenance of traditional farming systems in environmentally sensitive zones.

Table 9 shows that the measures targeted against agricultural pollution (Group I) are the least important, both in terms of the initial forecast as well as in terms of the actual achievement. Apart from the fact that measure 1 (protection of water from agricultural pollution) has not been implemented and that this measure could be better directed to the zones where more worrying concerns have been detected, the impact attributable to the remaining measures is mainly in the the disclosure of the existence these problems so as to sensitize technical staff and farmers to the problems of agricultural pollution.

It is the measures targeted at the maintenance and preservation of traditional farming systems (Group II) which have the greatest territorial impact. Notwithstanding their relative success and contribution to the maintenance of nature friendly farming systems and the conservation of the landscape, these measures, owing to the limited size and nature of the incentives available, are not expected to represent an alternative in zones in which intensification is possible. In marginal zones it also appears unlikely that the reconstruction of rural society can be based on the continued preservation of traditional farming systems. Rather than a structuring of a new relationship with the territory which takes environmental impacts into account, these measures have been a means to transfer funds. Added to the rents received from land, this transfer allows the continuation of farming systems whose viability was undermined some time ago by the major decline in the agricultural population.

Table 9   Initial forecasts and implementation of agri-environmental measures

MEASURES	Initial forecasts for 1994/1997 period (ha)	Implementation 1994/1995 (%)	Location of Zone of application
Group I	47,931	17	
1	720	0	Vulnerable Zones to be defined by the Nitrates Directive
2	36,630	5	Whole of the country
3	3,731	50	Whole of the country
4	700	8	Whole of the country
5	6,150	73	Whole of the country
Group II	531,085	79	
6	62,100	167	North of the country
7	48,420	59	Northern Interior and Alentejo
8	24,165	95	Northern Interior
9	175,050	57	Whole of the country, excluding Lisbon and Tagus Valley Region
10	18,330	135	Northern Interior
11	2,700	15	Small zone on northern bank of Tagus
12	9,180	78	Upper Rio Douro Valley
13	1,350	302	Whole of the country
14	11,250	145	Algarve
15	5,940	183	Northern Interior
16	112,500	54	Alentejo
17	28,600	0	Whole of the country excluding Lisbon and Tagus Valley Region
18	49,739 a)	87	Whole of the country
19	1,500	0	Northern coastal area
Group III	46,640	177	
20	25,600	39	North of the country
21	0	(42,047 ha)	North of the country
22	16,800	9	Whole of the country
23	4,240	29	Whole of the country
TOTAL	625,656	77	

a) Total number of Livestock Units (LUs).
Source: prepared using MADRP and Diniz (1997) databases.

As regards Group III measures, the objectives of measure 20 are similar to those envisaged by measure 3 of PAMAF albeit with less production and more geared towards fire prevention. In any event, its rate of realization has been well below forecast which has meant that its impact has not been significant.

12.4.4 Forestry measures in agriculture

The national forestry plan arising from the application of Regulation 2080/92 involves between 100,000 and 125,000 ha of new plantations on agricultural surfaces and improvements to 30,000 ha of existing plantations. It was prepared in the search for a balanced distribution of the programme between the north and south of the country, and encourages slow growth species. 11 zones and chosen species have therefore been selected and which, adapted to the ecological specifications of each zone, have received higher compensation. The programme, in addition to providing investment assistance and covering maintenance costs over a 5 year period, also attributes a bonus for loss of earnings which, based on the species in question and the type of plantation, can remain in force for 10 or 20 years. As applications normally refer to marginal land, the loss of income is very small or even nil. This programme has recorded a high application rate and is estimated to have covered more than 95,000 ha over the 1994/95 period in accordance with the regional distribution set out in Table 10.

Table 10   Regional distribution of forested areas according to Regulation 2080/92 in 1994/1995

Region	Forested area (%)
Northern Coast	
- Entre Douro e Minho	1.7
- Beira Litoral	1.0
Northern Interior	
- Trás-os-Montes	15.3
- Beira Interior	15.5
Lisbon and Tagus Valley	7.6
Alentejo	47.8
Algarve	11.1
Total	100

Source: MADRP.

If it is true that the objectives of the total area to be forested and the priority afforded to slow growth species have been achieved, the same cannot be stated as regards the balanced distribution of the programme between the north and south of the country. The programme has been effec-

Madureira, L. (1994) *A olivicultura nos sistemas de produção agrícola de Trás-os-Montes e Alto Douro. Um contributo para o estudo das relações entre a agricultura e o ambiente,* Masters Thesis in Agrarian Economy and Rural Sociology, Lisbon, Instituto Superior de Agronomia

Ministry of Agriculture (1992) *Reforma da PAC. Síntese dos principais aspectos;* Lisbon

Ministry of Agriculture (1994) *Medidas agro-ambientais;* Lisbon, IEADR

Pereira, J.D. (1997), *Modelização da produção de girassol em Portugal no período de 1971 a 1995;* End of Agronomical Engineering Course report, Lisbon, Instituto Superior de Agronomia

Quintela, C. (1995) *Poluição da água pela suinicultura, legislação e economia do ambiente. O concelho de Leiria como caso de estudo;* Masters Thesis in Agrarian Economy and Rural Sociology, Lisbon, Instituto Superior de Agronomia

Rolo, J.C. (1996) *Imagens de meio século da agricultura portuguesa;* In: Brito, J.P. et al. (coord.) *O voo do arado,* Lisbon, Museu Nacional de Etnologia, pp. 77-157

PART 4

NEW MEMBER STATES

13. AUSTRIA
Werner Kleinhanss

13.1 Introduction

Agriculture in Austria is recognized as having a number of objectives including the production of food, feed and other raw-materials, as well as the conservation of the environment and landscape, both as ends in themselves and for their importance to tourism. These objectives were expressed in the so-called Öko-Soziale Agrarpolitik (ecological-social agricultural policy) which included market and price policy measures with direct payments to compensate for structural disadvantages and for agri-environmental actions.

Until the 1980s agricultural policy was dominated by price policy. High levels of price support and market protection induced intensification, specialization and regional concentration of production, which contributed to surplus production and some environmental damage. Command and control measures were introduced to reduce surpluses, including quotas for milk, stocking limits and producer co-responsibility levies. In addition, a levy on mineral fertilizer and maize seed was introduced at that time. Direct payments for farms in mountainous and less favoured areas were given to support incomes and to prevent depopulation. Measures to support environmental friendly agricultural practices were established at the beginning of the 1990s.

The main causes of pollution problems in agriculture are assessed to be (Wytrzens and Reichsthaler, 1990): intensification, the separation of crop and livestock production, specialization of crop rotations, increase of erosion risky crops, soil improvement and the mechanization of crop production. Environmental problems are diagnosed as being mainly due to changes in the political, economic and social situation of agriculture, and only incidentally to lack of knowledge, ignorance or harmful actions on the part of farmers. Large farms are seen to pose a greater threat to the environment than small ones. Small farms have thus been favoured by structural as well as agri-environmental policy measures, in a policy that has deliberately retarded structural change in agriculture.

In 1995 Austria became a member of the EU. The principles of the Common Agricultural Policy were applied immediately and completely (Pirringer, 1997). On the one hand, this shock has been judged to be positive from the point of market orientation, structural improvement and international competitiveness (Schneider, 1996). On the other hand, CAP measures are judged to be less favourable with regard to the environment: they are thought to favour large farms and are considered to be too production oriented (Hovorka, 1996). However, the accompanying measures of the CAP are fully in accordance with the policy objective of the former Öko-Soziale Agrarpolitik.

The chapter describes the state of the agricultural environment before EU accession, explores the impacts of EU accession and of the CAP reform, and reviews the main results from the evaluation of the Austrian Programme of Environmental Friendly Agriculture (ÖPUL).

13.2 Interactions between agriculture and environment

The polluting pressures from agriculture seem rather small, compared to other EU Member States (Pirringer, 1997):

-    total consumption of commercial fertilizer (nitrogen, phosphorus and potassium) in 1993/ 94 at 76 kg/ha of UAA (or 107 kg/ha of fertilizer-worthy area), including 36 (or 51) kg of nitrogen per hectare;
-    consumption of pesticides at 0.96 kg of active agents per hectare; and
-    livestock density at 0.65 LU/ha of UAA (or 0.9 LU/ha of fertilized land).

The main developments and the state of the environment are summarized in accordance with Wytrzens and Reichsthaler (1990) and Pirringer (1997).

Erosion and soil compression

The risk of erosion is particularly high in intensively used arable land in the lower hilly areas and alpine zones. About 20% of UAA are threatened by erosion, especially in areas where maize, sugar beet and vines are grown, and on alpine grasslands with a slope of more than 30 degrees. The area of maize has increased drastically since the 1950s. Erosion in the mountain areas is mainly caused by higher grazing livestock numbers, but also partly through abandonment of alpine pasture, deforestation and skiing in the high mountains. The structural development of agriculture in the past led to the diminution of alpine pasture, which not only increased the risks of erosion, but also often detracted from the landscape.

Pollution of waters by minerals and pesticides

Because of the low level of fertilizer use and low livestock densities, mineral surpluses are relatively low by comparison with other EU countries. Assessments for 1994, based on the farmgate method, came up with a level of 22 kg of nitrogen surpluses per hectare of fertilized land and 14 kg of phosphorus per hectare (Bundesamt und Forschungszentrum für Landwirtschaft, 1996). Sampling of groundwater shows large regional variation in nitrate concentrations, ranging from about 50 mg/l in Burgenland and Vienna to 4 mg/l in Vorarlberg (Table 1). Six thousand square km are considerably polluted with nitrogen compounds. In most of the intensively used agricultural areas (valleys and basins), the nitrate content is higher than the Austrian standard of 45 mg per litre. Empirical analysis shows that statistically significant relations exist between erosion-endangering crops, farm size, density of population and nitrate stress on groundwater (Wagner, 1997).

Concerning pesticides, the relatively low level of 0.96 kg of active agents per hectare UAA has to be mentioned. Overall, 3,404 tonnes of active agents were used in 1995, of which 47% were herbicides, 41% fungicides, 4% insecticides and 0.5% growth regulators. The groundwater monitoring programme, which started in 1991, covers 47 pesticide agents. Of the total of more than 300,000 sample results, 6% were found to be contaminated with pesticides, mainly atrazine and its two metabolites. In 1996 about 6,800 square km of groundwater areas were threatened by atrazine and its metabolites, in particular in Upper Austria and Styria. It is expected that, due to the atrazine ban imposed in 1994, the contamination will fall below the target level (Pirringer, 1997).

Table 1   Overview of nitrate concentration in groundwater by region

Province	Number of sampling sites	1992/93 mg/l	1993/94 mg/l
Burgenland	121	52.53	54.44
Carinthia	173	23.08	19.70
Lower Austria	238	35.97	35.42
Upper Austria	159	26.98	24.40
Salzburg	70	13.32	10.64
Styria	234	30.49	26.05
Tyrol	115	9.10	6.37
Vorarlberg	61	5.34	4.11
Vienna	45	52.62	50.71
Austria	1,316	29.75	26.82

Source: Federal Ministry for Agriculture and Forestry, Wassergüte in Österreich, Jahresbericht, 1994.
Source: Pirringer, 1997.

Air pollution and emissions of greenhouse gases

There is not much information available on the links between agriculture and air pollution. The Austrian Research Centre of Seibersdorf and the Technical University of Vienna estimate ammonia emissions to be about 150,000 tonnes, of which 80% are from agriculture, as well as 220,000 tonnes of methane originating from agriculture and 12,000 tonnes of N_2O emissions. Livestock production and the use of mineral fertilizers have been identified as the most important sources of emissions (Pirringer, 1997).

Endangered species

Due to its natural conditions (alpine, basin and steppe areas) Austria has a broad spectrum of species. There are approximately 30,000 animal species. About 10,000 animals and about 2,800 ferns and flowering plants were judged according to whether they faced a threat to their existence (Table 2). The main threats to species and biological diversity are identified as agriculture, tourism, the exploitation of natural resources, urbanization, water use and forestry (Wytrzens and Reichsthaler, 1990).

Table 2   Endangered animal and plant species in Austria

	Animals		Ferns and flowering plants
	total	of which mammals	
Total number	10,882	83	2,873
of which:			
- not endangered	8,078	40	1,792
- potentially endangered	580	13	171
- endangered	794	18	401
- highly endangered	819	3	300
- threatened with extinction	425	4	156
- extinct, exterminated or disappeared	186	5	53

Source: State of the Environment in Austria, Federal Ministry of Environment.
          Youth and Family Affairs (1994).
Source: Pirringer (1997).

Destruction of landscape elements

Beside the factors mentioned above, the improvement of soils and consolidation of farmland have an impact particularly on the landscape. Since 1948 about 1 million ha have been included in land consolidation programmes (Wytrzens and Reichsthaler, 1990). Quantitative assessments of the environmental impacts are not available but the sorts of changes that have been prevalent include destruction of hedgerows and less diversified crop-rotations.

13.3 Impacts of agricultural and environmental policies

Economic as well as command-and-control instruments were introduced in the past for environmental policy objectives. The economic instruments were the fertilizer levy, a duty on maize seed, the restitution of fuel tax and subsidies for environmentally friendly agricultural practices. The restrictions on livestock size, restrictions on animal fertilizer use in water protected areas and standards for the quality of drinking water can be characterized as control measures.

Fertilizer levy and duty for maize seed

The fertilizer levy was introduced in 1986 as a soil protection levy within the market structural regulation; in 1990 the levy was about 50% of the fertilizer price and applied to all mineral fertilizers (Hofreither et al., 1996). Reductions of fertilizer use and yield reductions were less than expected; it is not apparent whether the slight decline of fertilizer use was induced mainly by the levy, technical progress or other policy measures. The effect of the levy was considerably offset by a cut in fertilizer prices by the fertilizer industry. Based on assessments it can be concluded

308

that the tax level was not high enough to induce significant changes. Indeed, the levy was mainly designed to generate revenues for other policy objectives (export enhancement, subsidising environmental measures). There also was a duty on maize seed. Even though the duty was quite high, farmers did not get enough incentive to switch to other crops. Both instruments were removed during the accession into the EU mainly because of concerns to maintain the competitiveness of Austrian agriculture within the EU.

Fuel tax restitution

Most EU countries have a system of fuel tax restitution for the agricultural sector. The former Austrian system is of special interest because restitution was given independent of actual use. The restitutions paid were mainly based on fixed amounts per hectare and differentiated between cereals, sugar beet and grassland. In principle, a lower fuel use could be expected, because farmers had to pay the full price of what they bought including taxes. The system was abandoned, however, in 1991 and the budget transferred to a programme for stabilising crop-rotations. Since 1994 the budget has been included in the ÖPUL programme. It is not clear whether these changes were guided by energy policy objectives or by the budget requirements of environmentally orientated subsidy programmes.

Restrictions on fertilizer use in water protected areas

The rules for fertilizer use in water protected areas are defined in the Wasserschutzgesetz from 1990 (Hofreither et al., 1996). The maximum livestock density is 3.5 LU/ha; beyond that level special permissions are required. The maximum level of (total) nitrogen use is 175 kg/ha on arable land and 210 kg per hectare on grassland. Within the ÖPUL programme nitrogen use from animal manure is limited to 150 or 120 kg/ha respectively. These limits are more restrictive than the EU Nitrate Directive.

Restrictions of livestock size

Restrictions on livestock numbers per farm are one of the options to reduce pollution problems proposed by environmental groups. The objectives are mainly linked with subsidies, because poultry production (which is not subsidized) is as concentrated as in other EU countries. Based on the assumption that small farms pollute less than large ones, the subsidies previously available included a fixed amount for all participating farms. Also, the subsidies were reduced with increasing farm size and there were upper limits. Small farms generally were favoured by these subsidies.

13.3.1   EU accession

Several investigations have been made to assess the impacts of EU integration. The studies mainly focus on economic and market impacts (Neunteufel and Ortner, 1989; Hofreither, 1996a;

Neunteufel, 1996; Schneider, 1994 and 1995). Simulations based on an Input-Output-Model (Köppl et al., 1995) showed that emissions of SO_2 and CO_2 from agriculture would not change. Assessments concerning environmental impacts are more or less on a qualitative basis and closely linked with price induced changes, restrictions on production, factor use, subsidies and programmes to stimulate environmentally friendly farming practices.

Price induced changes

After EU accession agricultural market prices were reduced by 23% (Schneider, 1994). Cereals and starch potatoes were affected most because market prices were reduced by 53 and 51% respectively. From a theoretical viewpoint lower intensities should have been induced and therefore the use of fertilizer and pesticides reduced. This effect was more or less neutralized, however, by the removal of the fertilizer levy, which meant a reduction of fertilizer prices by one third. For all those products where the effect of price reductions was lower than the impact of the levy (sugar beet, potatoes, vegetables and fruits) higher intensities may have been achieved. Overall, though, the decreasing tendency of fertilizer use since 1990 continues. This may be induced by better agricultural practices (Hofreither et al., 1996) and the introduction of environmentally orientated measures (ÖPUL).

The oilseed area drastically increased up to the guarantee of 127,000 ha 1). Winter rape is judged to have positive environmental effects (mainly in reducing nitrate pollution) because of its capacity to accumulate nitrogen during the winter.

Restrictions on production and factor use

As with lower intensities, the reduction of agricultural production is also assessed to have had positive impacts on water and soil (Köppl and Pichl, 1994). The following measures are of importance:

-   the set-aside of arable land increased from 2 to about 8% mainly owing to the requirements of the arable crop regime. Due to the small producer scheme the set-aside share is higher in regions with favourable farm structure and sometimes more intensive agriculture. Therefore mineral fertilizer use may be reduced more than in mountain and less favoured areas. In the latter case set-aside may be induced by the lower profitability of land use under CAP conditions;
-   the reduction of sugar quota may favour oilseed production; dairy cows might be substituted by suckler cows due to the reduction in milk quota. Extensive beef production is one opportunity to reduce livestock concentration and subsequently to contribute to a reduction of nitrogen surpluses;
-   the removal of restrictions on the maximum livestock number per holding may induce concentration and intensification of livestock production (Hovorka, 1996).

---

1)   Until the 1980s oilseed production was restricted to about 10,000 ha due to GATT agreements. Later on it has been increased because of bio-fuel production.

Agri-environmental programmes and structural policy

The Austrian Programme of Environmental Friendly Agriculture (ÖPUL) was accepted in 1994 and was modified according to the requirements of EC Regulation 2078/92. A recent evaluation of the programme indicates positive environmental effects (see below, section 3.2). ÖPUL is also important from an economic viewpoint because it allows income transfers to the agricultural sector which partially compensate for income losses due to EU-accession. The programme is announced to be a central element of existing and future agricultural policy (Hovorka, 1996). Compared with the former national programme of stabilising crop rotations, a larger budget is available for environmentally orientated incentives.

In the case of structural policy the level of subsidies before EU-accession is guaranteed for the next ten years. Beyond this period subsidization of small farms will be reduced which will, most probably induce structural changes. This may be particularly true for alpine and other less favoured regions. In some 'Objective 1' areas (Burgenland) major structural changes started some years ago, resulting in set-aside, the abandonment of agricultural land and a drastic decline in land prices. If that were to happen also in alpine regions, it would result in deterioration of the landscape.

According to Hovorka (1996) the modified programmes are much more in favour of production and large farms than the former ones. On the assumption that small farms pollute less than large ones, the changes mentioned above may induce negative environmental impacts.

13.3.2   The CAP

On account of the short time since Austria's accession into the EU the environmental impacts of the CAP are still difficult to assess. This is especially true for the market and price policy, although initial assessments have been made of the environmental effects of the ÖPUL programme. The description of environmental impacts will therefore be differentiated into two parts. In addition the impacts of structural policy measures will be explained.

13.3.2.1   Market and price support measures

The arable regime

The impact of price reductions on the intensity of farming are difficult to assess because of the lower fertilizer prices without fertilizer levies (see Section 3.1 above). The following aspects have to be mentioned with regard to environmental impacts:
-   increase of set-aside of arable land, mainly in regions with more favourable farm structures and partially intensively used soils. This may lead to a reduction of land to grow cereals, and reduced fertilizer and pesticide use;
-   increase of oilseed areas, especially rapeseed, within the limits of the base area for oilseeds. The available evidence to date suggests that lower nitrogen losses during winter can be expected;

Table 4   Participation in ÖPUL-measures 1995 c)

ÖPUL measure		Area (ha)	Farms (number)	Support 1995	
				Mill. ATS	%
1 a)	elementary supports	2,298,163	165,868	1,538.9	21.19
2 a)	organic farming	199,113	15,917	663.3	9.13
3 a)	renunciation of yield-increasing inputs on the whole farm	310,541	37,742	652.1	8.98
4	integrated fruit cropping	7,800	2,459	54.8	0.75
5	integrated viniculture	34,719	12,267	277.7	3.82
6	integrated floriculture	478	73	2.4	0.03
7	extensive grassland farming in traditional areas	114,553	11,118	271.5	3.74
8 a)	stabilization of crop rotation patterns	911,500	54,123	1,286.5	17.71
9 a)	extensive cereal growing	249,345	30,911	598.4	8.24
10 a)	renunciation of certain yield-increasing inputs on arable land	330,385	78,156	321.8	4.43
11 a)	renunciation of fertilizers and pesticides on grassland areas	248,668	46,119	443.3	6.10
12	restraints on cutting times of hay meadows	5,639	2,759	11.3	0.16
13	protection against erosion in fruit cultures	6,145	2,384	9.7	0.13
14	protection against erosion in vinicultures	3,350	2,727	7.6	0.10
15	protection against erosion in arable farming	716	261	0.5	0.01
16	raising endangered animal species	-	3,285	21.4	0.29
17	mowing of steep slopes and upland meadows	232,755	58,322	606.5	8.35
18	premium for alpine farming and sheep holding	274,427	8,734	265.3	3.65
19	maintenance of areas with ecological values	35,778	41,696	142.9	1.97
20	cultivation of endangered crop plants	18	11	0.1	0.00
21	maintenance of abandoned forest areas	638	162	2.6	0.04
22	20-years set-aside of areas for biotopes	147	168	1.2	0.02
23	allotment of areas for ecological objectives	542	777	3.1	0.04
24	allotment of areas for ecological objectives under set-aside	4,621	1,880	5.5	0.08
		b)	b)	7,263.6	100.00

a) Measures available nationwide; b) Because of the possibility to combine some measures on the same plot the total sum cannot be calculated; c) Subsidies for educational measures for control on organic farms are not included in the table.
Source: BMLF (1996).
Source: Pirringer (1997).

should not exeed 2.5 LU/ha in 1997; landscape features and the share of grassland should be maintained; and the rules should be followed for appropriate fertilising practises.

Measures applied for the whole farm (Measures 2-8) are organic farming, renunciation of specific yield increasing inputs, integrated production methods for permanent crops, extensive grassland farming in traditional areas, reduction of livestock density (which is included in extensive grassland farming) and stabilization of crop rotations. The premium varies from 900 to 3,000 ATS/ha for arable land, and between 200 and 2,500 ATS/ha for grassland; it is considerably higher for organic farming. Criteria for most of these measures are rather easily met, and therefore the measures can be applied by a large number of farms.

Single field orientated measures (Measures 9-12) are extensive cereal growing, renunciation of specific yield increasing inputs on arable land, renunciation of mineral fertilizer and pesticides on grassland and constraints on the cutting time for hay meadows. Premia are defined according to the type of measure (e.g. pesticide renunciation) and a regional classification of the meadows.

The so-called regional measures are orientated to erosion protection in vineyards, fruit growing, maintenance of areas with a high ecological value, conservation and support of landscape elements and long term set-aside for environmental purposes.

Acceptance

The programme is well accepted by farmers. In the first year of implementation in 1995, 179,000 farms took part in the programme, covering 2.6 million ha. Referring to the total budget, the following single measures have the greatest acceptance (see Table 4): elementary support; stabilization of crop rotation patterns; organic farming; whole farm renunciation of yield increasing inputs; extensive cereal growing, and mowing and sheep holding in alpine areas. Together, they account for 77% of the budget.

The measures for elementary support, organic farming and lower intensities (Measures 1, 2, 3, 10, 11) are widely applied in all regions (Wagner, 1996). The measure for organic farming (covering 8% of UAA) is widely applied on dairy and beef farms and therefore largely concen-

Table 5   Distribution of environmental support in 1995

Farm types	Partici-pating farms (number)	Average farm size (ha)	Support total (Mill. ATS)	Average support per ha (ATS)	Average support per farm (1,000 ATS)
Commercial farms	16,633	30.4	1,332	2,640	80.1
Forage farms Zone 0	21,800	15.3	810	2,400	37.1
Forage farms Zone 1-4	35,957	14.7	1,824	3,360	50.7
Granivore farms	7,108	18.3	1,824	3,360	50.7
Permanent crop farms	6,953	10.1	327	4,670	47.0
Mixed farms	11,183	18.0	479	2,380	42.9
Farms 25-75% forestry areas	19,325	12.5	1,022	4,050	52.9
Forestry farms	1,581	10.5	70	4,010	44.0
Large farms	321	130.7	112	2,640	348.9
Small farms	41,033	4.2	585	3,350	14.3
Other farms	9,365	-	444	-	47.4

Source: Federal Ministry of Agriculture and Forestry, Evaluation of economic aspects of the Austrian environmental programme (1997).
Source: Pirringer (1997).

trated in the Western part of Austria. Accession to the programme is relatively easy because of the low intensities and stocking rates in mountain farming. Because specialized distribution and marketing outlets are not always available, products from organic farming are often sold through conventional channels where price advantages cannot be realized. Although most of the budget is allocated to the most intensive agricultural regions, there are indicators that the programme is less accepted by the most intensive farms (Table 5).

The high acceptance can be explained partly by the generally low requirements and the relatively high premia. However, compensatory payments for some measures (e.g. for landscape protection or erosion) are rather small compared to the costs of farm adjustments (BMLF, 1996). Some of the measures and the level of premia may be modified during the ongoing revision of the programme.

Environmental impacts

An ecological evaluation of ÖPUL has been commissioned by the Agricultural Ministry; this project has been co-ordinated by the Bundesamt und Forschungszentrum für Landwirtschaft (BMLF, 1996). First evaluations of ÖPUL indicate positive environmental effects (Dietrich, 1997). The individual measures as well as the programme as a whole were assessed with regard to different environmental (abiotic and biotic) objectives. The following is a summary of the main results.

Maintenance of soil

- Fertilizer use: The level of fertilizer use is quite low (nitrogen - phosphorus - potassium. 47-22-25 kg/ha fertilizer-worthy area in 1994/95). ÖPUL may have contributed to the decreasing tendency of fertilizer use. The measure 'renunciation of easily soluble (nitrogen) fertilizers' (Measures 2, 10, 11) is only applied to 6% of arable land but to 70% of fertilizer-worthy grassland (involving substitution of mineral fertilizer by organic manure). Through better use of manure, nitrogen losses and therefore mineral surpluses will be reduced in the long term. The measure reduction of organic nitrogen to 120 or 150 kg/ha (Measures 1, 2, 3, 7, 10, 11) is to curb over-fertilization. However, farms with a high livestock density generally did not apply for the programme, and the measure to reduce livestock numbers has not been widely taken up. Therefore ÖPUL has little influence on the location, intensity and concentration of livestock production and related regional pollution problems.
- Concerning pesticides use, an overall reduction of 6%, including 9.6% for fungicides, 10% for insecticides and 58% for growth regulators, was achieved between 1994 and 1995. The major reductions are related to ÖPUL measures (the large reduction in growth regulators is not very important from a national point of view, because their share of total pesticide use is only 0.5%).
- Measures to reduce intensity (lower fertilizer and pesticide use, extensive cereal production, e.g. durum wheat, and organic farming) are applied on 0.64 million ha. Positive environmental effects are assumed. Erosion protecting measures (Measures 13-15) are only applied on 750 ha. Erosion protection is also a consideration in the measure for stabilization of crop

316

rotations (Measure 8) in limiting cereals (including maize) to 75% of arable area and in certain obligations for winter cover crops.
- There are positive signs with regard to water protection and to reduce pollution in the long term. Specific measures are available with regard to the reduction of pollution from intensive cropping and livestock production. They have not been well taken up, because the premia are too small compared with the costs of adjusting farm practices.

Impacts on climate influencing gases

The share of agriculture in total emissions of greenhouse gases is assessed at 9.1%, comprising 3.7% CO_2, 4.3% methane and 1.2% N_2O. ÖPUL impacts on fuel use are not significant, while emissions of N_2O may be reduced by 5%. Through lower fertilizer use, CO_2 emission will be reduced by 1.2%. Thus, problems of global warming will not be significantly reduced by ÖPUL. Other options, such as producing biomass for energy, could have greater impacts.

Biodiversity

Owing to the short experience with the programme it is difficult to assess impacts on biodiversity. ÖPUL measures should have positive impacts because environmentally damaging pressures (destruction of landscape, pollution through chemical inputs and mechanization) will be reduced. Organic farming is also stimulated. It has to be mentioned, though, that some of the measures with high ecological potential have elicited little response. Production of 'rare' crop varieties (Measure 20), for example, has only been carried out by a small number of farms. The reasons are the relatively low subsidy level, limited compatibility with other measures and the applicability of the measure to only certain provinces.

Concerning cultural landscape protection ÖPUL includes specific measures (Measures 19, 21-24), but only 2% of the budget is used for these items. The reasons are the low level of subsidies in relation to the costs of adaptation, and the incompatibility with local landscape protecting activities organized on a contractual basis. It is recommended that ÖPUL measures should be embedded in regional projects and that the measures should be extended towards other subjects of landscape protection and biodiversity.

Economic impacts

ÖPUL has significant income effects because of the relatively favourable premia, as well as the modest criteria for eligibility which have ensured a high level of participation. The total budget of 7 billion ATS considerably contributes to the support and stabilization of farm incomes. Indeed, the creation of additional farm income has been a secondary objective of the programme. At least 50% of the budget generates additional income, because the costs of the adaptations for some of the measures, especially the elementary support, are much lower than the subsidies (Schneider, 1996).

Farm level assessments by Schneeberger and Eder (1997) show that participation in ÖPUL has financial advantages for holdings. There are also distributional effects, because acceptance by farms with a high share of grassland is greater than by arable farms. Also, the acceptance in less favoured areas is higher than in regions with intensive agriculture (Wagner, 1996). Therefore ÖPUL is also important from the viewpoint of regional policy regarding subsidization of less favoured regions and the protection of the landscape by agricultural activities.

13.3.2.3 Structural policy

Structural policy is of importance with regard to regional economic development and compensation for economic burdens due to less favourable natural conditions. The share of mountain and less favoured areas is 68.6% of total UAA. Compensatory allowances are paid to 126,000 farms and the total budget in 1995 was 2.9 billion ATS (Hovorka, 1996). Payments are differentiated by the degree of regional natural disadvantages: There is a so-called Wahrungsregel which guarantees that, if subsidies under EU regional funds are lower than in the former national programme (mountain farming scheme), the difference will be covered by national funds for each of the farms participating in the former programme (Schneider, 1996; Amsing, 1996; Ortner, 1996). Compared with the former regulation, EU structural policy measures are much more in favour of production (payments per hectare or LU) and therefore show less preference concerning environmental policy objectives (Hovorka, 1996). Although environmental assessments are not available the following general conclusions can be drawn:
- agriculture in ecologically sensitive areas, mainly in mountainous regions, will be maintained;
- a minimum level of active population will be ensured in less favoured areas, which seems to be a necessary condition for tourism (Klasz, 1996a).

13.4 Concluding observations

Austrian agricultural policy before EU-accession was characterized by:
- price protection and supply control through quota and limitation of livestock numbers;
- the support and stabilization of farm income by direct payments to compensate for structural disadvantages;
- regulations and subsidies to stimulate environmental-friendly agricultural production.

Measures were taken in such a way that small farms and less favoured regions were favoured in comparison with large farms, which were thought to be mainly responsible for environmental damages (Wytrzens and Reichsthaler, 1990). Although the intensity level and concentration of livestock production is less than the EU average, there are regional environmental problems induced by agriculture: nitrate contamination of groundwater in basins and intensively used agricultural areas and soil erosion because of the rise of maize producing areas. Agriculture with its side effect of landscape protection is of special importance for alpine regions and tourism. Positive external effects of agriculture are commonly regarded as a reason for subsidies (direct payments) to holdings for the provision of environmental benefits (Pevetz et al., 1990; Puwein, 1994; Hofreither, 1996b; Neunteufel, 1994; Wörgötter, 1994).

In 1995 Austria entered the EU. The principles of the CAP were applied immediately and completely. Price reductions will be compensated by factor bounded degressive payments for a period of five years. Compared to the former agricultural policy, CAP may have positive and negative environmental implications:

(+) intensity of production will be reduced due to significant reductions of market prices; this effect is partly compensated by the removal of the fertilizer levy;
(+) set-aside of arable land will reduce usage of agrochemicals, but may induce higher intensity on cropping areas;
(+) a greater budget is available to support programmes of environmental-friendly agriculture (ÖPUL);
(-) suspension of the fertilizer levy for the reason of competitiveness;
(-) CAP compensation payments, as well as payments within agri-environmental measures and for less favoured areas are much more orientated towards production and may induce the structural development of farms with probable negative environmental effects.

Agri-environmental measures will become a central part of agricultural policy in Austria. At present ÖPUL measures are characterized by the way that a large number of farms is able to participate without high costs of adaptation. Therefore the programme creates additional income and compensates for income losses resulting from EU-accession. The programme is much more orientated towards the conservation of the actual state of environment than on special objectives of environment and landscape protection. A revision of the programme is presently being discussed. With regard to further negotiations within the WTO it has to be mentioned that payments for environmental preservation of agricultural practices should be related to the costs of adaptation; therefore payments within the agri-environmental schemes cannot at all be used for income transfers.

The accession to the EU is regarded as a necessary step to overcome the structural conservatism of the past decades and to move towards a market orientated agriculture. Sharp structural changes will be induced especially in less favoured regions. Negative impacts on landscape values may be induced on account of the withdrawal of agriculture in some regions. Therefore direct payments are necessary to compensate for the functioning of agriculture.

References

Amsing, S. (1996) *European Agricultural Structural Policy in the New Member States, Finland, Sweden and Austria;* LEI-DLO, The Hague

Bundesministerium für Land- und Forstwirtschaft (BMLF) (1994) *Wassergüte in Österreich;* Jahresbericht 1994, Wien

Bundesministerium für Land- und Forstwirtschaft (BMLF) (1996) *Ökologische Evaluierung des Umweltprogrammes (ÖPUL);* Band 1 und 2, Bericht des Bundesminsteriums für Land- und Forstwirtschaft an die Europäische Kommission gemäß Artikel 16 der VO (EG), Nr. 746/96, Wien, Dezember

Dietrich, M. (1997) *Ökologische Evaluierung des Umweltprogrammes;* Der Förderdienst, 1, Wien, 10-11

Fiegl, T. (1994) *Agrarförderungen und landeskulturelle Leistungen der Landwirtschaft - Projekt Lanersbach;* In: Schneeberger, W. und. Wytrzens, H.K. (eds.), Naturschutz und Landschaftspflege als Agrar- und Forstpolitische Herausforderung, Wien, 143-155

Hofreither, M.F. (1996a) *Macroeconomic development after Austria's EU-accession - some selected observations;* In: Ortner, K.M., Neunteufel, M.G., Jumah, A., Hofreither, M.F. (eds.)' Agriculture after joining the EU - Sectoral Analyses for Austria'; In: Schriftenreihe der Bundesanstalt für Agrarwirtschaft, 78, Wien, 55-63

Hofreither, M.F. (1996b) *Bewertung von Umweltleistungen der Land- und Forstwirtschaft;* Der Förderdienst, 1, 9-15, Wien

Hofreither, M.F., K. Pardeller, E. Schmid and F. Sinabell (1996) *Austrian country report for the inventory on mineral emission of agriculture;* In: Simonsen, J.W. (ed.) Inventory on Mineral pollution from agriculture, Part I: Country reports. (AIR 3 CT93-1164). Norwegian Agricultural Economics Research Institute, Oslo, January 25, 65-80

Hovorka, G. (1996) *Das Direktzahlungssystem in Österreich nach dem EU-Beitritt;* In:Bundes anstalt für Bergbauernfragen, Forschungsbericht Nr. 37, Wien

Klasz, W. (1996a) *Die Multifunktionalität der Landwirtschaft als Motor regionaler Entwicklung;* Der Förderdienst, 10, Wien, 301-302

Klasz, W. (1996) *Reform der EU-Strukturfonds-Ergebnisse der Bergbauern;* Der Förderdienst, 12, Wien, 381-389

Köppl, A., K. Kratena and C. Pichl (1994) *Umweltpaket und umweltpolitischer Gestaltungsspielraum;* WIFO Monatsberichte, Wien, 95 - 101

Köppl, A. and C. Pichl (1995) *Anreizorientierte Instrumente der Umweltpolitik;* WIFO Monatsberichte, 11, Wien, 697-707

Neunteufel, M.G. (1994) *Über die Internalisierung der sogenannten externen Effekte;* In: Schneeberger, W. und Wytrzens, H.K. (eds.), Naturschutz und Landschaftspflege als Agrar- und Forstpolitische Herausforderung, Wien, 61-70

Neunteufel, M.G. (1996) *Environmental aspects of EU-integration of Austrian agriculture;* In: Ortner, K.M., Neunteufel, M.G., Jumah, A., Hofreither, M.F. (eds.) Agriculture after joining the EU - Sectoral Analyses for Austria, Schriftenreihe der Bundesanstalt für Agrarwirtschaft, 78, Wien, 29-41

Neunteufel, M.G. and K.M. Ortner (1989) *Auswirkungen eines EG-Beitrittes auf die österreichische Landwirtschaft;* Schriftenreihe der Bundesanstalt für Agrarwirtschaft, 54, Wien

Ortner, K.M. (1996) *The Austrian farm sector's adjustment to CAP in 1995;* In: Ortner, K.M., Neunteufel, M.G., Jumah, A., Hofreither, M.F. (eds.) Agriculture after joining the EU - Sectoral Analyses for Austria, Schriftenreihe der Bundesanstalt für Agrarwirtschaft, 78, Wien, 11-27

Pevetz, W., O. Hofer and H. Pirringer (1990) *Quantifizierung von Umweltleistungen der österreichischen Landwirtschaft;* Schriftenreihe der Bundesanstalt für Agrarwirtschaft, Wien, 60

Pirringer, H. (1997) *Economic and Structural Impact of Changing (Higher or Lower) Intensity in Agriculture in Pursuance of the Goal of Sustainable Agriculture;* Paper presented at Regional FAO Workshop (2.-4. April), Gödöllö, Hungary

Puwein, W. (1994) *Abgeltung der Landschaftspflege - Subvention oder Produktionsentgeld;* In: Schneeberger, W. und Wytrzens, H.K. (eds.), Naturschutz und Landschaftspflege als Agrar- und Forstpolitische Herausforderung, Wien, 17-23

Riegler, J. (1996) *Weiterentwicklung der Gemeinsamen Agrarpolitik;* Der Förderdienst, 12, Wien, 390-393

Rupprechter, A. (1995) *Die vier großen Herausforderungen für die österreichische Landwirtschaft im Rahmen der gemeinsamen Agrarpolitik;* Der Förderdienst, 11, Wien, 338-344

Schneeberger, W. and M. Eder (1997) *Modellrechnungen zu betriebswirtschaftlichen Grundsatzentscheidungen unter den neuen wirtschaftlichen Rahmenbedingungen;* Institut für Agrarökonomik, Universität für Bodenkultur, Wien, http://www.boku.ac.at/i.../Boku 4-5, Boku 4-95.html

Schneider, M. (1994) *Chancen und Risiken der Landwirtschaft im EU-Binnenmarkt;* WIFO Monatsberichte, Wien, 46-61

Schneider, M. (1995) *Agrarsektor 1994: Kräftige Erholung;* WIFO Monatsberichte, 8, Wien, 525-532

Schneider, M. (1995) *Bilanz der ersten Erfahrungen mit der Gemeinsamen Agrarpolitik;* WIFO Monatsberichte, 5, Wien, 333-335

Schneider, M. (1996) *Agrarsektor 1995 im Zeichen der EU-Integration;* Der Förderdienst, 10, Wien, 306-308

Schneider, M. (1997) *Österreichische Landwirtschaft unter EU-Bedingungen;* WIFO Monatsberichte, 3, 155-170, Wien

Wagner, K. (1996) *Regional differenzierte Wirkungen des ÖPUL;* Der Förderdienst, 7, Wien, 207-214

Wagner, K. (1997) *Ökonomische Auswirkungen der Grundwassersanierung auf die Landwirtschaft;.* In: Schriftenreihe der Bundesanstalt für Agrarwirtschaft, 80, Wien

Wörgötter, A. (1994) *Landschaftspflege durch Landwirtschaft - empirische Schätzungen und agrarpolitische Konsequenzen;* In: Schneeberger, W. und Wytrzens, H.K. (eds); Naturschutz und Landschaftspflege als Agrar- und Forstpolitische Herausforderung, Wien, 25-38

Wytrzens, H.K. and R. Reichsthaler (1990) *Agrar - Umweltpolitik, Agrarrelevante Konzeptionen der Umweltpolitik;* Wien

14. FINLAND
Asko Miettinen

14.1 Introduction

Finland is one of the northernmost countries in the world where agriculture is practised. This means that the growing season is short, although the days are long in summer. The total area of the country is 337,000 km² of which the land area is 305,000 km². That is mainly covered with forests (77%); only 9% is agricultural land. The most consolidated areas of cultivation are located in the southwestern and western part of the country. The share of agriculture in the gross domestic product (GDP) in 1995 was 1.6%, and its share in the employed labour force was 6.7% (MMM, 1996a).

14.1.1 Interactions between agriculture and environment

Agriculture has both beneficial and harmful effects on the environment. Agricultural landscapes for all to see and enjoy can be regarded as a positive externality of agricultural production, and environmental degradation by nutrient run-off as a negative one. It is often difficult to measure the extent of these external effects or attribute them to a particular farm or farming practice.

Mineral emissions

Eutrophication of surface waters due to nitrogen and phosphorus emissions is the most significant environmental problem caused by agriculture in Finland. Nitrogen is the limiting factor in the eutrophication of coastal waters (i.e. the Baltic Sea), and phosphorus in the case of inland lakes. Agriculture is relatively intensive in the southern and southwestern parts of Finland. Such intensive production systems have contributed to the eutrophication of coastal waters. Some problems also exist in inland areas where the livestock farming is more intensive. It has been estimated that 20% of the lakes and about 55% of the rivers are eutrophic, with the problems concentrated in the southern and western part of the country. By comparison, the quality of the groundwater is quite good: nitrate pollution is only a localized problem; in most cases the nitrate levels in groundwater supplies remain below 25 mg/l.

The use of fertilizers in Finland has decreased from when they peaked in 1990 (Figure 1). One reason for this reduction are the fertilizer taxes, which were first introduced in 1990. Later on the measures for the reduction of overproduction have also affected the use of fertilizers and, moreover, the composition of fertilizers has changed due to the implementation of the Finnish Agri-Environmental Programme (reg. EEC 2078/92) which includes upper limits for the application of nitrogen and phosphorus in inorganic fertilizers and manure. The fertilizer industry has responded to these requirements by reducing the share of phosphorus in fertilizers and by adjusting the recommended application levels.

The livestock density in Finland is generally low (0.61 LU/ha in 1995). Across regions it ranges between 0.27 LU/ha in Uusimaa to 0.97 LU/ha in Keski-Pohjanmaa (Table 1). There are

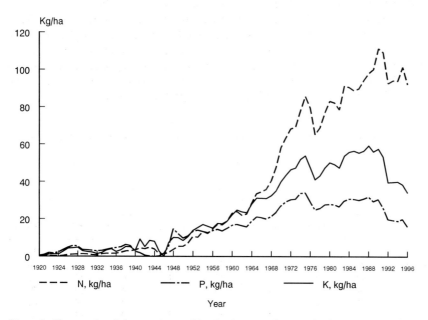

Figure 1   The quantity of plant nutrients sold in fertilizers per hectare of cultivated land (MMM 1996a)

Table 1   Livestock density (LU/ha) by region in 1995

Uusimaa	0.27
Turku	0.45
Satakunta	0.50
Häme	0.47
Kymi	0.61
Mikkeli	0.84
Pirkanmaa	0.56
Etelä-Pohjanmaa	0.66
Vaasa	0.64
Keski-Pohjanmaa	0.97
Keski-Suomi	0.72
Kuopio	0.86
Pohjois-Karjala	0.84
Kainuu	0.87
Oulu	0.64
Lappi	0.89
Ahvenanmaa	0.61
Whole country	0.61

also regional differences in the distribution of production, with pig production concentrated in the southwest, and dairy production mainly practised in the central part of Finland.

Nutrient balances can be used in order to assess farm specific nutrient surpluses or deficits. In 1994 the average surplus for nitrogen was 47 kg/ha and for phosphorus 12 kg/ha (Table 2). There has been a downward trend in the nitrogen surplus in Finland during the period 1985-95 (Pirttijärvi, 1997b). The nitrogen surplus per hectare has been decreasing, on average around 3 kg/ha per year, although during the past few years nutrient surpluses have actually increased. Similar trends can be observed in most other OECD countries (Pirttijärvi, 1997b).

Table 2   Nitrogen and phosphorus surpluses (kg/ha) and the efficiency ratio of the fertilizer use (%) by region in 1993 and 1994

| | 1993 | | | | 1994 | | | |
| | N | | P | | N | | P | |
	surplus	efficiency	surplus	efficiency	surplus	efficiency	surplus	efficiency
Uudenm. ja Nylands sv.	40	56	7	57	37	58	7	58
Farma ja Finska Hush.	44	55	9	53	50	51	10	50
Satakunta	35	59	11	44	40	55	12	43
Häme	48	53	11	45	55	48	12	42
Pirkanmaa	36	55	9	44	42	48	9	41
Päijät-Häme	37	58	10	47	52	48	12	41
Kymeenlaakso	44	54	10	47	48	52	10	46
Etelä-Karjala	43	52	11	40	50	46	12	36
Mikkeli	43	52	13	33	47	50	13	33
Pohjois-Savo	59	47	15	31	65	44	16	31
Pohjois-Karjala	48	49	13	32	50	48	13	32
Keski-Suomi	41	52	11	35	43	48	11	35
Etelä-Pohjanmaa	43	55	14	38	49	51	14	37
Österbotten Sv.	24	68	10	46	24	66	9	48
Keski-Pohjanmaa	60	53	18	32	66	49	18	32
Oulu	43	51	13	33	48	47	13	32
Kainuu	44	46	12	28	48	41	12	26
Lappi	33	56	11	31	41	47	12	26
Total	43	54	11	41	47	51	12	40

Source: Pirttijärvi, 1996.

Ammonia emissions

Emissions of ammonia from agriculture contribute to the occurrence of acidification and eutrophication of the soil and water courses. Most of the ammonia emissions in Finland, about 80%, are caused by agriculture, mainly livestock manure (Pipatti 1990). In the crop year 1985/86 ammonia emissions in Finland were about 43,000 tonnes of which the share of agriculture was around 90%

(about 38,000 tonnes). Although more than half of the ammonia deposited on Finland originates from abroad, local emission sources are significant. Other emissions from agriculture include methane and nitric oxide, which are greenhouse ...

The use of pesticides

The use of pesticides in Finland (Figure 2) increased until the early 1980s but since then the amount of active ingredients has decreased (MMM, 1996b). This is partly due to production policy, particularly the increase in set-aside, and partly due to technical improvements in plant protection products which mean that smaller doses are needed. Generally, the use of pesticides in Finland is very low compared to other European countries (Miettinen, 1996). This is due to climatic conditions - particularly the harsh winters - which limit the growth of the pest population.

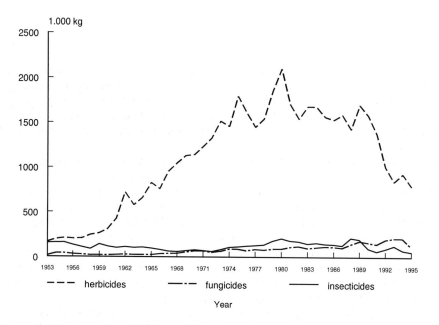

Figure 2   Sales of pesticides (1,000 kg) in Finland during the period 1953-1995 (MMM, 1996b)

Landscape and biodiversity

During recent decades the visual and natural diversity of the cultivated environment has declined in Finland due to developments in agricultural production methods, including decreased use of permanent pasture, and the general use of subsurface drainage and pesticides. At the same time society increasingly appreciated the value of diversity. Such aspects contributed to the value of

326

the final products from rural areas and the pleasantness of the rural environment. The size of Finnish farms is still very small, and open cultivation areas comprise the fields of several farms. It is feared that the decreasing number of farms may in the future induce a reduction of biodiversity through the loss of landscape features as well as a further concentration of production.

The rural landscape is highly appreciated, and there is concern about the decline of the cultivated landscape. Presently, meadows, fields, and grazing grounds are the habitats of about one-fifth of Finland's endangered plant and animal species. About 17% of the plant species are endangered and about a third of them are threatened by agriculture (Ryttäri and Kettunen, 1997). The development targets for the near future call for the annual afforestation of 20,000 ha of arable land, which mainly concerns the eastern and northern part of the country. Intensified production methods have also had strong impacts on the landscape. The pressure for economies of scale requires large scale production units and leads to uniform and monotonous landscapes. The greatest threat to the rural landscapes, though, is the discontinuation of farming and the depopulation of rural areas.

14.1.2 Evolution of environmental policy

The relationship between nature and agriculture has always been very delicate and farmers have traditionally been seen as guardians of the rural environment. Finnish agriculture is facing changes as it adapts to the CAP. There are difficulties in this process, but there are also possibilities. The EU Agri-Environmental Regulation is regarded as a very important way to promote environmental protection, both by preventing environmental degradation and by producing environmental benefits.

Recognition of environmental problems caused by agriculture began in Finland in the early 1970s although the magnitude of such problems was not considered very significant until the mid-1980s when a committee was set up to consider water protection measures for agriculture (Jokinen, 1995). The pollution load from industry and municipalities was still considered to be the biggest problem for water courses. At the beginning of the 1990s, however, new research results showed agriculture to be the largest single source of phosphorus and nitrogen to surface waters (Rekolainen et al., 1992). These research results were used in the formation of the environmental programme for rural areas (Ministry of the Environment, 1992). Thus the perceived environmental effects of agriculture changed from the minor pesticide problem of the 1970s to the significant nutrient loading problems of the 90s (Jokinen, 1995).

At the beginning of the 1990s new measures were launched in order to reduce the detrimental effects caused by agriculture. In 1990 a phosphorus tax of FIM 1) 0.50/kg P was set, and it was gradually raised to FIM 1.70/kg P by 1992. A nitrogen tax of FIM 2.90/kg N also came into effect in 1992. Although intended partly to finance the export of surplus production and set-aside (Miettinen, 1994), these taxes did cause a slight reduction in the use of fertilizers and this extensification of production may have also had a positive effect on the environment. The fertilizer

1)   FIM 1 = ECU 0.17.

taxes were abolished, though, on EU membership to help offset the additional cost burden farmers faced.

Sumelius (1994), having assessed the effects of fertilizer taxation, suggests that a nitrogen tax may be more cost-effective than other measures such as a nitrogen quota in reducing nitrogen leakage. The most cost-effective measures are likely to be those designed to reduce leakage from areas with the highest nutrient losses (Sumelius, 1994). According to Miettinen (1993), the nitrogen tax would have had to be considerably higher in order to achieve a significant reduction in input use.

14.1.3   Agriculture and society

The abandonment of agricultural land linked to rural depopulation and the economic difficulties of agriculture have been a major concern of rural areas in Finland. The socio-economic structure of the Finnish countryside consists of agriculture together with other related occupations and small-scale industries. A substantial share of the incomes generated in the rural areas comes from agriculture, and agriculture has also been considered an important agent in maintaining overall population levels in the rural areas (Kettunen and Niemi, 1994). This structure is likely to be weakened considerably by the decline in agricultural incomes.

The environmental awareness of consumers mainly concerns the healthiness of food. When people were asked the reason for buying organic products, 40% stressed the purity and healthiness of food and only about 10% of them gave environmental reasons (Väisänen and Pohjalainen, 1995). This result is quite different compared to other Nordic countries where environmental reasons are stressed more. The reason for the opinion of Finnish consumers might be that Finnish food is generally considered pure. Consumers consider the greatest risk in food to be the added preservatives and pesticide residues. When consumers were asked whether EU membership had affected food safety in Finland, about 57% considered that food safety had decreased (Siikamäki, 1997). According to the opinion of the majority of consumers, agriculture should be developed in a way that takes environmental considerations better into account.

Participation in the Finnish Agri-Environmental Programme was relatively high in the first year in the EU. Some 80% of farmers joined the programme, and an even bigger share of the cultivated land was included (Siikamäki, 1996). However, the attitudes of farmers towards environmentally sound production depend on personal factors such as age and education. Less educated and older farmers do not pay much attention to the environmental effects of production (Tiilikainen, 1997).

14.2 Impacts of market and price support measures

14.2.1   Agricultural policy before the EU membership

The main factors which shaped agricultural policy before EU membership were the desire to guarantee food supply in all conditions, to develop farmers' incomes, and to maintain population levels in the rural areas. Agriculture was protected against foreign competition, and the income objective could be achieved by regulating prices. Already in the 1950s production exceeded domestic consumption and since then restricting and reducing overproduction have been matters of ongoing political debate (Kettunen, 1992a).

Agricultural policy was divided into the following measures:
- production policy;
- structural policy;
- income and price policy;
- employment in the countryside and maintaining the rural population level.

Production policy was based on the achievement of self-sufficiency, i.e. to balance production and consumption in the long run. This was also a strategic objective and, moreover, agricultural production was considered to be important for reasons of employment, regional policy, and the inhabitants of the rural areas. The effect of the production policy on the environment was negative as regards the nutrient load to waters, but was generally positive in terms of the preservation of agricultural landscapes. Structural policy had to support the self-sufficiency objective. The objective of structural policy was to increase productivity but, at the same time, to limit the increase of farm size, in order to reduce overproduction and to maintain the rural population level. These rather controversial objectives of structural policy were set in order to maintain the so-called family farm structure, and through this, to secure the basic population in the rural areas.

The objective of the income policy was to guarantee the agricultural population a fair income level compared with other population groups. Disparities due to the location of farms and farm size were equalized through the price policy. The Price Act was the most important means of price policy. It was used to guarantee compensation for the increase in costs from rises in input prices as well as to keep farm incomes in line with the growth in incomes in other sectors (Kettunen, 1992a). The cost compensation of price policy favoured intensive production methods (i.e. increased use of inputs like fertilizers and pesticides), and its objectives were generally in conflict with the goals of the production restriction measures.

Strong criticism was directed at the subventions required for the export of surplus production. In the early 1990s pressures on the independence of agricultural policy came from the GATT negotiations. Efforts to liberalize foreign trade in agricultural products were supported by Finnish consumers who demanded lower food prices (Kettunen, 1992a).

14.2.2 The CAP

Through the accession of Finland to the European Union the national agricultural policy was replaced by the CAP. Finnish agriculture and agricultural policy have not faced the CAP reform itself but the reformed CAP, which represents a remarkable change from the earlier national agricultural policy.

The objective of Finland in the entry negotiations was to achieve a package of supports that would guarantee the profitability of agriculture even if the producer prices dropped by half when Finnish prices were to be adjusted to the EU level immediately following accession. Finland applied for the normal LFA aid to the whole country, and so-called Nordic aid on the basis of its

329

unfavourable climatic conditions. According to the Accession Treaty, 85% of the total area was included in the mountain support of the LFA. The remaining 15% - all in Southern Finland - was excluded on the basis that it produced wheat. Finland is also allowed to implement its own Northern aid (to areas north of the 62nd parallel) during the transitional period of five years.

In the beginning of EU membership the agricultural prices fell, on average, by 40% (Table 3): cereal prices fell by 60-70%; fodder cereals by around 60%, and milk prices by about 25% (Kettunen, 1996a). The reduction in farm income was compensated by direct support, consisting of the CAP support and national forms of support which are to decline during the transitional period. Agricultural income, taking into account all support, has dropped by 14% or FIM 1.1 billion. The major part of the farm income depends on support; in Finland, on average, the share of support of the total income was about 45% in 1996 (AERI, 1997). In connection with the accession negotiations, the agri-environmental programme was also prepared according to Regulation EC 2078/92. This programme replaced all earlier environmental action plans for agriculture, and it was adopted widely right from the beginning of EU membership.

Table 3   The most important market prices in 1993 and 1995 (FIM/kg)

	1993	1995	Change (%)
Wheat	2.26	0.87	-61.5
Rye	2.85	0.89	-68.8
Fodder cereal	1.77	0.72	-59.3
Turnip rape	3.66	1.20	-67.2
Sugar beet	0.41	0.30	-26.8
Pig meat	16.18	8.00	50.6
Eggs	8.74	2.83	-67.6
Milk	2.73	2.04	-25.3
Beef	23.5	15.75	-33.0

Source: Kettunen (1996a).

14.2.3   The arable regime

The CAP-reform base area for arable crops in Finland was set at 1.6 million ha. Before the accession, the price support levels were clearly higher than in the EU. High overproduction of agricultural products was a key problem of Finnish agriculture, and several so-called 'balancing measures' had been applied. To reduce the overproduction of fodder cereals set-aside was used extensively. There was a mandatory system whereby farmers had to set aside at least 15% of the arable land, with a premium paid for the area taken out of cultivation (Kettunen, 1992a). This reduced the arable area by about 500,000 ha, or 20%, in the years 1991-94.

After the accession to the EU, the area under cultivation increased by about 125,000 ha from the year 1994 as the area under set-aside decreased from 505,100 hectares in 1994 to 223,200 ha in 1995. However, the total area of arable land (i.e. cultivated area and set-aside together) decreased by some 150,000 ha. In 1995 the area under cereals increased especially as a result of the growth in the cultivation of spring wheat and rye (MMM, 1996a). Set-aside was no longer mandatory to the same extent as before, and the 92-tonne limit for set-aside was relatively high for small Finnish farms. The set-aside area has continued to decrease, down to 179,300 ha in 1996 (MMM, 1996a).

The reduction of prices had been expected to lead to more extensive production in which the use of fertilizers and pesticides would be reduced. According to Sumelius (1993), the 50% reduction in the market prices of cereals would have caused a reduction in nitrogen application of about 25%. However, the reduction in fertilizer use has not been that large because of the abolition of the fertilizer taxes and, possibly, the direct income support which is used to compensate the income loss. Another issue concerning the price changes is that they may affect the relative cultivated areas of different crops - for example a shift from grassland to cereals - which may have a bigger effect on, say, nutrient loading than a change in the intensity.

The area of grassland increased due to greater use of grass-based fodder in livestock production. The effect of the increase in the cultivation of cereals and the reduction of the set-aside area may be twofold: with an increased total nutrient load due to the crop production being offset, to some extent, by the decrease in the previous practice of bare set-aside. According to Rekolainen et al. (1992), set-aside with green cover reduces the nutrient load by some 10% compared to bare set-aside.

Consequently, the overall effect of the arable crop regime on the environment might have resulted either in an increase in N- and P-load or at least in the status quo. Assessments of the exact effects have not been made, and they also are difficult to make. In coming years the area of crops cultivated is expected to decrease due to the reduction in the national transitional aid. This may lead to set-aside or afforestation or even land abandonment, which will have very negative effects on the agricultural landscape.

14.2.4   The dairy regime

In the negotiations, the milk quota for Finland was set at 2,342 million kg, including a direct sales quota of 10 million kg. This corresponded to the amount of milk delivered to dairies in Finland in 1992. The production of milk fell only slightly after accession although the change in the market price was considerable. This was probably due to the uncertainty concerning the price formation among farmers, and moreover, the price incentive due to the income support may have been higher than the actual market price for farmers. In 1996 milk production decreased further by 1.5%, compared to the previous year.

The total of the reference quantities of all individual holdings exceeded the national total guaranteed quantity (i.e. national quota) of milk production, and therefore in the beginning of 1997 reference quantities were cut by 4.5% for every dairy farm. This adjustment will be compensated by reallocation of national aid so that an additional FIM 0.04/litre is paid for the produced milk. At the same time the connection of milk production to the national Northern aid was abolished, which may increase milk production in areas in Central Finland where it has traditionally been concentrated. This may increase the livestock density regionally and thus may increase the risk of nitrate leaching to groundwater as well as problems in the handling of manure.

Area payments under CAP may stimulate fodder cereals compared to grassland farming (Table 4) because the opportunity cost for silage, for example, is now defined by the price of cereals (Ryhänen and Sipiläinen, 1996). Dairy production based on coarse fodder has lost some of its competitiveness against fodder cereals. The allocation of arable land to different crops is determined by relative prices, production technology and support policy. Therefore, the potential possibility for increased nutrient losses due to the reduced grass area may increase with rising support levels for fodder cereals. The increased use of fodder cereals may also cause instability in the regional nutrient balances through transportation of fodder cereals from the crop production areas of Southern Finland to Central Finland (see also Ryhänen and Sipiläinen, 1996).

Table 4   Production costs without and with the aid for fodder (FIM/fodder unit)

	Production cost without aid	Production cost with aid
Pasture	0.90	0.29
Feed grain	1.68	0.86
Silage	1.53	1.03
Hay	1.77	1.07

Source: Käytännön Maamies, 1997.

14.2.5   The beef regime

In the Accession Treaty the special beef premium quota was set at 250,000 animals, which was equal to the existing production. Compensatory payments are provided to farmers, up to the maximum livestock density of 2.0 LU/ha. This condition is expected to limit very intensive production methods and, therefore should be beneficial for the environment, even if such high livestock densities are not that common in Finland. After the accession, the production of beef decreased by some 9% to a level of 96 million kg in 1995 due to the increased slaughtering at the end of 1994 as well as the reduction in the number of cows (Kettunen, 1996b). The development of beef production in Finland is partly dependent on the milk quota. In the long run a reduction in milk production will subsequently reduce beef production.

In general, beef is a byproduct of milk production in Finland and, therefore, the production is located in the same areas as milk production. Compared to other European countries, the problems of beef production in Finland arise from its non-specialized nature and the relatively small size and high production costs of beef production units. The profitability of Finnish beef production has been estimated to decline 30-60% during the transitional period of five years and to be especially marked in Southern Finland (Rantala, 1996). The Northern aid, in particular, improves the relative profitability of beef production in the central and northern parts of Finland. This may have a positive impact on the preservation of landscape in those areas.

Similar to milk production, the proportion of fodder cereals is also projected to increase in the feeding of beef cattle, because the relative price of cereals is projected to be reduced compared to hay and silage (Ryhänen et al., 1996). This will decrease the grass production of dairy and beef cattle farms. This type of change in feeding practises and intensification of production in general will cause increased ammonia emissions and enhance the risk of nitrate leaching due to increased amounts of livestock manure. The intensified livestock production may also increase the transportation of fodder from other regions, like Southern Finland, which will cause imbalance in regional nutrient balances.

Moreover, the changes in the international beef markets aim at liberalising and increasing the trade which implies intensified competition that will lead to a shift of production to regions with low production costs (Hemmilä, 1996). This will weaken the competitiveness of beef production in Finland and reallocate beef production more closely to milk production, which will have detrimental effects, at least for the maintenance of pastoral landscapes due to the reduction in grazing herds.

14.3 The accompanying measures

14.3.1 Agri-environment policies

14.3.1.1 Structure of the FAEP

The Finnish Agri-Environmental Programme for 1995-99 (FAEP) was prepared according to EC Regulation 2078/92. The total amount of environmental aid is ECU 270 million annually which is 50% financed by the EU (Table 5). The preparation of the Finnish programme under the Regulation was done in co-operation between the Ministry of Agriculture and Forestry and the Ministry of the Environment. The five-year programme was finally approved by the Commission in October 1995.

Table 5   The payments of Agri-Environmental aid in Finland

General Agricultural Environment Protection Scheme:			
	farmers	ha.	paid support ECU million
1995	78,475	1,765,000	226
1996	79,635	1,838,000	232

Supplementary Protection Scheme:				
	contracts	ha	LU	paid support ECU million
1995	6,809	90,000	5,310	16.4
1996	4,112	47,500	1,790	10.4
Total	10,921	137,500	7,100	26.8

The programme (MMM, 1994) is intended to ensure that agriculture is practised in a sustainable manner, and the objectives are defined as follows:
- to reduce pressure on the environment, especially on surface waters, groundwater and air, and to reduce hazards caused by the use of pesticides;
- to preserve biodiversity and manage agricultural landscapes;
- to protect wildlife habitats and endangered species of flora and fauna;
- to produce agricultural commodities in an extensive and environmentally friendly manner.

The environmental aid programme is mainly oriented towards the cultivation of field crops, as well as the preservation of the landscape related to agriculture. It is expected to induce a major shift towards environmentally sound production methods. There are connections to forestry, mainly for traditional biotopes, forest pastures, and the distinctive habitat and wildlife of the field-forest margin. The FAEP aims at compensating the farmers for the costs or income losses they incur in fulfilling the criteria set for joining the programme (Pirttijärvi, 1997a).

General Agricultural Environment Protection Scheme

The FAEP consists of four elements: the General Agricultural Environment Protection Scheme (GAEPS), the Supplementary Protection Scheme (SPS), Advisory Services, and Training and Demonstration Projects (MMM, 1994). The GAEPS is the most significant of the elements; it is available in the whole country and is intended to cover as large a share of the total agricultural area as possible (Siikamäki, 1996). In order to obtain aid under GAEPS, the following requirements have to be fulfilled:
- a farm environmental management plan (FEMP) must be prepared;
- a certain base level of fertilising must not be exceeded (Table 6);
- manure must be appropriately stored for 8-12 months and may not be spread on frozen soil or snow;
- stocking density must be below 1.5 LU/ha;
- buffer strips, between 1-3 m wide, must be left on the sides of main ditches or water courses;
- at least 30% of arable land must have plant cover during the winter in areas A and B (see Annex of this chapter on agricultural support areas in Finland);
- landscape and biodiversity must be appropriately maintained on the farm;
- spraying devices must be tested by an authorized agency and pesticides may be applied only by a person who has completed training on pesticide use.

Farmers who meet these requirements are paid an annual premium per hectare. Compensatory payments vary between 40-275 ECU/ha, and they are highest in area A (for support areas see: Annex) and decrease towards the north. According to the agri-environmental protection scheme, farmers must be compensated for any additional costs or income losses due to undertaking environmentally beneficial measures (EEC reg. 2078/92). In area A (i.e. Southern Finland) the aid based on the GAEPS also partly compensates income losses due to the adoption of EU prices and the lower national aids allowed by the Commission.

Table 6   Base levels for fertilizing (incl. inorganic fertilizers and manure) (kg/ha)

Crop	Nitrogen	Phosphorus
Barley	90	15
Oats	90	15
Spring wheat	100	15
Winter wheat	120	15
Rye	120	15
Oil-seed plants	100	15
Sugar beets	120	30
Potatoes	60	40
Starch potatoes	80	40
Silage	180	30
Hay	90	15
Pasture	180	30

Supplementary Protection Scheme

The Supplementary Protection Scheme (SPS), which is directed only to a certain limited number of farms, is aimed at promoting special measures to improve the environment. Compensatory payments provided under the SPS are based on the actual costs of the implementation of the measure, which can be:
-    organic conversion and production;
-    establishment of riparian zones and treatment of run-off waters from arable land;
-    commitment to storing, handling and using manure from other farms;
-    landscape management and enhancing biodiversity;
-    extensification of agricultural production;
-    rearing animals of local breeds in danger of extinction.

The most important of these in monetary terms is the organic measure (see Table 7). The conversion period to organic production lasts for three years, and during this period the farmer receives, on average, ECU 270/ha/year to compensate for the lower yields and lower income during the conversion period. From the third year onwards the crops can be labelled organic and sold at a possible premium price. The fixed organic production aid is then about ECU 120/ha/year (Miettinen et al., 1997). In 1995 over ECU 10 million was spent on the organic measure. Prior to the programme, about 30,000 ha of arable land was cultivated organically, whereas by 1996 the area under conversion was about 75,000 ha.

Scheme for Advisory Services and Training and Scheme for Demonstration Projects

The Finnish Agri-Environmental Programme puts emphasis on advising and educating farmers to improve their environmental management. Advisory services and consultants can get funding to organize courses under this scheme. In general, training, education, and demonstration projects promote the better understanding of the FAEP, and also give practical directions on how to meet the GAEPS criteria.

14.3.1.2 Environmental effects

After the first year of the implementation of the FAEP some 80,000 farms had joined the programme (Siikamäki, 1996). This is over 80% of the total number of farms and about 90% of the total cultivated area on these farms. Participation was generally higher in Southern Finland than in other parts of the country, and this is mainly due to higher premiums paid in area A.

Differences in the participation between different types of production were generally not very big, except for the lower response of pig and poultry farmers. The main reasons not to participate in the GAEPS on animal husbandry farms were, for pig and poultry producers, the criterion for the base level of fertilizing and, for dairy farms, the investment needed for manure storage facilities (Siikamäki, 1996).

Some *a priori* estimates of the effects of the FAEP on the environment were made when setting the goals of the programme. It was expected that the programme would result in more extensive agriculture and thus have a positive impact on the environment. It was estimated that both erosion and phosphorus losses into watercourses would decrease by about 40% (~25% dissolved phosphorus) and nitrogen losses by about 30% in the long run (MMM, 1996b). This reduction will be achieved only if a large majority of the farmers join the GAEPS and if SPS-measures can be focused on the designated areas where their effectiveness will be the greatest.

Monitoring of the impacts of the FAEP on the environment has begun, including participation levels, changes in agricultural practises, and the environmental impacts of the FAEP (mainly GAEPS measures). Four areas representing the range in natural conditions and the structure of agriculture have been selected for the study. The environmental impacts are assessed using mathematical models to calculate nutrient loss estimates for all possible combinations of physio-biological and management conditions (Grönroos et al., 1997).

According to the results of the first round of monitoring covering the period 1994-1995 there seems to be a slight reduction in the use of fertilizers, but the data available are insufficient for exact assessment (Grönroos et al., 1997). The reduction of fertilization indicates extensification of production, which should have positive effects on the environment as it reduces the nutrient leaching in the long run. However, it is quite difficult to separate the effects of the Agri-Environmental Programme from the effects of the market and price regimes given the considerable changes that have occurred in the prices of agricultural products.

As regards the other requirements of the GAEPS, the animal densities in the study areas decreased, but the limit of the programme (1.5 LU/ha) was still exceeded on some farms (Grönroos et al., 1997). In general the livestock density, which averages 0.61 LU/ha in Finland, is only a local problem in areas with intensive agricultural production. The extensification of livestock production will reduce the leaching of nitrates as well as reduce ammonia emission from manure. Ammonia emissions can also be reduced by ordinary management of manure, e.g. covered storage and tillage immediately after the spreading of manure.

The increase of the plant cover in winter was significant in Southern Finland where cereal production is concentrated. This will reduce soil erosion and thereby also reduce phosphorus losses, which are closely connected to erosion.

A slight change in manure application from the autumn towards the summer was also observed (Grönroos et al., 1997). Application during the summer months is projected to substantially reduce the potential for nutrient leaching but requires investments in manure storage for livestock farms in order to reach storage capacity of 8-12 months. According to the Farm Environmental Management Plans (FEMP), 67% of livestock farms need to rebuild or expand their manure storage facilities (MKL, 1997).

So far no further assessments of the environmental impacts of the changes in agricultural practices due to the GAEPS have been made. The changes in agricultural practices from 1994 to 1995 were made before the actual conditions of the FAEP were available. Even so, the trend seems to be towards more environmentally sound agricultural practices. The environmental awareness of farmers has increased and attitudes towards a better environment have improved (Tamminen, 1997). This has been beneficial for the adoption of better management practices, which are essential in order to achieve any improvements in the state of the environment.

SPS measures aim at producing further positive environmental impacts, such as increasing biodiversity and maintaining the landscape. To be eligible for the measures under SPS, participation in the GAEPS is also required. Especially measures like riparian zones, treatment of run-off waters, and extensification of production are limited to focal areas of water protection in order to obtain the best possible result. The most important of the measures under the SPS is the aid for organic production.

Table 7   Contracts under the Supplementary Protection Scheme in 1996.

	ha (or LU)	Number of contracts
Aid for conversion to organic farming	57,917	2,954
Aid for organic farming	18,705	1,150
Riparian zones	929	410
Sedimentation ponds	1,327	100
Wetlands	278	16
Drainage systems 1	1,940	162
Drainage systems 2	235	22
Neutralization of sulphate soils	40,211	1,989
Management of traditional biotopes	7,215	1,041
Landscape development 5 years	1,419	304
Landscape development 20 years	44	23
Biodiversity 5 years	519	145
Biodiversity 20 years	22	6
Balanced use of nutrients in manure	6,582	624
Extensification of production	271	34
Maintenance of local breeds	(7,104)	1,941

Source: Ministry of Agriculture and Forestry.

The use of mineral fertilizers and pesticides is not allowed in organic farming. Instead of these chemicals, the organic farmer relies on animal manure, biological pest control, etc. Since the nutrients are a scarce input, it is important to create a cycle, where the nitrogen and phosphorus are recycled in the farm ecosystem. As a consequence, most organic farms emit a smaller nutrient load into the surface waters. According to various nutrient leaching studies, the nitrogen load of organic farms is, on average, 50% smaller than that of conventional farms. Organic farming practices also improve the overall condition of the soil, thereby reducing soil erosion and phosphorus load. Naturally, problems related to pesticide use also diminish (Miettinen et al., 1997).

14.3.2   Early retirement measures

An increasing share of Finnish farmers belong to the older age groups. In 1990 their average age was 51.2 years. In the case of full-time farmers it was somewhat lower at 47.6 years (MMM, 1992). Before EU membership there was a scheme for a pension for a farmer 55 years of age or over transferring a farm to a successor who was obliged to stay on the farm for a period of at least five years after the transfer. The objective of the measure was to improve the structure of agriculture. The number of the pensions granted annually was about 1,500. This scheme was abolished after the accession to the EU, but a national scheme for the support of a successor continues.

Regulation 2079/92 allows for compensation to be paid to full-time farmers aged 55-64 years for stopping agricultural work. The objectives of the measure are the lowering of the average age of farmers, the improvement of the production structure, the control of production, and support for older farmers. The successor must be a skilled farmer who will continue full-time farming for at least 10 years after the transfer and, moreover, the successor must increase the total arable area of the farm by at least 2 ha or 10%.

The operation of the scheme in Finland has an environmental requirement concerning the farming land transferred. The successor has to comply with the conditions of the General Agricultural Environment Protection Scheme (GAEPS), and the early retirement scheme needs to be joined for a period of at least five years. The beneficial effects on the environment however may be marginal compared to the FAEP scheme in general.

The premia paid for the retiring farmer depend on the former income level and, on average, they have been ECU 543/farmer/year and ECU 883/farm/year. It has been estimated that some 2,500 farms will join the early retirement scheme and by the end of 1999 there will be about 15,000 farms in the scheme (MMM, 1997).

14.3.3   Forestry measures in agriculture

An afforestation scheme applied in Finland before EU membership, and the premium for the afforestation of agricultural land was paid according to agricultural income for the first five years and ECU 1,190-1,800/ha for the next five years depending on the district (Kettunen, 1992b). The objective of this scheme was simply to reduce agricultural overproduction.

The EU likewise allows the payment of aid for the afforestation of agricultural land. This scheme is obligatory for each member state. The regulation is also aimed at contributing to kinds of management which serve the conservation of nature. At least 2 ha of arable land have to be afforested in order to obtain the premium in Finland, and this area may comprise several parcels (>0.1 ha). The area to be afforested must have been cultivated (including set-aside) after 1991. Premiums according to Regulation 2080/92 are 3,000-4,000 ECU/ha for the cost of planting, and the compensation for income losses for a professional farmer is 600 ECU/ha for the maximum of 20 years. If over 60% of the arable land area of the farm is afforested the compensation for income losses will be 1.5 times higher.

The incentive for afforestation under Regulation 2080/92 is lower than the former national scheme. Therefore it is not expected to have a significant effect. Primarily the areas to be afforested will be the least valued agricultural land. Forests already occupy 77% of the Finnish land area and, therefore, the possible increase of the forest area due to the afforestation scheme for agricultural land is likely to be modest. A more important aspect is the good management of the existing forests. The major effect will be the reduction in the open, agricultural landscape.

14.4 Structural measures

In Finland especially Objectives 5 and 6 concern the rural areas. Objective 5 aims at facilitating the adjustment to the changes in the Common Agricultural Policy by either promoting the structural change of agriculture (5a) or by granting aid to developing the rural areas (5b). Objective 6 is regional aid aimed at sparsely populated rural areas in the north. The LFA payments are included in the aid of Objectives 5 and 6.

In mountain areas classified as unfavourable, compensation for natural disadvantage is paid on the basis of animal units or hectares. According to the Accession Treaty, the LFA aid covers about 85% of the agricultural land area in Finland, which indicates the importance of the aid for the viability of Finnish agriculture and the possible impact on the preservation of the rural population. The area covered is defined according to difficult climatic conditions which result in a shortened growing season. The amount of aid is 180 ECU per unit of cattle or hectare of agricultural land. On cattle and sheep farms the aid is mainly paid on the basis of the livestock. On pig, poultry, and cereal farms the aid may be paid on the basis of the area only. There are also certain national investment aid schemes for the LFA area.

The LFA aid has a positive effect on the viability of agriculture and thereby on the environment by maintaining the agricultural landscape and rural infrastructure to some extent. The short growing season is the major problem in Finnish agriculture, and this problem is not removed through any system of aid. The LFA can be seen as a means to slow down the decline in agricultural production.

14.5 Conditions in environmental policies

Nitrates Directive (91/676/EEC)

The Nitrates Directive (91/676/EEC) concerns the protection of waters against pollution caused by nitrates from agricultural sources. This directive has not yet been implemented in Finland, but has been considered by an administrative working group (Ministry of the Environment, 1997). The working group was not unanimous about the conditions for the implementation of the directive. The main reason for this was that similar rules already exist in the Finnish Agri-Environmental Programme (Reg. EC 2078/92).

The measures proposed in the report are to be allocated to vulnerable zones where the risk of damage of ground and surface waters is notable. In the proposal of the government decision for the implementation of the directive there are conditions describing so-called good management practices including rules for the management of livestock manure, such as the size of storage facilities and limits for the spreading time of manure, and the amounts to be applied. Moreover, in the proposal there are rules for livestock density limits (<1.5 LU/ha), the plant cover of arable land during winter, and maximum limits for the use of nitrogen and phosphorus (i.e. from fertilizers and manure) (Ministry of the Environment, 1997).

Water Act (L 264/61)

The Water Act, which is a national act, contains very specific provisions on a permission procedure based on prohibitions against the closing off, altering, or polluting of waters. Permission from the Water Rights Court is required for any activities resulting in non-compliance with the prohibitions. This type of permission procedure must be followed, for example, if a new production unit is to be established (Miettinen, 1994).

Pesticides Act (L 327/69)

The Pesticides Act, which is also a national act, includes legislation for testing the quality and properties of pesticides, the inspection and supervision of their production, and the trade of pesticides. The producer or the importing company has to pay the registration charge of pesticides for the inspecting authorities. This is indirectly reflected in the prices of pesticides (Miettinen, 1994).

Fertilizer Act (L 377/86)

The Fertilizer Act, which is also a national act, controls the quality of fertilizers concerning heavy metal contamination in a similar way as the Pesticides Act in order to prevent the possible damage caused by these substances. Moreover, in the Fertilizer Act there are regulations for the production, marketing, and import of fertilizers or their raw material. The Fertilizer Act is aimed at the producers and importers of fertilizers, and therefore its effect on farmers is indirect (Miettinen, 1994).

14.6 Concluding observations

The adoption of the Common Agricultural Policy has changed remarkably the agricultural price and support policy in Finland. Agricultural prices fell on average 40 % and this reduction was compensated by direct support in different forms. Moreover, several other measures were adopted simultaneously including the accompanying measures and structural measures. It is therefore difficult to separate very clearly the effects of market and price support measures and agri-environmental measures on the environment. Moreover, the adoption of the CAP in Finland began in 1995 and therefore the actual effects on the environment cannot be indicated yet.

The allocation of support on different products - e.g. fodder cereals - seems to be more important in defining the allocation of production than the actual prices. In milk and cattle production the relative price (including support) of fodder cereals is lower than the price of coarse fodder which causes the increased use of fodder cereal in animal husbandry and thereby changes in regional nutrient balances. Therefore the reallocation of direct support concerning the natural production conditions e.g. for livestock feeding, could also be very effective from an environmental point of view.

The agri-environmental programme (FAEP) is the most important scheme included in the CAP to improve the agricultural environment. It replaced all the former programmes aimed at the environmental protection of agriculture, and its significance is enhanced by the absence, as yet, of a far reaching environmental policy affecting agriculture. The greatest influence of the FAEP should be on the reduction of nutrient losses due to several restrictions included in the conditions of the FAEP e.g. in manure usage and fertilizer application. Already, there are signs of progress, e.g. the use of phosphorus in fertilizers and manure has declined rapidly during the first two years of the application of the FAEP.

However, the implementation of the FAEP is not very specific concerning the regional variation in environmental conditions. The main emphasis in the FAEP is laid on the General Agricultural Environment Protection Scheme GAEPS which is the horizontal measure for the whole country. The Supplementary Protection Scheme has a more detailed approach on specific environmental targets although the support for organic farming is more like a horizontal measure.

One way to summarize is that the CAP is likely to have reduced, or more exactly is likely to reduce, leakages of N and P although separating the effects of different measures is difficult. Regional specialization may counteract this development to a certain extent. The adverse effects of the CAP on the environment could be dealt with through targeted measures within the Supplementary Protection Scheme (SPS) of the agri-environmental programme.

References

AERI (1997) *Maatalouden kokonaislaskelmat;* Moniste
Grönroos, J., S. Rekolainen and A. Nikander (1997) *Maatalouden ympäristötuen toimenpiteiden toteutuminen v. 1995;* Suomen ympäristö: 81. Suomen ympäristökeskus. 84 p. Helsinki
Hemmilä, T. (1996) *Naudanlihan kansainväliset markkinat;* Vaikutukset EU:ssa ja Suomessa. PTT:n raportteja ja artikkeleja 142. 81 p. Espoo
Jokinen, P. (1995) *Tuotannon muutokset ja ympäristöpolitiikka. Ympäristösosiologinen tutkimus suomalaisesta maatalouden ympäristöpolitiikasta vuosina 1970-1994;* Turun yliopiston - julkaisuja, sarja C - osa 116. Turku
Käytännön Maamies (1997) *Nurmirehu tarvitsee tehoa ja lisätukea;* Vol 6. Helsinki

Kettunen, L. (1992a) *Finnish agriculture in 1991;* Research Publications 65a. AERI. 59 p. Helsinki

Kettunen, L. (1992b) *Suomen maatalouspolitiikka;* Research Reports 185. AERI. 148 p. Helsinki

Kettunen, L. and J. Niemi (1994) *The EU Settlement of Finnish Agriculture and National Support;* Research Publications 75a. AERI. 91p. Helsinki

Kettunen, L. (1996a) *The adjustment of Finnish agriculture in 1995 in 'First experiences of Finland in the CAP';* Research Publications 81: 7-25. AERI. Helsinki

Kettunen, L. (1996b) *Finnish agriculture in 1995. Research Publications 79a;* AERI. 61 p. Helsinki

Miettinen, A. (1993) *The effectiveness and feasibility of economic incentives of input control in the mitigation of agricultural water pollution;* Agric. Sci. Finl 2:453-563

Miettinen, A. (1994) *Maatalouden ympäristönsuojelu - toimenpiteet ja niiden kehittyminen Suomessa. Vesi- ja ympäristöhallituksen monistesarja Nro 553;* 33 p. Helsinki

Miettinen, A. (1996) *Herbisidien käytön vähentämisen vaikutus viljelyn tuottoon;* Research reports 205: 53-71, AERI, Helsinki

Miettinen, A., K. Koikkalainen, V. Vehkasalo and J. Sumelius (1997) *Luomu-Suomi? Maatalouden tuotantovaihtoehtojen ympäristötaloudelliset vaikutukset -projektin loppuraportti;* Publications 83, AERI, 124 p. Helsinki

Ministry of the Environment (1992) *Ehdotus maaseudun ympäristöohjelmaksi;* Maaseudun ympäristöohjelmatyöryhmän muistio. 1992:68. 48 p. Helsinki

Ministry of the Environment (1997) *Nitraattidirektiivityöryhmän muistio 12.3.1997. The memorandum of the Nitrates directive working group;* Ympäristöministeriö, Ministry of the Environment, Helsinki

MKL (1997) *Maatilojen ympäristönhoito-ohjelmat 1995-1996;* The Rural Advisory Centre, 15 p. Helsinki

MMM (1992) *Maatalouslaskenta 1990;* National Board of Agriculture. Maa- ja metsätalous 1992:1. 151 p. Helsinki

MMM (1994) *Ehdotus Suomen maatalouden ympäristöohjelmaksi;* The proposal for Finnish agri environmental programme. Ministry of Agriculture and Forestry. 1994:4

MMM (1996a) *Maatilatilastollinen vuosikirja;* Maa- ja metsätalous 1996:5. MMM: nietopal\-velukeskus, Ministry of Agriculture and Forestry, Helsinki

MMM (1996b) *Maatalouden ympäristöohjelma 1995-1999;* Seurantatyöryhmän väliraportti 2.9.1996, Maa- ja metsätalousministeriö, Ministry of Agriculture and Forestry. 61 p. Helsinki

MMM (1997) *Maaseudun selviytymisopas;* Ministry of Agriculture and Forestry, K-Mediat, Peltoirkan Päiväntieto. 80 p. Helsinki

Pipatti, R. (1990) *Ammoniakkipäästöt ja -laskeuma Suomessa;* Valtion teknillisen tutkimuslaitok sen tutkimuksia 711. 41 p. Espoo

Pirttijärvi, R. (1996) *Maatalouden ravinneongelmat Hollannissa, Saksassa ja Suomessa;* Research reports no. 205: 5-36. AERI. Helsinki

Pirttijärvi, R. (1997a) *The impact of the Finnish agriculture on environment and the Finnish agri-environmental programme. Regional workshop on economic and structural impact of changing intensity in agriculture in pursuance of the goal of sustainable agriculture;* OECD, Gödöllö, Hungary

Pirttijärvi, R. (1997b) *Nutrient balances in Agri-Environmental Policy. Proceedings of the workshop towards operationalisation of the effects of CAP on environment, landscape and nature: expiration of indicator needs;* Wageningen, April 17-19, The Netherlands

Rantala, J. (1996) *Naudanlihan tuotannon kannattavuus siirtymäkaudella 1995-2000;* PTT:n raportteja ja artikkeleita 143. 49 s. Espoo

Rekolainen, S., L. Kauppi and E. Turtola (1992) *Maatalous ja vesien tila. Luonnnonvaraineuvosto, maa- ja metsätalousministeriö;* Luonnonvarainjulkaisuja 15. Helsinki

Ryhänen, M. and T. Sipiläinen (1996) *EU-jäsenyyden vaikutuksen kasvintuotantoon;* In: Maatalousyritysten sopeutuminen EU:ssa vallitseviin hintasuhteisiin' edited by Matti Ylätalo. Publication no. 12. University of Helsinki. Department of Economics

Ryhänen, M., P. Huhtanen, S. Jaakkola and S. Ahvenjärvi (1996) *EU-jäsenyyden vaikutus aidontuotantoon;* In: 'Maatalousyritysten sopeutuminen EU:ssa vallitseviin hintasuhteisiin' edited by Matti Ylätalo. Publication no. 12. University of Helsinki. Department of Economics

Ryttäri, T. and T. Kettunen (1997) *Uhanalaiset kasvimme.* Kirjayhtymä. 335 p. Helsinki

Siikamäki, J. (1996) *Finnish Agri-Environmental Programme in practice - participation and farmlevel impacts in 1995;* Research Publications 81: 83-98. AERI. Helsinki

Siikamäki, J. (1997) *Torjunta-aineden käytön vähentämisen arvo?* Contingent valuation - tutkimus kuluttjien maksuhalukkuudesta. Research reports 217. AERI. Helsinki

Sumelius, J. (1993) *A response analysis of wheat and barley to nitrogen in Finland;* Agric. Sci. Finl. 2:465-479

Sumelius, J. (1994) *Controlling nonpoint source pollution of nitrogen from agriculture through economic instruments in Finland;* Research publication 74. AERI. 66 p. Helsinki

Tamminen, A. (1997) *Tiedollinen ohjaus maatalouden ympäristönsuojelussa;* Manuscript, AERI, Finland

Tiilikainen, A. (1997) *Maidontuottajien asenteen ympäristöä, eläimiä ja kuluttajia kohtaa;* Publications no 14, Marketing. University of Helsinki. Department of Economics

Väisänen, J. and L. Pohjalainen (1995) *Kiinnostus luomutuotteisiin ja niihin liittyvä maksuhalukkuus. Luonnonmukaisen viljelyn liitto;* Mikkeli

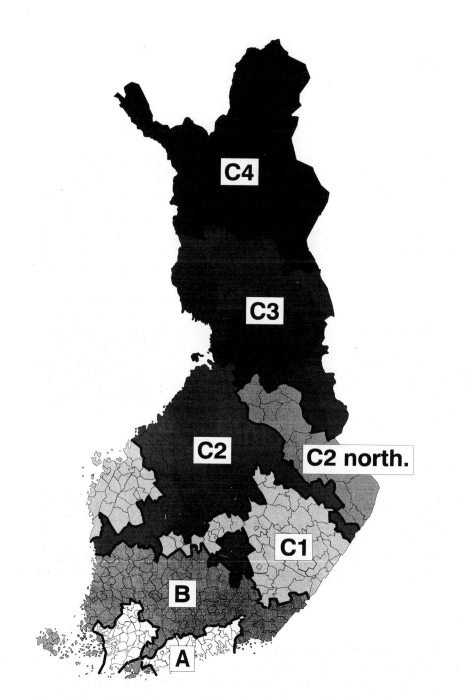

Figure   Agricultural support areas in Finland

344

15. SWEDEN

John Sumelius and Aage Walter-Jørgensen

15.1 Introduction

Sweden is a sparsely populated country. The total area is 449,964 km², of which 39,030 km² are lakes and water courses. More than half of the land area is covered by forest while a quarter of the country is mountains and marshes. Only about 9% of the land is in agricultural use. Structural change has led to a considerable reduction in the number of holdings with at present around 91,000 and an average size of 30 ha. The farms in the north of the country tend to be smaller (OECD, 1997). The major part of the agricultural area is situated in the southern and central part of the country. In northern Sweden, which represents 60% of the total area and 15% of the total population (Svedsäter, 1996), only 1.5% of the area is covered by agricultural land (Drake and Gustafson, 1996).

Among the themes related to environment, health and animal ethics that have been debated in Sweden, nitrogen leaching is the most important. Agriculture is estimated to account for some 30% of what is leached. Agriculture also accounts for approximately 10% of phosphorus leaching but this has received less attention (OECD, 1997). There is some concern over the leaching of pesticides and cadmium in fertilizers too. More emphasis, though, has been laid on the importance of marginal agricultural landscapes and farmland with high natural value. Among health-related topics, nitrates and pesticides in groundwater are matters of public concern as well as antibiotic and hormonal residues in food although mainly in relation to imported products. The freedom of movement of laying hens and of pigs also arouses concern.

15.2 Impacts of market and price support measures

Sweden became a member of the EU at the start of 1995. However, a reform package for agriculture had already been introduced in 1991, which was intended to take account of expected requirements of the Uruguay Round in GATT but also to prepare Swedish agriculture for EU entry. Following the reform of the Common Agricultural Policy in 1993, further adjustments were undertaken to assist in the adaptation to the CAP. To investigate the impact of the CAP on the environment in Sweden, therefore, it is necessary also to take into account recent changes in national policies.

15.2.1 The 1991 reform

The primary objectives of Swedish agricultural policy are to satisfy national security requirements, to safeguard the environment, and to promote regional development (Swedish Ministry of Agriculture, 1989). During the post-war period, these objectives were pursued by a protective line of policy involving high levels of support at extensive costs to society. By the end of the 1980s, increasing concern about structural problems and rigidities in the economy as a whole led to comprehensive reforms from which agriculture was not excluded.

The general direction of agricultural reform was towards a deregulation of domestic markets for the major commodities. It was intended that guaranteed intervention prices and export subsidies be abolished over a five year period. The internal deregulation was to be followed by reductions in import protection at a pace set by the Uruguay Round. Dairy quotas had already been removed and direct payments were introduced in 1989 in order to compensate for price reductions. The reform resulted in a unilateral decrease - on average 10% from 1990 to 1991 - in protection at the border (OECD, 1995, p. 10).

Following Sweden's application for EU membership in 1991, some changes to the reform programme were introduced with effect from 1993-94 to align national policies to the CAP. This included abolition of import levies on protein feedstuffs, introduction of direct compensatory payments in the crop sector, reduction of export subsidies for several crop and livestock products and re-introduction of administered prices within the crop sector. Milk quotas were not re-introduced until the actual accession to the EU in 1995. Taxes on fertilizers and pesticides have been retained throughout the reform process.

Measured by OECD's aggregate PSE value, support to the agricultural sector in Sweden fell from 64% in 1991 to 52% in 1993 (OECD, 1995, p. 12), which compared with an overall aggregate PSE value of 48% in the EU in 1993.

15.2.2 The implementation of the CAP

Sweden's accession to the EU implied full membership by January 1, 1995. National price policy was replaced overnight by the price policy of the Common Agricultural Policy allowing free trade between Sweden and the EU and full price alignment immediately. An analysis of the impact of EU accession (SJFI, 1994) found that cereal prices were reduced by nearly 20% from 1987-90 to 1995, and those of milk and beef by about 10%. However, much of the price adjustment took place before 1995 as a result of the 1991 reform. Quotas for farm land eligible for EU compensatory payments, sugar quotas and milk quotas were set, by and large, according to historical production. Quotas for male beef cattle eligible for support were, however, reduced slightly relative to prior production levels.

A considerable part of Sweden is classified as disadvantaged areas which are eligible for special income compensation to farmers. Of a total base acreage of 1.8 million ha at the time of entry, 60,000 ha were classified as 'Northern agricultural areas' and 360,000 ha as Less Favoured Areas. Special support is crucial for the maintenance of agricultural production in disadvantaged areas.

15.2.3 Impacts of policy reforms

The impact of the CAP on the environment is primarily reflected through the effect on production intensity in agriculture, notably the application of nutrients and chemicals in plant production, and through land use. In Sweden, most concern is about nitrogen leaching (Drake and Gustafson, 1996) but leaching of pesticides to groundwater and disposal of heavy metals in the soil are also of major concern. High support for agricultural production encourages farmers to use more fertilizers and pesticides and is thereby increasing the potential for pollution of soils and waters.

Support for agriculture also affects the landscape by bringing more land into production. In high productive areas this may be a threat to wildlife and to the diversity of the landscape whereas, in less productive areas, cultivation may be beneficial to the environment by keeping the landscape open and free from unwanted vegetation (Svedsäter, 1996). In such areas, there can be conflicting interests between the maintenance of agricultural activities for landscape purposes and restricting agricultural production for environmental reasons.

Most Swedish studies of the CAP have dealt with the impact on the agricultural sector and the Swedish economy of entering the EU. The basis for investigation has been a mathematical linear programming model (SASM) for Swedish agriculture developed at the Swedish Agricultural University (Jonasson, 1996). A few studies have dealt with the environmental impact of EU membership compared with the effect of the previous national policy.

In a report prepared prior to Sweden's entry into the EU, Hasund and Jonasson (1993) investigated the projected impact of EU membership on the environment. EU entry is compared with full implementation of the 1991 reform. According to the authors, this implies that farmers would have received lower support if the 1991 reform had been implemented than they actually receive as members of the EU. In judging the likely effect on the landscape and pollution from agriculture, it was assumed that deregulation under the 1991 reform would ultimately have involved lower support to Swedish agriculture than would EU membership. It was found that:

- EU membership would increase the cultivated area as compared with the eventual situation under the 1991 reform. It is estimated that the acreage of reform crops (cereals, rape seeds, etc.) might be increased by 0.5 - 1.0 million ha. The increase is measured from a lower level than what is actually observed at the time of entry. Furthermore, the number of beef cattle could increase by 150,000 - 350,000 heads;
- the change in land use would affect, in particular, natural grassland areas where the 1991 reform tended to reduce agricultural activities. The accession to the European Union is therefore considered to have a positive effect on the landscape in such areas. In the northern regions, open farming areas are considered to be an amenity in themselves. A reduction of support, following the 1991 reform, would have reduced the incentive to keep farm land free of scrub and thereby would have made it more difficult to maintain an open landscape. It needs to be mentioned that Sweden maintains special support programs for landscape preservation;
- the price of fertilizers would fall, meaning that the use of fertilizers would decline at lower rates than anticipated under the national policy reform. The larger area under cultivation would further increase the pressure on the environment;
- the use of pesticides might decline by about 15% as compared with an expected fall of 40% under the 1991 reform;
- the supply of heavy metals (cadmium) to the soil would increase relative to the situation under the national reform.

Due to special support programmes, northern regions would be more or less unaffected by EU membership.

Rabinowicz (1993) has estimated the short-term changes in farm production due to EU membership (Table 1). A comparison is made between internal market policy and EU policy, and it should be noted that the figures presented show the ultimate effect expected from the full implementation of the policies. It is found that cereal production will increase whereas livestock production, including production of milk, is expected to decline. The production of pigs and poultry would be affected most by entry.

Table 1   EU membership compared with internal deregulation by products sold (in 1,000 tonnes)

	Internal market	EU policy
Cereals	690	4,520
Rape seeds	257	472
Milk	3,212	3,189
Beef	137	108
Pigmeat	333	131
Poultrymeat	39	0

Source: Rabinowicz (1993), p. 50.

The impact of EU membership on Swedish agriculture was investigated in a recent report (LES, 1996). The basis for calculation was the above mentioned sector model using data for 1995 (the year that Sweden entered the EU). Three scenarios were investigated (Table 2): baseline, which builds on the actual use of farm land in 1995; short-term equilibrium, which illustrates adjustment within fixed resources of land and buildings; and medium-term equilibrium illustrating adjustment if land and building capacity can vary. The impact of integration was measured by the difference between the baseline and the short-term and medium-term outcomes respectively, assuming that 1995 price relations will prevail (i.e. no account was taken of subsequent changes in prices and compensatory payments). The results are very much in line with the findings in previous investigations. Livestock production will fall as a result of the change of policy whereas the

Table 2   Total production of selected products (in 1,000 tonnes)

	Baseline-1995	Short-term	Medium-term
Bread grain	1,700	2,050	2,010
Coarse grain	3,370	4,470	2,580
Oil seeds	280	480	470
Milk	3,250	3,230	3,300
Beef	137	130	126
Pigmeat	311	312	234
Eggs	92	92	69

Source: LES, 1996.

production of cereals and oil seeds will increase in the short run. In the medium term, when land use and building capacity can vary, cereal production is expected to fall and livestock production will be further reduced. The increase in cereal production in the short run is explained by higher cropping intensity due to higher prices following EU entry.

The study suggests a wide range of medium-term effects of EU entry:
- land use will be affected considerably;
- cereal acreage will decrease and the area of grassland will increase;
- the intensity of plant production will fall;
- the pressure on the milk quota may increase, which subsequently may increase the value of the quota to farmers;
- milk production will tend to move from cropping areas in the South to natural grassland areas in Mid- and Northern Sweden. The regional distribution of quotas, however, sets a limit on a reallocation of milk production;
- the economy of beef production will improve in Northern Sweden;
- the economy of pig and poultry production will deteriorate.

The overall picture arising from the study is that the accession to the EU will promote regional specialization of production. Cattle production will tend to be concentrated in areas receiving special support whereas the production of commercial crops will take place on the better soil.

15.2.4   Impacts on the environment

As indicated above, the change of market policy in connection with Sweden's entry into the EU is having a significant impact on agricultural production and land use which, in turn, will also effect the environment. Leaching of nitrogen and pesticides is closely linked to intensity of production in agriculture, and changes in the economic conditions for agricultural production will therefore have effects for the environment. Higher support for agriculture will enhance the detrimental effect on the environment and vice versa.

Support for agricultural production will also influence land use and thereby the landscape. As described above, this is of particular importance to less productive farming areas where agriculture has beneficial effects by keeping the landscape open and free from unwanted growth. Clearly, such objectives cannot be achieved through general support for agriculture but must be dealt with locally through individual guidance and intervention at the local level. Sweden has in this respect shown the way by targeting support for the maintenance of the agricultural landscape in selected areas where natural amenities are to be protected.

The arable regime

The 1991 reform, combined with Sweden's entry into the EU and the reformed CAP have resulted in a considerable reduction of producer prices for cereals, pulses and rape seeds. In compensation, Swedish farmers receive acreage support. As a result, the intensity of production in Swedish crop production is expected to fall and so is production in the longer term following a considerable fall in the total acreage of cereals. Part of the latter fall is explained by set aside. Consequently, the total acreage of grassland is expected to increase.

Taken as a whole, the change of market policy is expected to lower the use of nutrients and pesticides in Swedish agriculture which will reduce the detrimental effects on the environment. The main cause of this development is lower production intensity and change of land use towards more grassland. Were acreage payments to cereals and set aside to be reduced, this trend could be further enhanced. It is also important in this respect that Sweden has maintained taxes on nitrogen fertilizers and pesticides.

The impacts of policy reforms show major regional variations because of the combined effect of acreage support and regional support for livestock production. The expected trend of development is towards regional specialization with cereal production concentrating on the better land (primarily in mid- and southern Sweden) and milk and beef production moving towards less productive areas which are more suitable for grass production. The possibilities to apply for various support programmes (acreage payments, payments for set aside, special support for agriculture in northern Sweden, regional differentiated rules for number of animal units per ha, regional support for beef production, special support for maintenance of farming landscape, etc.) provide ample scope for combinations of support in less favoured areas. It is therefore difficult to assess the aggregate effect of the reform of the arable crop regime in isolation from other CAP measures. The special support for maintaining grass-based livestock production in less productive areas could easily make up for the fall in production intensity in crop production.

The dairy regime

So far the production of milk in Sweden has remained slightly below the total milk quota of 3.3 million tonnes adopted in 1995. As indicated above, however, it is expected that milk production will be enhanced and that the value of the quota to farmers will increase in the coming years. This is based on the assumption that structural adjustments in the dairy sector will proceed and milk quotas be made transferable.

The change of policy implies a break in the regional trend in milk production in Sweden. Prior to EU entry, milk production increasingly moved into crop production areas. This development was supported by the 1991 reform. After the EU entry, with more emphasis to be given towards regional development, this trend is expected to be reversed, meaning that milk production will have its stronghold in grassland regions. Milk production in Northern Sweden will be less affected by the change of policy as regional support has been applied to that region already before EU entry and, even more important, the region has been allocated a fixed milk quota.

The main impact of the dairy regime is therefore towards concentration of milk production in grassland areas which may enhance the risk of nitrogen leaching to groundwater and surface waters in such regions. On the other hand, the fall in livestock intensity may alleviate the pressure on the environment in cropping areas.

The beef regime

Beef production in Sweden is primarily a by-product of milk production. In 1995, the ratio of beef cows to dairy cows was about 1:3 but the number of beef cows is expected to decline considerably in the coming years. The stock of dairy cows, on the other hand, is set by the milk quota and the development in milk yield per cow. Thus, the production of beef will increasingly depend on the trend in milk production.

The profitability of beef production in Sweden is closely linked to regional support. Without such support, the production of beef cattle would cease in most regions. Support for the agricultural landscape and enhancing biodiversity may temporarily also maintain the number of beef cattle. However, in the longer run, this task will probably be taken over by milk producers as beef production declines and Sweden becomes more dependent on the import of beef. Thus, the beef regime provides little incentive in itself to ensure landscape preservation but may help in reducing the pressure on the environment.

15.3 Agri-environment policies

15.3.1 National programmes before the accession

Prior to EU membership two national programmes of landscape conservation existed. Nature Conservation Measures in the Agricultural Landscape was set up in 1986. In 1990 a similar programme, Measures for Landscape Conservation, was launched (Svedsäter, 1996). The objective of these programmes was to find new forms of direct payments for the provision of public services (protection and enhancement of biological diversity, and conservation of landscape and the natural and cultural heritage) instead of trying to achieve a number of objectives through price support. The second of the two programmes was tied to the Swedish agricultural reform of 1990 which terminated when Sweden joined the EU. The Swedish agri-environmental programme based on EC Regulation 2078/92 partly replaced and partly extended these older national programmes (Rundqvist, 1996).

The Swedish agricultural landscape is of great importance for biological diversity and nature conservation. Over two thirds of the more than 400 species of vascular plants that exist in Sweden occur in the agricultural landscape. Because of changes in land use, cultivation methods and afforestation, landscape and biological diversity had been decreasing for decades (Rundqvist, 1996). There has been concern that the long-term transformation of the agricultural landscape in the northern part of the country, where agricultural land represents such a tiny but vital part, and in the forest regions in the central and southern parts will lead to land abandonment, afforestation and losses of agricultural landscapes and biological variety. According to a contingent valuation study by Drake (1992) the Swedes seem willing to pay SEK 541/person/year or SEK 975/ha/year to preserve the agricultural landscape.

The Swedish regional programmes for conservation emphasise both nature conservation value and historical and cultural values. The programme has classified the natural heritage of the whole country. For each area in the regional programmes the *nature conservation value* is classified into three categories according to the following criteria:
- biological qualities representative of the region;
- biological variety, specifically a rich variety of habitats and areas with a mosaic structure used in a traditional way for grazing, mowing and cultivation;

- variety and density of species;
- presence of semi-natural grassland; and
- presence of threatened habitats and species.

Cultural or historical values are classified according to the representativeness of different historical periods, presence of landscape elements with historical value like fences, ancient remains, roads, ditches and ancient cultivated fields, and continuity of agricultural activity for a long time. Recreational, educational and scientific values are also considered (Rundqvist, 1996).

15.3.2 Agri-environmental programmes

The Swedish agri-environmental programme is expected to enhance biodiversity and limit nitrate leaching by setting up protective zones around lakes and streams. However, no quantitative assessment is yet available regarding impacts on the environment and the costs involved.

When Sweden joined the EU the on-going programmes for landscape conservation were modified somewhat in accordance with EC Regulation 2078/92. The Swedish agri-environmental programme was planned for five years with an evaluation after two years.

The objectives of the programme are that agriculture shall be practised in a way which protects human health, ensures a sustainable use of natural resources, maintains biodiversity and protects the natural and cultural landscape. Pressure on the environment from agriculture through the use of fertilizers and pesticides shall be minimized. In particular the programme shall:
- conserve the biological diversity and the cultural remains of high value in the agricultural landscape, created by traditional farming practices, through appropriate maintenance of meadows, semi-natural pastures, landscape elements and other cultural heritage values. The purpose is to maintain the traditional and representative values of different regions for the future;
- conserve the genetic resources in local animal breeds threatened by extinction;
- restore and establish habitats to enhance biological diversity;
- reduce leaching of nutrients and the use of pesticides in order to avoid health risks and provide suitable conditions for flora and fauna (Svedsäter, 1996).

The programme consists of three parts:
1. conservation of biodiversity and cultural heritage values in the agricultural landscape, as well as maintenance of an open agricultural landscape in the forest region in northern Sweden;
2. protection of ecologically sensitive areas;
3. promotion of organic production.

The first part of the programme includes conservation of mowed meadows, semi-natural grazing land, habitats and cultural heritage environments and open landscape; while the second part concerns the establishment of wetlands and of permanent grassland to counter nitrogen leaching, and support for catch crops and for endangered farm animal breeds.

In terms of realized budget in 1996, the second and the third parts of the programme -the protection of ecologically sensitive areas and the promotion of organic farming - with SEK 17 millions and SEK 178 millions respectively - have received much less support than the first part aiming at conservation of biodiversity and cultural heritage values in the agricultural landscape for which SEK 1,045 million (ecu 110 million) was appropriated (Swedish Board of Agriculture 1997). It is therefore quite clear that the Swedish agri-environmental programme is focusing much more on landscape issues than on ecological issues such as the reduction of nutrient leaching.

In order to realize the first part of the programme farmers can make agreements with the government to manage the landscape. Prior to Swedish entry into the EU the Landscape Conservation Programme covered 15,000 agreements and 376,000 ha throughout Sweden. Approximately half of the area covered was under arable crops; the rest was semi-natural grassland (Rundqvist, 1996). About 15% of farmers were involved in the Landscape Conservation Programme. The objective was to extend the coverage to include the remaining 370,000 ha of semi-natural land. The measures available are summarized in Table 3.

Table 3   Summary of the scheme 'Conservation of Biodiversity and Cultural Heritage Values'

Programme	Acreage (ha)	Compensation (ecu/ha/year)	Cost (MECU)	Cost (MSEK)
Mowed meadows	3,000	182	0.7	6.6
scything	1,000	182	0.3	2.5
pollarding	400	36	0.01	0.2
Semi-natural grazing lands				
'highest value'	80,000	146	11.5	112
'high value'	32,000	64	2.3	22.4
Cultural heritage features				
on arable land	20,000	34	0.8	7.5
pollarding	300	36	0.01	0.1
Open agricultural landscape				
support area 1-3	141,000	187	34.4	326
support area 4	70,000	128	4.6	44
forest areas	690,000	109	40.5	384
Total estimated costs				
for acreage compensation			95.1	905.3
Information, education				
and demonstration projects			7.9	75

Source: Svedsäter (1996).

The agri-environmental programme covers all the remaining traditionally managed (mowed) meadows in Sweden. They total 3,000 ha and are divided into two groups of different value. A farmer who receives compensation is obliged to remove each year the leaves and twigs from the meadows, mow the sward-growth annually, and remove the cuttings within two weeks after mow-

353

ing as well as keep a record of management activities. He may not spread fertilizers or pesticides on the meadow or use the ground for the supplementary feeding of grazing animals.

The semi-natural land is classified into four categories A ('highest value'), B ('high value'), C ('certain biological and cultural value') and D ('high cultural value'). The areas are subject to similar types of management as meadows. However, the areas can be used for the supplementary feeding of grazing animals. The scheme on cultural heritage features includes a wide variety of elements valuable for the landscape, which are eligible for management agreements. The scheme, which aims towards maintenance of an open landscape in north Sweden and in forest regions, is intended to support semi-natural grazing land and animal husbandry production based on forage from ley and permanent grassland.

The second part of the agri-environmental programme - the protection of ecologically sensitive areas - involves the following measures (Rundqvist, 1996):

1. restoration and establishment of wetlands and ponds on arable lands. It covers 4,200 ha, and compensation is provided for a maximum of 20 years. The maximum payment is SEK 4,000/ha during the first five years; it should not exceed SEK 2,500/ha during the remaining 15 years;
2. establishment of permanent grassland (13,000 ha) in areas sensitive to nitrogen leaching. Compensation is SEK 1,000/ha for extensive leys and 3,300 for riparian zones;
3. establishment of 40,000 ha of catch crops in south Sweden with SEK 500/ha as compensation;
4. improvement of local Swedish breeds of cows, pigs, goats and sheep and to preserve the genetic diversity of domestic animals for a compensation of SEK 1,000 per livestock unit.

The smaller amount available to support these measures suggests that they are not given as high a priority as the conservation of landscapes, biodiversity and cultural heritage.

15.4 Concluding observations

Sweden became a member of the EU in 1995. However, a reform package for agriculture had already been introduced in 1991 in order to make Swedish agriculture ready for entering the EU, and further adjustments were undertaken between 1993 and the time of entry to adapt to the CAP. The impact of the CAP may therefore be viewed in relation to the situation prior to the 1991 reform, although part of the adjustment took place before entry.

The change in market policy in connection with Sweden becoming a member of the EU has had, and is still having, a significant impact on land use and the intensity of production in agriculture. The support for commercial crops was reduced considerably already before entry and, although support was enhanced at the time of entry, a resulting fall in production intensity is believed to have reduced nitrogen leaching. Furthermore, it is expected that the cereal acreage will decrease in the medium term. Support for milk production was affected only to a limited extent by entry. However, due to regional support and allocation of quotas to less favoured areas, milk production is expected to move from cropping areas in the south to natural grassland areas in mid- and northern Sweden. Support for pigs and poultry, being situated largely in the better farming areas, has been reduced significantly with entry, thus causing production to fall. It is therefore believed that the change in policy in connection with entry to the EU will alleviate nitrogen leaching in productive farming areas but with only limited environmental effects in less favoured areas.

There is also a potential negative effect on the environment speeded up by the CAP which concerns the biological and cultural values connected with the agricultural landscape. Land with a high biological or historical value is in danger of becoming abandoned. The reasons for land abandonment are several: price changes, change in technology and rationalization. The agricultural reform in connection with EU membership would have accelerated the process. In order to avoid these negative effects on the landscape two national programmes for preserving the landscape were initiated before EU membership. The programmes were continued after Sweden joined the EU on the basis of EC Regulation 2078/92. Through the programme some of the potentially negative effects on landscape and biodiversity of agricultural reform have been avoided. While it may be too early to evaluate the programme, Sweden has shown one possible way forward in targeting support for the maintenance of agricultural landscapes in selected areas.

References

Drake, L. (1992) *The non-market value of the Swedish agricultural landscape;* European Review of Agricultural Economics 19(3): pp. 351-364

Drake, Lars and Arne Gustafson (1996) *Inventory on Mineral pollution from agriculture, Part I: Countryreports;* Sweden. pp. 199-206; In: Simonsen, Jesper W. (ed.) EU concerted action Policy measures to control environmental impacts from agriculture (AIR3 CT93-1164), Oslo. Ed. Simonsen, J. Norwegian Agricultural Economics Research Institute, Oslo, January 25, 1996

Hasund, Knut Per and Lars Jonasson (1993) *'Jordbruket och Miljön' (Agriculture and the Environment);* In: Sverige och den Europeiska Miljöpolitiken (Sweden and the European Environmental Policy), Naturvårdverkets Forlag, Solna, pp. 142-171

Jonasson, L. (1996) *Mathematical programming for sector analysis - some applications, evaluations and methodological proposals;* Avhandling 18. Sveriges Lantbruksuniversitet, Inst för Ekonomi, Uppsala

LES (1996) *Jordbrukets anpassning til EU - Modellberäkningar av optimal anpassning på kort och medellång sikt (The adjustment of agriculture to EU - Model calculations of optimal adjustment at short and medium terms);* Livsmedelsekonomiska samarbetsmnänden, Stocholm, 34 p.

OECD (1995) *Agricultural Policy Reform and Adjustment - The Swedish Experience;* Paris, Organisation for Economic Co-operation and Development

OECD (1997) *Sweden: Incentives for Environmental Management of Farmland with High Biological Values;* pp. 129-132, in the Country Case Studies report from the Helsinki Seminar on Environmental Benefits from Agriculture. OCDE/GD(97)110. 257 p. Paris

Rabinowicz, Ewa (1993) *Konsekvenser av EG-medlemskapet för jordbruket och livsmedels-sektorn (Consequences for Agriculture and the Food Sector of EC-membership);* Bilaga 3 till EG-Konsekvensutredningen, Samhällsekonomi, Sveriges Lantbruksuniversitet, Uppsala, 80 p.

Rundqvist, Bengt (1996) *Sweden;* chapter 10; In: Whitby, M. (ed.); The European Environment and CAP Reform Policies and Prospects for Conservation Wellingford, CAB International, 271 p.

SJFI (1994) *EU-udvidelsen og markedsbalancen (Enlargement of the EU and Market Balance);* Chapter III; In: Danish Agricultural Economy - Autumn 1994, Institute of Agricultural Economics, Copenhagen

Swedish Board of Agriculture, (1997) *Report 1997:10 (Jordbruksverket, rapport 1997:10);* 163 pp. + annexes. Jönköping

Swedish Ministry of Agriculture (1989) *A New Food Policy - Summary of a report from a Parliamentary Working Group to the Swedish Government;* presented to the Minister of Agriculture on October 25th, 1989

Svedsäter, H. (1996) *Swedish Country Report;* pp. 203-220; In: J. Umsätter and S. Dabbert (eds.), Policies for Landscape and Nature Conservation in Europe, An inventory to accompany the workshop on 'Landscape and Nature Conservation held on 26th-29th September, 1996 at the University of Hohenheim, Germany